T0251235

HERBICIDES

HERBICIDES

CHEMISTRY, DEGRADATION, AND MODE OF ACTION

Volume 3

edited by

PHILIP C. KEARNEY and
DONALD D. KAUFMAN

Pesticide Degradation Laboratory
United States Department of Agriculture
Agricultural Research Center
Beltsville, Maryland

CRC Press
Taylor & Francis Group
Boca Raton London New York

CRC Press is an imprint of the
Taylor & Francis Group, an **informa** business

First published 1988 by Marcel Dekker, INC.

Published 2022 by CRC Press
Taylor & Francis Group
6000 Broken Sound Parkway NW, Suite 300
Boca Raton, FL 33487-2742

ISBN 13: 978-0-8247-6175-2 (hbk) (Volume 1)
ISBN 13: 978-0-8247-7804-0 (hbk) (Volume 3)

Visit the Taylor & Francis Web site at
http://www.taylorandfrancis.com

and the CRC Press Web site at
http://www.crcpress.com

Library of Congress Cataloging-in-Publication Data
(Revised for vol 3)

Herbicides : chemistry, degradation, and mode of
    action.

    Published in 1969 under title: Degradation of
herbicides.
    Includes bibliographical references and indexes.
    1. Herbicides--Collected works. I. Kaufman, Donald
DeVere. II. Kearney, P. C. (Philip C.).
Degradation of herbicides.
SB951.4.K4   1975   668'.654   75-158

PREFACE

The present volume is the third in a series entitled Herbicides, Chemistry, Degradation, Mode of Action first published in 1975, which encompassed 19 chapters. The two-volume 1975 publication was preceded, in 1969, by a single volume entitled *Degradation of Herbicides* covering twelve major classes of herbicides. The format of the current volume resembles the 1975 volumes, which provided a broad coverage of herbicide technology on a compound by compound basis. Specifically, it contains an introductory section that covers the history of the compound, its physical and chemical properties, toxicology, synthesis and analytical methods, chemical reactions, and formulations. The second part of each chapter examines the metabolism of the herbicide in a number of environmental niches, specifically: plants, animals, soils, and microorganisms, and environmental processes such as persistence, movement, and photodecomposition. The third section considers the mode of action of the herbicide along with its absorption, translocation, selectivity, and the physiological/biochemical effects. Finally, a comprehensive contemporary review of the literature is provided.

There has been a tremendous growth in the herbicide industry over the last decade. One market projection estimates that herbicide sales will be about $2.4 billion in 1987, or slightly over 70% of all pesticide sales estimated at about $3.5 billion in the United States.

Alachlor is the leading herbicide in current usage. Alachlor, and the structurally related metolachlor, are reviewed in the current volume. Herbicides are used on over 90% of the corn, soybeans, and rice grown in the United States. As conservative tillage practices are more widely used, it is anticipated that the volume of herbicide usage will also increase.

Weed control is one major obstacle to full implementation of any minimum tillage scheme. Currently about a third of the 300 million acres of our tilled soil is under some form of minimum tillage. During this period of unprecedented usage, certain environmental problems have emerged that have occupied a major portion of the research effort devoted to herbicides. The most serious problem facing the industry is the detection of minute residues, in the parts per billion range, of several major herbicides and their metabolites in groundwater. Although less than 1% of the applied material finds its way into both shallow or deep aquifers, concerted efforts are underway to determine the toxicological significance of these residues. One development that may alleviate much of the current concern about groundwater residues is the development of some extremely active new classes of herbicides that control weeds in the grams per acre range. The sufonylureas are one such class of new herbicides that hold considerable promise for reducing the environmental burden. Volume 3 contains the most comprehensive coverage of this new class of herbicides yet published.

One unique problem that has surfaced in recent years is the occurrence of enhanced metabolism of certain soil-applied herbicides, primarily in the carbamothioates family, that reduce the effectiveness of these compounds against their target pests. Evidence is emerging that soil microorganisms are becoming adapted to degrade each successive addition of herbicides like EPTC at ever-faster rates. Literature on the carbamothioates was reviewed in a previous volume. The phenomenon of problem soils has generated such interest in the scientific community, as well as new insights on their mode of action, that they are updated in this volume.

Biotechnology promises to play an ever-increasing role in crop protection programs in the future. The broad-spectrum herbicidal activity and environmental safety of glyphosate offers an ideal compound to develop engineered resistance in selected crop species. Glyphosate does not move in soils, and therefore poses no threat to groundwater. To date, no resistant weed species has developed to glyphosate. Because of its potential use in future biotechnology-oriented crop protection programs and its current widescale usage, an extensive review of glyphosate is featured in Chapter 1.

Triazine chemistry has been important in herbicide chemistry since the introduction of the widely used compound atrazine and several related commercial symmetrical triazines. These herbicides were covered in previous volumes. The asymmetrical triazines have also made a contribution to weed control with the introduction of metribuzin, metametron, and isomethiozin and are discussed in Chapter 4.

The volume is intended for scientists in government, industry, and universities engaged in pest control research. Policymakers and regulatory personnel have also found previous volumes a good learning guide and reference source. It is also intended for classroom use in both advanced undergraduate and graduate level coursework. Some previous coursework in organic chemistry and plant physiology are obvious prerequisites. Public interest groups may find the complete series a useful source in developing background information or just general awareness of the direction of this area of pest control research.

Philip C. Kearney

CONTRIBUTORS

Elmo M. Beyer, Jr.   Discovery Research, Agricultural Products Department, E. I. du Pont de Nemours & Company, Inc., Wilmington, Delaware

Michael J. Duffy   Agricultural Products Department, E. I. du Pont de Nemours & Company, Inc., Wilmington, Delaware

Stephen O. Duke   United States Department of Agriculture, Agricultural Research Service, Southern Weed Science Laboratory, Stoneville, Mississippi

E. Ebert   Agricultural Division, CIBA-GEIGY Limited, CH-4002, Basel, Switzerland

Kriton K. Hatzios   Department of Plant Pathology, Physiology, and Weed Science, Virginia Polytechnic Institute and State University, Blacksburg, Virginia

James V. Hay   Agricultural Products Department, E. I. du Pont de Nemours & Company, Inc., Wilmington, Delaware

William H. Kenyon*   MSU-DOE Plant Research Laboratory, Michigan
   State University, East Lansing, Michigan

Homer M. LeBaron   New Technology and Basic Research, CIBA-GEIGY
   Corporation, Greensboro, North Carolina

Janis E. McFarland   Agricultural Division, CIBA-GEIGY Corporation,
   Greensboro, North Carolina

Donald Penner   Department of Crop and Soil Sciences, Michigan
   State University, East Lansing, Michigan

David D. Schlueter   Agricultural Products Department, E. I.
   du Pont de Nemours & Company, Inc., Wilmington, Delaware

Dexter B. Sharp[†]   Environmental and Formulations Technology,
   Technology Division, Monsanto Agricultural Company, St. Louis,
   Missouri

B. J. Simoneaux   Agricultural Division, CIBA-GEIGY Corporation,
   Greensboro, North Carolina

Robert E. Wilkinson   Department of Agronomy, Georgia Station,
   University of Georgia Agricultural Experiment Station, Experiment, Georgia

_____

Current affiliations:
*Agricultural Products Department, E. I. du Pont de Nemours &
Company, Inc., Wilmington, Delaware.
[†]Retired; Consultant.

CONTENTS

CONTENTS OF VOLUME 1

CONTENTS OF VOLUME 2

HERBICIDES

Chapter 1
GLYPHOSATE

STEPHEN O. DUKE

United States Department of Agriculture
Agricultural Research Service
Southern Weed Science Laboratory
Stoneville, Mississippi

I. INTRODUCTION

A. History

A vast literature of thousands of publications on the herbicide
glyphosate, N-(phosphonomethyl)glycine (1), has developed [1]
since it was first described as a herbicide in the scientific literature
[2]. This literature includes a book [3] and an extensive bibliography [4]. A major reason for this high level of interest in glyphosate is that it is expected to be the first herbicide to reach
the billion dollar sales mark by 1986 from markets in 119 countries
[3]. This nonselective, broad spectrum, postemergence herbicide
is registered for use in more than 50 crops and is used extensively
for vegetation control and management in many nonagricultural
settings.

$$\underset{\text{OH}}{\underset{|}{\text{HO}-\overset{\overset{\text{O}}{\|}}{\text{C}}-\text{CH}_2-\overset{\overset{\text{H}}{|}}{\text{N}}-\text{CH}_2-\overset{\overset{\text{O}}{\|}}{\text{P}}-\text{OH}}} \tag{1}$$

In this review, the term glyphosate refers to the free acid form
of N-(phosphonomethyl)glycine, even though much of the literature
on glyphosate uses the term for the isopropylamine salt of glyphosate.
This salt is the active ingredient of Roundup herbicide.

Phosphonic acids are a relatively new class of herbicides first
patented in 1969 [5] by Monsanto Company. An earlier patent [6]
was issued to Stauffer Chemical Co. that included a process by
which phosphinic acids could be converted to phosphonic acids. The
compounds described in this patent were described primarily as
chelating agents and no mention was made of herbicidal activities of
phosphonic acids. After patenting glyphosate as a herbicide in 1969,
Monsanto Company issued several other patents for various analogues
and salts of phosphonic acids [e.g., Refs. 7–12], including the

patent for the salt of glyphosate used in the commercial formulation Roundup [8].

The phosphonic acids and their salts patented by Monsanto in their original patent [5] are tertiary amines of the types shown below (2,3), prepared by methods published earlier by Monsanto scientists [13]. Of these earlier compounds, only glyphosine (4) was developed commercially, as a sugar cane ripener with a trade name of Polaris. The patent that specifically covers glyphosate as a herbicide includes secondary amines of the general structure below (5). The R groups are OH or OR; the OR group containing a cation (metal, ammonia, or certain organic-substituted ammonia) to form a salt.

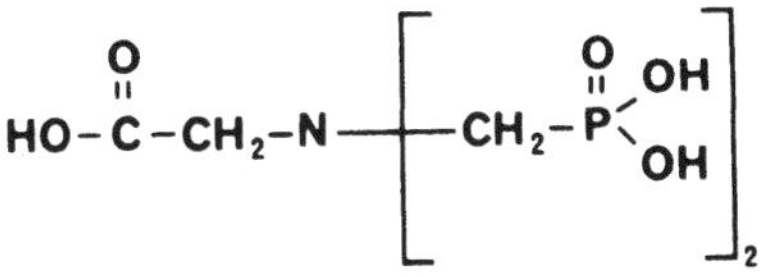

$$HO-\overset{\overset{\displaystyle O}{\|}}{C}-CH_2-N \left[-CH_2-\overset{\overset{\displaystyle O}{\|}}{P} \overset{OH}{\underset{OH}{<}} \right]_2$$

(2)

$$\left[HO-\overset{\overset{\displaystyle O}{\|}}{C}-CH_2 \right]_2 -N-CH_2-\overset{\overset{\displaystyle O}{\|}}{P} \overset{OH}{\underset{OH}{<}}$$

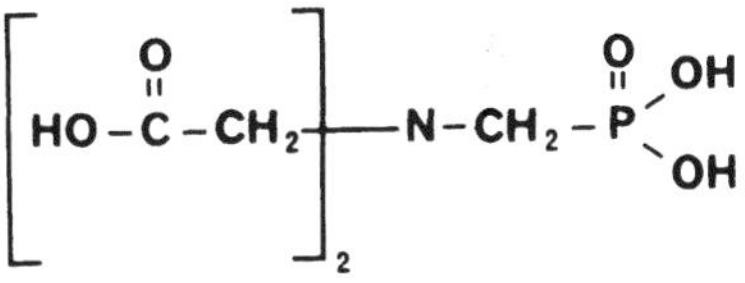

(3)

$$HOOC-CH_2-N \left(CH_2-PO_3H_2 \right)_2$$

(4)

$$R-\overset{\overset{\displaystyle O}{\|}}{C}-CH_2-\overset{\overset{\displaystyle H}{|}}{N}-CH_2-\overset{\overset{\displaystyle O}{\|}}{P} \overset{R_1}{\underset{R_2}{<}}$$

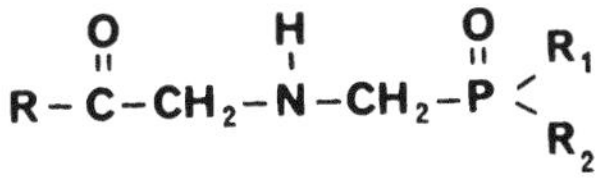

(5)

Stauffer Chemical Co. has subsequently applied for several patents [14,15] for various salts (sulfonium, sulfoxonium, and phosphonium) of glyphosate.

Several of these salts, such as the trimethylsulfonium [SC-0224 (6) and the trimethylsulfoxonium (SC-0545) (7)] salts, are being developed as herbicides. These salts have phytotoxicities similar to the isopropylamine salt of glyphosate [16,17]. Although phosphinates are also under development as herbicides (e.g., HOE-00661 or HOE-39866), [Refs. 16–18], this review will not deal with them. Many aminophosphonate analogues of glyphosate have been screened for herbicidal activity; however, none has as much activity as glyphosate [19–21]. Some of these compounds, however, are more selective than glyphosate [19].

The general area of phosphonates in living systems and phosphonate pesticides has been reviewed recently [22].

$$\left[\begin{array}{c} \text{O} \\ \parallel \\ \text{HO} - \text{C} - \text{CH}_2 - \underset{\underset{\text{H}}{|}}{\text{N}} - \text{CH}_2 - \underset{\underset{\text{O}}{\parallel}}{\overset{\overset{\text{OH}}{|}}{\text{P}}} - \text{O} \end{array} \right]$$

$$\left[\begin{array}{c} \text{CH}_3 - \underset{\underset{\text{CH}_3}{|}}{\text{S}} - \text{CH}_3 \end{array} \right]_n^+ \qquad (\underline{6})$$

$$\text{SC} - 0224$$

$$\left[\begin{array}{c} \text{O} \\ \parallel \\ \text{CH}_3 - \underset{\underset{\text{CH}_3}{|}}{\text{S}} - \text{CH}_3 \end{array} \right]_n^+ \qquad (\underline{7})$$

$$\text{SC} - 0545$$

B. Physical and Chemical Properties

Chemical properties of glyphosate have been discussed in two recent reviews [1,23]. In the free acid form glyphosate is a white, crystalline solid with zwitterionic properties (Fig. 1). It does not dissolve well in water (Table 1), indicating strong intermolecular hydrogen bonding in a crystalline lattice [1] (Fig. 1). These hydrogen bonds are between OH and NH groups and phosphono oxygens [23]. With the isopropylamine salt of glyphosate, hydrogen bonds form only between the NH group of the isopropyl amine and the PO_3^{-2} group of the glyphosate cation [25], making it much more water soluble. Details of the crystalline structure [23,26,27] and infared spectra [25] of glyphosate are published.

Glyphosate is very stable at room temperature. When heated to 200–230°C it softens, loses water, and resolidifies to form N,N'-diphosphonomethyl-2,5-diketopiperazine [1]. Glyphosate can be regenerated by refluxing with strong mineral acids, such as hydrobromic acid [1].

At physiological pHs glyphosate exists as both mono- and dianions (Fig. 2). Glyphosate is a relatively strong acid and will thus readily form metal and onium (organic ammonium and isologues, such as sulfonium, arsonium, phosphonium, etc.) salts [1]. At physiological pHs copper and zinc are strongly complexed with glyphosate, whereas manganese, calcium, and magnesium are complexed to lesser degrees (Fig. 3). The stability constants (log K_{m1}) of the metals are: Cu(II) 11.92, Zn 8.4, Mn(II) 5.53, Ca 3.25, and Mg 3.25 [28].

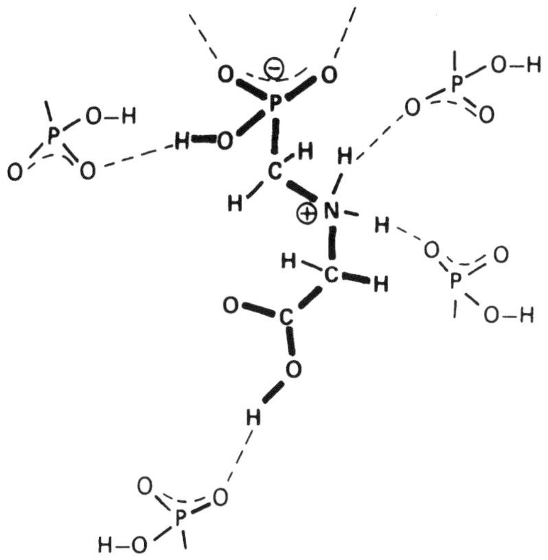

FIG. 1   Zwitterionic character and hydrogen bond system of glyphosate.   Reproduced from *The Herbicide Glyphosate* [3] by permission of the publishers, Butterworths & Co., Ltd.

TABLE 1

Physical and Chemical Properties of
Glyphosate Acid

Molecular weight	169.1
Solubility in water	1.2–8% at 25–100°C
Melting point	200°C
Vapor pressure	Negligible
Physical state	Solid
Color	White
Odor	Odorless
Density	0.5 g/cc
pK_1, pK_2, pK_3	2.27, 5.58, 10.25

Source: From Refs. 23, 24.

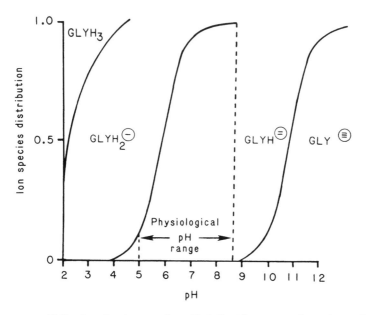

FIG. 2   Ionic species distribution as a function of pH [29].
Reproduced by permission of The American Chemical Society, © 1976.

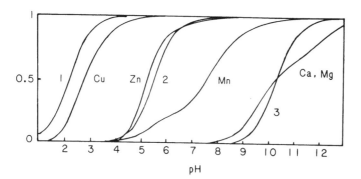

FIG. 3   Fractions of metal complexed as functions of pH at
equimolar concentrations (2.5 mM) glyphosate and metal ion in 0.1 M
KNO_3 [28].   The curves marked 1, 2, and 3 represent the base
fractions at the three dissociation steps of glyphosate.

C. Toxicological Properties

Only one previous review of toxicological properties of glyphosate exists [30]. Literature on this subject is extremely sparse, considering the importance of the herbicide. The paucity of published research in this area may be the consequence of the general lack of toxicity of glyphosate compared to most other pesticides.

Acute toxicological properties of glyphosate are summarized in Table 2. Neither glyphosate nor its commercial formulation are very acutely toxic compared to other pesticides. In some cases the formulated product Roundup is more toxic than technical-grade glyphosate, apparently due to additives and or adjuvants [36,39]. The LD_{50} value for ingestion by rabbits is about threefold greater than that for aspirin. The LD_{50} value for humans is unknown; however, it is almost certainly less than for rats or rabbits on a weight basis because of scaling considerations [e.g., Ref. 40]. Bentonite clay caused greater toxicity of glyphosate to *Daphnia pulex* [37], indicating that metal ion chelation may alter toxicity of glyphosate to animals.

Glyphosate has no remarkable subacute or chronic toxicity, nor does it have mutagenic or teratogenic effects at doses well above those expected to be received by ingestion of crops treated with glyphosate [30] (Table 3). Vigfusson and Vyse [42] found glyphosate to be weakly mutagenic, based on induction of sister-chromatid exchanges in human lymphocytes in vitro; however, their study involved lymphocytes from only two blood donors, which may be an insufficient sampling [30].

The basis for the low level of toxicity of glyphosate in mammals is not known. Although, glyphosate has been reported to inhibit oxidative phosphorylation of mammalian mitochondria by uncoupling [43,44], it is possible that this observation was a pH effect, just as an early report of glyphosate effects on photosynthesis [45] was found to be a pH effect [46]. These results [43,44] could not be confirmed by other studies [47], in which it was suggested that the isopropylamine moiety of the salt caused the effects. Effects of the glyphosate anion were found, however, on protein synthesis, oxygen uptake, and passive swelling of maize mitochondria [47].

Glyphosate has been found to reduce the hepatic levels of cytochrome P-450 and monooxygenase and the intestinal levels of aryl hydrocarbon hydroxylase in rats [48]; however, this work has not been confirmed and the isopropyl moiety of the salt used may have been the responsible agent.

D. Synthesis and Analytical Methods

One of the earliest synthetic schemes found to yield glyphosate was that of Westerback et al. [49], in which chloromethylphosphonic

TABLE 2

Acute Toxicological Properties of Glyphosate

Species	Test	Dose (effect)	References
Quail	Acute oral LD_{50}	>3.8 g/kg	24
Rabbit	Acute oral LD_{50}	3.8 g/kg	31, 32
Rat	Acute oral LD_{50}	4.3 g/kg Isopropylamine salt of glyphosate, 4.9 g/kg	32, 33
Rat	Acute oral LD_{50}	5.6 g/kg Roundup, 5.4 g/kg	24
Rabbit	Acute dermal LD_{50}	7.9 g/kg Roundup, >5.0 g/kg	34
Rabbit	Eye irritation	6.9 of FHSA scale of 100 (slightly irritating) Roundup, moderately irritating	24
Rabbit	Skin irritation	0.1 on FHSA scale of 8.0 (nonirritating) Roundup, 4.3, moderately irritating	24
Rat	Acute vapor inhalation	Roundup, no effect 4 hr at 12.2 mg/L air	24

Species	Test	Value	Ref.
Atlantic Oyster	96 hr TL_{50}	>10 mg/kg (slightly toxic)	35, 36
Blue gill	96 hr LC_{50}	120 ppm Roundup, 14 ppm (almost nontoxic)	24
Carp	96 hr TL_{50}	115 ppm (almost nontoxic)	24
Daphnia	48 hr LC_{50}	780 ppm (almost nontoxic) Roundup, 5.3 ppm	24
	48 hr EC_{50}	<8 ppm ai of Roundup	37
Duck	8 day dietary LC_{50}	>4640 ppm (almost nontoxic)	24
Fiddler crab	96 hr TL_{50}	934 mg/L (almost nontoxic)	35
Harlequin	96 hr LC_{50}	163 ppm (almost nontoxic)	35
Quail	8 day dietary LC_{50}	>4640 ppm (almost nontoxic)	24
Shrimp	96 hr TL_{50}	281 ppm (almost nontoxic)	35
Trout	96 hr LC_{50}	86 ppm (slightly toxic) Roundup, 11 ppm (slightly toxic)	24
Rat	Acute intraperitoneal LD_{50}	190–280 mg/kg	38

Source: Adapted from *The Herbicide Glyphosate* [3] by permission of the publishers, Butterworths & Co., Ltd.

TABLE 3

Effects of Glyphosate in Long-Term Toxicity Studies

Species	Test	Dose/effect	References
	Subacute toxicology		
Rat, dog	Subacute toxicity: body weight food consumption behavior mortality hematology blood chemistry urinalyses gross pathology histopathology	Dietary level of 2000 ppm for 90 day, no difference from control	24, 35
Rat, dog	Chronic toxicity: 2 year feeding study	300 ppm, no adverse effect	24
Hen	Neurotoxicity test	No effect 15 g/kg (1.25 g twice daily for 3 day), with dosage regimen repeated 21 days later)	
Rat	Subacute toxicity: 6 mo blood chemistry	Dietary level of 245 ppm no significant effects	41

Teratogenic effects

Oral			
Rabbit	Bodyweight gain Death Administration day 6–27 of gestation	Negative effect 350 mg/kg per day (highest dose tested) NOEL[a] for fetotoxicity 175 mg/kg per day	31, 33
Effects on fetal toxicity and birth defects			
Rat	3-generation reproductive study: parental and pup bodyweight gain, behavior, survival	NOEL 10 mg/kg per day	31, 33
	Teratology study day 6–19 of gestation Fetotoxic study	Negative 3.5 g/kg per day (highest dose tested) NOEL 100 mg/kg per day for fetotoxicity	31, 33
Carcinogenic and Mutagenic Effects			
Mouse	18-month feeding study	NCP[b] 300 ppm (highest dose tested)	31
	Dominant lethal	negative 2 g/kg	33
	Host-mediated mutagenicity test	negative	33

TABLE 3

(continued)

Species	Test	Dose/effect	References
Salmonella sp.	Ames test	negative	33
Bacillus subtilis	Rec-assay mutagenicity Cohn, Prazmowski test	negative up to 2 mg per disk	33
Rat	Dominant lethal test	negative 2 g/kg	33
	26-month feeding study	NOEL (31 mg/kg per day) (no oncogenic effects, highest dose tested)	33

[a]NOEL, no observable effect level.

[b]NCP, no carcinogenic potential.

Source: Adapted from *The Herbicide Glyphosate* [3] by permission of the publishers, Butterworths & Co., Ltd.

acid was reacted with glycine in basic media [1,8]. Moderate yields were obtained by carefully controlling reaction conditions and using excess glycine; however, glyphosate was difficult to separate from the other products [1]. More efficient syntheses of glyphosate have since been published [50,51].

Detection of glyphosate by physical [52] and bioassay [53] techniques has been reviewed recently. The present review will not discuss bioassays, because this author does not feel that bioassays are appropriate for glyphosate. Many other compounds, both natural and synthetic, have similar effects that are difficult to separate from those of glyphosate in bioassay. There are now a considerable number of sophisticated instrumental assays for glyphosate from which laboratories can choose. These are listed in Table 4.

As discussed by Bardalaye et al. [52], glyphosate's three polar functional groups (phosphonic acid, carboxylic acid, and secondary amine) complicate analytical techniques for glyphosates. Its high degree of polarity necessitates derivatization for many separatory methods and derivatization is not always easy because of the lack of suitable solvents for glyphosate, the derivatization reagents, and the derivatives. Derivatization methods are being improved [e.g., Ref. 74] to simplify analysis by gas chromatography (GC). The simplest methods, however, are those that separate glyphosate in aqueous solution (e.g., thin-layer chromatography or ion exchange chromatography) and then allow for detection by reaction of a func- tional group (e.g., ninhydrin reaction with the amine).

E. Chemical Reactions

Glyphosate is a very stable compound that will undergo most chemical reactions that any compound with a phosphonic acid, a carboxylic acid, or a secondary amine will undergo. Its low solubil- ity in organic solvents limits some chemical transformations, making derivatization for analytical procedures difficult (see Section I. D). Derivatizations include N-alkylation, N-sulfonylation, N-acylation, esterification, amination, dehydration, and production of phosphonyl and acyl halides. Details of several of these derivatizations are discussed by Franz [1].

F. Formulations

Glyphosate formulations have been reviewed previously by Turner [75]. A herbicide formulation is the herbicide prepared for practical use. Exact formulations are usually trade secrets and often vary from year to year and(or) from one geographical location to another. Glyphosate is no exception. Thus, this discussion will deal only with published studies of the effects of additives, ad- juvants, carriers, and different salts of glyphosate.

TABLE 4

Analytical Techniques for Detection of Glyphosate

Derivatization	Detection	Application	References
TLC			
None	Amine-specific reagent	Soil and water	54
None	0.5% Ninhydrin in butanol	Bindweed tissues	55
None	Ninhydrin, copper-nitrate, and rhodamine B	Soil and plant tissues	56
N-nitroso derivative	UV degradation, followed by fluorogenic labeling	*Cirsium, arvense* (L.) Scop. tissues	57
HPLC			
Labeled with fluorogenic compound (9-fluorenyl-methyl chloroformate)	Fluorescence	Soil and water	58
Post column labeling with o-phthalaldehyde	Fluorescence	Blackberry fruit	59
None	UV spectroscopy and refractive index	Formulated and technical glyphosate	60, 61

Labeling with fluorogenic compound	Fluorescence	Various fruit tissues	62–64
Labeling with fluorogenic compound	Fluorescence	Straw	65

GLC

Methyl 1-N-trifluoroacetylester	Flame photometry	Plant and animal tissues, soil, and water	66
2-Chloroethyl-N-heptafluoro-butyryl ester	Electron capture	Blueberry tissues	67

GC/MS

n-Butyl N-trifluoroacetyl ester	Flame ionization	Standards	68
Esterification and acylation with fluorinated alcohols and perfluorinated anhydrides	Flame photometry or electron capture	Standards	69

Ion Exchange Chromatography

Nitrosation	Polarography	Water, soil, and plant tissues	70, 71

TABLE 4

(continued)

Derivatization	Detection	Application	References
	Amino Acid Analyzer		
None	Spectrophotometric detection of post column ninhydrin reaction product	Standards	72
	HPLC/GC		
N-trifluoramethyl o,o-trimethyl derivative	Flame photometry	Fruit	73

Source: Adapted from *The Herbicide Glyphosate* [3] by permission of the publishers, Butterworths & Co., Ltd.

The most common commercial formulation of glyphosate, Round-up, contains the isopropylamine salt of glyphosate. This salt is considerably more soluble in water than the free acid form, making it much more convenient to use. During development of glyphosate, other highly soluble salts of the herbicide, such as monosodium and monodimethylamine salts, were tested, but the isopropylamine salt was chosen by Monsanto Co. for reasons which were not disclosed by them. Stauffer Chemical Co. has chosen to develop trimethyl-sulfonium (SC-0224) and trimethylsulfoxonium (SC-0545) salts of glyphosate as herbicides [16,17]. In studies comparing the efficacy of different salts of glyphosate, no marked differences have been noted between the potassium and isopropylamine salts [76] and the trimethylsulfonium, trimethysulfoxonium, and the isopropylamine salts [16,17,77]. The question arises as to whether or not glyphosate enters the plant in the ionized form. Since the salt used seems to make little difference in activity, provided it is nontoxic, it is likely that glyphosate enters the plant as an anion.

For this reason, glyphosate does not penetrate thick cuticles well. Activity can be enhanced in some cases by applying it in pure oils or surfactants, which form glyphosate micelles [75]. Generally, however, a small amount of surfactant in aqueous solution with the isopropylamine salt of glyphosate is more efficacious. This effect is generally due to increased glyphosate absorption. For instance, [14C]glyphosate absorption increased almost threefold when applied in Roundup, which contains a surfactant, when compared with up-take from aqueous solution of technical-grade isopropylamine salt [78]. Further enhancement of activity of Roundup by further sur-factants has often been reported, particularly when low doses are used on relatively resistant weed species [75]. Apparently, sur-factant-enhanced phytotoxicity is not entirely due to promotion of cuticular wetting.

Addition of certain salts to glyphosate formulations can sub-stantially increase phytotoxicity in some cases. Ammonium salts seem to be most consistently effective in increasing performance. Several examples of this are listed in Table 5. Generally, the pres-ence of ammonium ions doubles the phytotoxicity of glyphosate. In the comprehensive study of Wills and McWhorter [87], ammonium, potassium, and sodium ions significantly increased glyphosate activ-ity. Glyphosate absorption and translocation were approximately doubled by ammonium ion. Bicarbonate salts of potassium and sodium were more efficacious than salts of chloride, bisulfite, or nitrate.

Because di- and trivalent metal cations complex with and in-active glyphosate [88,89], compounds that complex or sequester these inactivating cations will enhance herbicidal activity. Among the ionic species that will enhance activity in this manner are phosphates [87] and organic acids, such as tartaric acid [90] and EDTA [91].

TABLE 5

Effects of Ammonium Salts on Efficacy of Glyphosate

Rate or concentration of ammonium salt	Glyphosate or Roundup rate or concentration	Target organism(s)	Effect	References
0.7–7% w/v Ammonium sulfate	?	Woody species	Increased phytotoxicity	79
10 kg/ha Ammonium sulfate	0.25–0.05 kg ai/ha	Rubber plantation weeds	Increased phytotoxicity	80
0.6–2.5% Ammonium sulfate	0.25 kg ai/ha	Cyperus rotundus L.	2-Fold increased effect	79
1 kg/ha Ammonium sulfate	1 kg ai/ha	C. rotundus Parthenium hysterophorus L. Digitaria, spp.	2-Fold increased effect	81

0.1–2.5 kg/ha Ammonium sulfate	Very low doses	Dwarf bean (Phaseolus, sp.)	Greatly enhanced activity	82
5 kg/ha Ammonium sulfate	1 kg ai/ha	Couchgrass (Agropyron repens, (L.) Beauv	2-Fold increased effect	83
2.5–5 kg/ha Ammonium sulfate	0.5–2.0 kg ai/ha	Potato (Solanum tuberosum L.)	Improved volunteer control	84
5 kg/ha Ammonium sulfate	5 L/ha Roundup	Aegopodium podagraria L.	2-Fold increased activity	85
1.25–10 kg/ha (0.5–4% w/v) Ammonium sulfate and nitrate	?	Agropyron repens (L.) Beauv	Increased activity	86
0.1 M in 150 L/ha Ammonium bisulfate, chloride, phosphate, nitrate, or bicarbonate	0.25 kg ai/ha	C. rotundus	1.5-2-Fold increased activity	87

Source: Adapted from *The Herbicide Glyphosate* [3] by permission of the publishers, Butterworths & Co., Ltd.

As commercially formulated, isopropylamine glyphosate is particularly vulnerable to being washed off of foilage by rainfall because of its high solubility in water. This is a particularly serious
problem with target species that absorb glyphosate slowly. Formulation of glyphosate with adjuvants that increase the speed of absorption for use with such species might be worth while in particularly
wet geographical locations [92]. Also, use of formulations, such as
gels [93] that may increase the "rainfastness" of the herbicide, may
help solve this problem. However, if it is the ionic form of glyphosate which is absorbed, decreasing solubility may decrease
absorption.

II. DEGRADATION PATHWAYS

A. Degradation in Plants

As pointed out by Coupland [94], in his review of metabolic
alteration of glyphosate by plants, little literature exists on this
subject. This is partly because of technical difficulties in analysis
of glyphosate and its potential metabolites (Fig. 4), but it may also
be largely due to the general lack of metabolic degradation of glyphosate in higher plants. Since the physiological basis for herbicide
selectivity is often differential metabolic capabilities [95], it is not
unreasonable to expect little metabolic capability in most species
toward nonselective herbicides such as glyphosate. This expectation
is supported by the available literature.

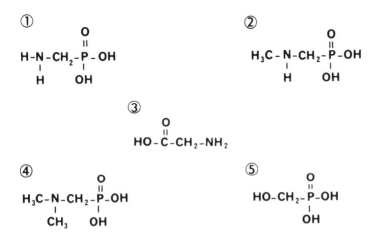

FIG. 4   Potential metabolites of glyphosate. (1) aminomethylphosphonic acid (AMPA), (2) N-methylphosphonic acid, (3) glycine,
(4) N,N-dimethylphosphonic acid, (5) hydroxymethylphosphonic
acid.

First, it should be pointed out that few studies of glyphosate metabolism have accounted for all of the ^{14}C label. In studies in which all ^{14}C found is in the glyphosate molecule, it is possible that some glyphosate has degraded to CO_2 if not all of the ^{14}C label applied is accounted for. Another possibility, however, is that loss of glyphosate has occurred through root exudations. Undegraded glyphosate has been reported to be exuded from roots of foliarly treated plants in several studies [e.g., Ref. 96].

A number of studies have reported no evidence of metabolic degradation of glyphosate in a variety of species in studies lasting as long as several months (Table 6). In many of these studies, not all of the applied [^{14}C]glyphosate was recovered. In the study by Wyrill and Burnside [104], although not all ^{14}C was recovered as glyphosate, more than 90% of recovered ^{14}C was as glyphosate and this proportion did not change during a period of 1–20 days, indicating that the less than 100% recovery was not due to metabolism. In the study by Schultz and Burnside [105], approximately 95% of the [^{14}C]glyphosate applied was recovered as intact glyphosate and less than 0.5% was determined to be converted to CO_2. No metabolites were identified.

Extraction and separation techniques can greatly affect results. For instance, Coupland [99] found considerably more glyphosate degradation in couchgrass tissues by changing the TLC solvent system. There is some evidence that both glyphosate and AMP can be conjugated with an unknown plant constituent [99].

In several studies metabolites have been identified (Table 6); however, the proportion of metabolites reported is generally low and proof that microbial contamination did not occur during the experiment is not usually available. Evolution of $^{14}CO_2$ from radiolabeled glyphosate-treated tissues in darkness [e.g., Ref. 99] could well be the result of microbial degradation [see below].

B. Degradation in Soil

The subject of degradation of glyphosate in soil has been reviewed recently by Torstennson [108]. He concluded that chemical degradation of glyphosate in soil is negligible and that virtually all degradation in soil is brought about by microflora. Thus, this discussion will deal with soil factors that influence degradation in soil, as well as the degradation by microorganisms occurring in the soil. Later sections will deal with degradation by microorganisms in nonsoil environments and persistence in the environment.

Glyphosate is rapidly inactivated as a herbicide in most soils by adsorption to clay particles through its phosphonic acid moiety [e.g., Refs. 109–111]. It has very low mobility in the soil [111, 112] and is not volatile. The only documented, nonbiological, chemical reaction of glyphosate in the soil is the formation of

TABLE 6

Results of Studies of Degradation of Glyphosate by Higher Plants

Species	Duration	Metabolic products (5 of 14C recovered)	References
Purple nutsedge (*Cyperus rotundus* L.)	16 days	None	97
Canada thistle (*Cirsium arvense* (L.) Scop.)	7 days	None	98
Leafy spurge (*Euphorbia escula* L.)	7 days	None	98
Couch grass (*Agropyron repens* (L.) Beauv.)	8 days	None	96
	21 days	1– 26% AMP 4– 15% Other	99
Wheat (*Triticum aestivum* L.)	10 days	None	100
	5 months		101

Creeping red fescue (*Festuca rubra* L.)	14 days	None	102
Reed canary grass (*Phalaris arundinacea* L.)	14 days	None	101
Barley (*Hordeum vulgare* L.)	3 days	None	103
Common milkweed (*Asclepias syriaca* L.)	20 days	None	104
Hemp dogbane (*Apocynum cannabinum* L.)	20 days	None	104
	12 days	None	105
Apple (*Malus domestica* Borkh.) tree	94 days	2– 8% AMP <1% Other	106
Pear (*Pyrus communis* L.) tree	80 days	2– 7% AMP <1% Other	106
Field bindweed (*Convolvulus arvensis* L.)	30 days	2– 4% AMP 2– 4% Glycine 2– 3% Sarcosine	55
Tall morning glory (*Ipomoea purpurea* (L.) Roth)	21 days	4– 6% AMP 1– 5% Glycine 2– 5% Sarcosine	107

N-nitrosoglyphosate in the presence of high levels of nitrite [112],
although formation is not observed with low levels of glyphosate and
nitrite [113,114].   Thus, loss of the glyphosate molecule from soil is
due almost entirely to microbial degradation.

Degradation of glyphosate occurs under both aerobic and anaerobic conditions [112].   Although there is evidence that most
microbial decomposition of glyphosate in soil is through cometabolism
(i.e., not using the herbicide as a food source) [111,115,116],
strains of *Pseudomonas* sp. and *Alcaligenes* sp. have been isolated
that use glyphosate as a sole phosphorus source [117–119].   The
major metabolic product of glyphosate by soil microorganisms is
aminomethylphosphonic acid (AMP) [107,112,116].   AMP is readily
decomposed by soil microorganisms [118].   The enzymatic activity
involved in degradation of glyphosate by these soil organisms is
intracellular, inducible, and requires the cofactors pyruvate and
pyridoxyl phosphate [118].

The half-life of glyphosate in soil varies considerably, ranging
from a few days to months or years [108].   There is no single,
overriding factor responsible for this variation.   In some soils, such
as sandy loam, strong biological effects on *Rhizobium* symbiosis with
legumes can occur for up to 120 days after application of as little as
2 µg of herbicide per gram of soil [120].   The soil microbial population and degree of adsorption of glyphosate to soil particles appear
to be the most important factors in affecting degradation.   Generally,
the amount of microbial activity in the soil is positively correlated with
the degradation rate of glyphosate [107,121], whereas other factors
such as pH or percent organic matter appear to have little influence
[108].   Addition of metal cations which strongly complex with glyphosate
(e.g., Fe^{3+} or Al^{3+}) reduces the degradation rate [122].   Addition of
phosphate to soil which competes for glyphosate binding sites, thus
releasing more glyphosate anion, stimulates degradation [111,122].

C. Degradation in Animals

A search of the literature has revealed no published reports of
metabolic degradation of glyphosate in vertebrate or invertebrate
animals.   Bioaccumulation in aquatic animals was concluded to be low
by Tooby [36].   It is also not accumulated by terrestrial animals
[123], apparently due to excretion or metabolism.

D. Degradation by Microorganisms

Although glyphosate is quite toxic to many microorganisms such
as *Escherichia coli* [124] and *Fusarium* [125], others degrade it and
can use it as a sole source of phosphorus [118–120] (see Section

II.D). Degradation of glyphosate by microorganisms apparently is the chief route of dissipation of this herbicide in the environment. Thus far, the only pure cultures demonstrated to metabolize glyphosate completely are *Pseudomonas aeruginosa* [117] *Pseudomonas* sp. [118,119,126], and *Alcaligenes* sp. [118]. Earlier, Cerol and Seguin [127] had reported growth of a *Pseudomonas* species was stimulated by glyphosate. All of these cultures can use glyphosate or AMP as a sole phosphorus source, but not as a sole carbon source. Talbot et al. [118] concluded that degradation by *Pseudomonas* sp. SG-1 is a stepwise degradation from the carboxyl end of the molecule with AMP as an intermediate, since an initial cleavage of the C-P bond would yield sarcosine, a product that the microorganism can utilize as a sole carbon source. They concluded that a transamination could be involved in degradation since pyridoxyl phosphate is a requirement for degradation. Recent studies utilizing solid-state nuclear magnetic resonance (NMR) determination of in vivo glyphosate metabolism in *Pseudomonas* sp. PG-2982 have, however, indicated that AMP is not an intermediate [126]. Rather, the phosphonomethyl carbon-nitrogen bond was cleaved, yielding glycine. The glycine was subsequently incorporated into protein (70%) and purines (20%). The phosphonomethyl carbon was incorporated into several molecules, including nucleic acids, methionine, serine, and thymidine.

Biotic degradation of glyphosate appears to be restricted to a limited number of microorganisms, since glyphosate inhibits the growth of most microorganisms tested [124]. Because degradation by plants and animals has not been demonstrated conclusively, dissipation of this herbicide in the environment may be almost entirely dependent on soil microbial metabolism.

E. Degradation by Photodecomposition

Whether there is any significant photodegradation of glyphosate in the environment is not clear from published information [128]. It is unlikely that significant photodegradation of glyphosate occurs in soil or on leaf surfaces, since it is shaded in the soil, and metabolism studies with plants have often revealed little or no degradation. In water, however, there is some evidence of photodegradation. Rueppel et al. [112] found only 2% degradation of glyphosate in a Crosby reactor over a period of 16 days with 8 hr of sunlight per day. This study was done in sterile, dionized water. Other studies, using sterile, nondeionized water or high ultraviolet (UV) light have indicated that glyphosate will photodegrade to AMP in water [128]. AMP appears to be less photolabile than glyphosate, although rigorous studies have not been published.

F.   Persistence Under Field Conditions

Persistence of glyphosate in terrestrial environments is not a
significant problem. In natural habitats, glyphosate has little or no
lasting effect on ecosystems within one or two years after application
[e.g., Ref. 129]. The compound is strongly bound by soil particles
and is rapidly degraded by apparently ubiquitous soil microflora
when released from soil particles. The persistence of glyphosate
under field conditions has not been well studied. It appears, however, to be quite variable [108], apparently due to differences in
soil type, microflora populations, and environmental conditions. In
a field study on six different forest soils, the period required for
90% loss of applied glyphosate ranged from 3 weeks (iron podsol) to
more than 45 weeks (weakly formed iron podsol) [108]. The half-
life of glyphosate in an Oregon forest sprayed with 3.3 kg/ha varied
considerably, depending on the location in the forest [123]. For
instance, the half-life in the shrubs, ground cover, litter, and litter-
covered soil were 26.6, 10.4, 14.0, and 29.2 days, respectively.
AMP was found at relatively low concentrations in the soil, but degraded rapidly. Others have found higher concentrations of AMP
than of glyphosate in the soil of irrigation ditchbacks approximately
6 months after glyphosate application [130].

G.   Movement and Persistence in Water

Glyphosate does not move readily through soil because of strong
complexing properties; therefore, it is unlikely to enter the aquatic
environment in significant amounts unless used for aquatic weed
control [see Refs. 131,132] or unless it washes from foilage into
streams or canals [e.g., Ref. 123]. Edwards et al. [133] found
that less than 1% of applied glyphosate was generally lost through
runoff in fields treated with recommended rates of the herbicide.
Despite glyphosate's extreme stability in sterile water in the laboratory, it is generally considered to dissipate rapidly in aquatic environments in the field [128]. This rapid dissipation apparently is
due to three factors: (a) photolysis to AMP [e.g., Ref. 112], (b)
microbial degradation to AMP and CO_2, and (c) adsorption to sediment, becoming soil residues. No published studies have rigorously
determined the relative contributions of each of these factors in a
field situation; however, it an aquatic environment with considerable
suspended material in the water, the latter factor is likely to be the
overriding mechanism of dissipation.

Glyphosate can move considerable distances in canal or stream
water. Comes et al. [130] accounted for 58% of applied glyphosate
at distances 8 and 14.4 km downstream from sites of introduction
into the canals. In this study, approximately 30% of the glyphosate
was lost in the first 1.6 km, presumably due to adsorption to

suspended material that settled out. Bowmer [131] found 63% of the glyphosate in irrigation channels to be lost to adsorption within 2 km. In a forest ecosystem sprayed with glyphosate, most of the glyphosate found in stream water was concluded to be introduced by direct spraying [123]. Initial levels of 0.3 mg/L decreased to trace amounts within about 1 week, whereas glyphosate in stream sediment increased over a period of 2 weeks to about 0.5 mg/L and then decreased to a level of about 0.1 mg/L at 58 days.

III. MODE OF ACTION

A. Absorption and Translocation

A large body of literature exists on absorption and translocation of glyphosate and the subjects have been reviewed by Caseley and Coupland [134]. In general, glyphosate is readily absorbed and translocated by vascular green plants. Several of its characteristics are responsible for its relatively high levels of absorption and translocation; (a) its high water solubility and anionic nature, (b) a lack of significant degradation in green plants, and (c) a slow mechanism of action.

Glyphosate is usually applied as a foliar spray with a surfactant. Its absorption is easily measured by removal of unabsorbed [14C] glyphosate by two or three rinses or dips in water or 1% ethanol in water [135]. Apparently the amount of glyphosate absorbed increases with concentration of glyphosate applied [136] and adjuvants or surfactants usually increase absorption of the herbicide [98,137–139]. Whether glyphosate enters the apoplast of the leaf from the cuticle as an ion or as a salt or complex is not clear from the literature.

As mentioned earlier, several metal ions reduce the herbicidal efficacy of foliar applications of glyphosate [e.g., Refs. 140–142]. Since chelating agents restore activity in the presence of metal ions [e.g., Refs. 90, 91, 143], one might conclude that metal ions complex glyphosate, preventing absorption. Nilsson [88,143], however, has shown that ferric citrate or $MnCl_2$ increased foliar absorption of glyphosate, while decreasing the phytotoxicity of the herbicide. Absorption of glyphosate by pea leaflets was delayed by Ca^{2+}, Mg^{2+}, Fe^{2+}, and Zn^{2+}. These data strongly suggest that glyphosate can be absorbed as a complex, but that if the complex is extremely stable under physiological conditions, it may not be active.

Plants can absorb glyphosate through their roots [143–145], however, this is not likely to be an important route of entry, even in aquatic plants. Although some herbicidal damage can occur following treatment of soil with glyphosate [e.g., Ref. 146], significant injury is not likely, probably due to the strong complexing capacity of soil for glyphosate. Furthermore, Hance [147] has suggested that roots have little capacity for absorption of glyphosate. Low, nonphytotoxic levels of glyphosate (50 μM) in hydroponic solution without iron can stimulate growth of wheat plants, presumably by complexing Zn^{2+} and Cu^{2+}, and allowing a more favorable ionic balance.

Foliar absorption of glyphosate is generally increased by any factor raising the water potential of the plant. Increased soil moisture [148,149] and high relative humidity [98,139,148,150,151] are reported to increase foilar glyphosate absorption. Plants with a higher water potential absorb more glyphosate than those that are more water stressed [150]. These factors may be involved by their contribution to cuticle hydration. A more hydrated cuticle more readily absorbs hydrophilic herbicides like glyphosate [152]. Also plants grown under high relative humidity have lower amounts of epicuticular wax in the cuticle [139]. Increased humidity results in slower drying of spray droplets, which can result in longer hydration of the droplet/cuticle interface for better penetration of a hydrophilic herbicide like glyphosate. Also, a higher water potential favors a more hydrated cuticle and absorption. Wiping the leaf cuticle with water- or chloroform-wetted paper will greatly increase absorption [153], presumably through cuticle removal. Wyrill and Burnside [104] explained tolerance of hemp dogbane to glyphosate to be due to decreased absorption because of a relatively thicker and waxier cuticle than is present in more susceptible species.

Plants grown under low light intensity apparently absorb more glyphosate than plants grown under high light intensities, because they have less epicuticular wax [139].

Once glyphosate has penetrated the cuticle, uptake by mesophyll cells is relatively much slower [154], perhaps because the negative charges of the cell wall and plasmalemma repel the anionic glyphosate [155]. Gougler and Geiger [156] concluded that uptake and loss of glyphosate from leaf cells is passive. Uptake was not affected by pH. These characteristics may explain why glyphosate is readily translocated.

Gougler and Geiger [156] have postulated that glyphosate accumulation and transport in the phloem is explained by the intermediate permeability mechanism [157,158]. This mechanism predicts that compounds toward which the plasmalemma has relatively low permeability are taken up slowly by phloem cells and that relatively little of the compound leaks from phloem to xylem before the compound is deposited in metabolic sink tissues by diffusion. More of the compound will diffuse into the phloem than into mesophyll cells because the diffusion gradient is higher because of the continual movement toward metabolic sinks. Such compounds are termed "ambimobile," and glyphosate anionic permeability characteristics are within the range of theoretical ambimobile compounds.

In fact, glyphosate is readily translocated from sites of foliar application to metabolic sink tissues, especially meristematic and storage tissues. Gougler and Geiger [156] found that translocation of radiolabeled sucrose and glyphosate applied to a sugar beet leaf is proportionally similar in virtually all parts of the plant (Fig. 5). McAllister and Haderlie [159] had similar results in studying the translocation distribution patterns of $^{14}CO_2$-derived assimilates and glyphosate in field-grown Canada thistle. In studies in which flow of ^{14}C assimilate was manipulated by stem girdling, leaf shading, and cytokinin application, only minor differences in glyphosate and assimilate translocation were found in tall morning glory [160]. These data indicate that anything that would affect assimilate production, movement, or partitioning would also affect glyphosate translocation. This conclusion is supported by the findings of Kells and Rieck [161] and Schultz and Burnside [105] that lighting conditions that favor photosynthesis result in greater translocation of glyphosate. Furthermore, photosynthetic inhibitors, applied prior to or at the time of glyphosate treatment, greatly reduce translocation [162]. Results of Gougler and Geiger [163] suggest that glyphosate stops its own translocation at the point at which it inhibits export of carbon from the glyphosate-treated leaf.

It follows from these findings that glyphosate transport should be greatest from application sites that are exporting more carbon and that glyphosate should accumulate in greater concentrations in the strongest metabolic "sinks." For this reason, storage organs, such as tubers, rhizomes, or stolons, accumulate relatively high levels of glyphosate [e.g., Refs. 150, 162, 164–168], making glyphosate an excellent herbicide for control of perennial weeds. In fact, glyphosate applied to one of several plants connected by perennating organs, such as rhizomes or stolons, can kill or injure other connected plants [169].

Translocation of glyphosate is both acropetal and basipetal from source leaves [e.g., Refs. 105, 144, 162], just as is export of carbon. Those sinks, both acropetal and basipetal, that are the most metabolically active (i.e., distal meristems) or are larger accumulate the highest concentrations of glyphosate [153,165,170,171]. Furthermore, the site of application can greatly influence translocation patterns [e.g., Ref. 105] because the metabolic sinks vary with photoassimilate source location.

Translocation of glyphosate generally is favored by high relative humidity [e.g., Refs. 148, 172, 173] and high water potential [148, 150,153,174], both factors which should increase photosynthesis and translocation of assimilate. In quackgrass, however, relative humidity had no effect on translocation [175]. Klevorn and Wyse [174] found that increased water availability increased both photosynthesis and glyphosate translocation to rhizome buds of quackgrass

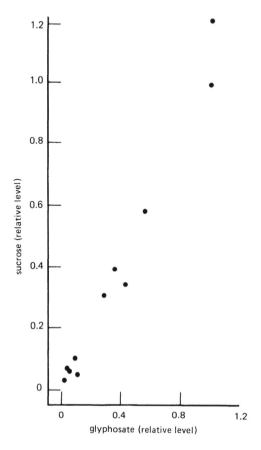

FIG. 5   Correlation between sucrose and glyphosate translocation in sugar beet (adapted from Ref. 156).

plants. Radiolabeled assimilate partitioning, however, did not correlate well with glyphosate translocation patterns.

The effect of temperature on translocation is more complicated,
probably because the optimum temperature for photosynthesis and
assimilate translocation varies considerably between species. Unfortunately, only one study [176] has been conducted in which dual
measurements of photosynthetic rates and translocation of assimilate
and glyphosate were made, as influenced by temperature. Translocation has been reported to be higher in hemp dogbane at 30°
than at 25°C [105] and in johnsongrass at 35° than at 24°C [148].
In soybean, however, translocation was dramatically higher at 24°
than at 35°C [148]. In quackgrass, there was no significant effect

of temperature (8–21°C) on translocation or photosynthesis [176]. Similarly, Coupland [175] found no significant effect of temperature on glyphosate translocation in quackgrass 48 hr after treatment; however, there were slightly significant increases in translocation at times earlier than this.

Soil temperature can also greatly influence translocation. Klevorn and Wyse [174] found that glyphosate was more readily translocated to rhizome buds of quackgrass plants with soil temperatures of 12° or 18°C than at 7°C. These results were probably primarily due to reduced metabolic activity of buds at the lower temperature.

In several species surfactant increases absorption but decreases translocation of glyphosate [139,148], possibly because the permeability of the mesophyll plasmalemma is increased. This could then lead to perferential mesophyll absorption, resulting in a more rapid cessation of assimilate production. A paradox in this study is that in some cases surfactant markedly increases toxicity of glyphosate. McWhorter et al. [148] speculated that a phytotoxic, nonradiolabeled metabolite of glyphosate or a phytotoxic secondary metabolite could have been translocated. In cotton, however, a surfactant increased translocation and phytotoxicity [173].

Potassium phosphate increased both translocation and phytotoxicity of glyphosate in soybeans [148]. This may be the result of phosphate competition for glyphosate anion binding sites, just as phosphate releases glyphosate from soil particles to facilitate degradation [111,122]. Phosphorus deficiency had no effect on glypohsate translocation in field bindweed, but nitrogen deficiency reduced acropetal transport while increasing basipetal movement [177]. Nitrogen deficiency may increase the relative activity of root meristems compared with shoot meristems.

Compounds with hormonal properties have been reported to influence glyphosate translocation greatly. The application of GAF 141 [(2-chloroethyl)phosphonic acid plus N-methylpyrrolidone], an experimental ethylene-releasing agent, to bean (Phaseolus vulgaris L.) plants 24 hr prior to glyphosate treatment inhibited acropetal translocation of glyphosate [178]. Both 2,4-D amine and 2,4-D ester greatly reduced glyphosate absorption and translocation when applied with glyphosate [103]. Exogenously applied cytokinin had no significant effect on translocation of glyphosate; however, it did slightly increase absorption [179].

Although, after foliar absorption, most glyphosate movement is considered to be symplastic, enough apoplastic movement occurs to consider it an ambimobile compound rather than a phloem-mobile compound [160]. In comparison with sucrose, there is slightly more apoplastic movement of glyphosate [180]. Studies in which phloem movement has been blocked by girdling have shown that glyphosate moves more readily in the apoplast than ^{14}C from photosynthate

[160]. Similarly, Klevorn and Wyse [167] found significant amounts
of glyphosate to be apoplastically transported in girdled leaves,
culms, and rhizomes of quackgrass. In this study apoplastic transport was greatest in water-stressed plants. Injection of glyphosate
into stems [160] or introduction through cut stems [155,181] has
demonstrated apoplastic movement into transpiring tissues as well as
considerable symplastic movement. A major criticism of studies involving girdling, stem injections, or cut stems is that such anatomical-
ly altered plants cannot accurately reflect movement of foliarly applied glyphosate in an intact plant. Certainly the amount of apoplastic translocation in plants in which phloem movement has been
blocked will be overestimated.

In summary, glyphosate penetrates the cuticle of most species
to enter the plant apoplast. It then slowly enters the symplast
passively and is translocated primarily via the phloem to metabolic
sinks. Since it is essentially not metabolized in higher plants,
relatively high levels can accumulate in meristematic areas at relatively long distances from the site(s) of application. Any factor,
plant-related or environmental, that influences phloem transport
from the site of application to other parts of the plant will similarly
influence translocation of glyphosate.

B. Enzymatic Studies of Amino Acid Synthesis

Glyphosate is one of the few herbicides for which a specific site
of action at the molecular level has been identified [182]. Glyphosate inhibits the activity of 5-enolpyruvyl shikimic acid-3-phosphate
[EPSP] synthase, an enzyme of the shikimic acid pathway (Fig. 6),
in a wide variety of plants and organisms (Table 7).

Inhibition is competitive with respect to phosphoenolpyruvate
(PEP) with a K_i near 1 µM in both higher plants and microorganisms
[183,195,196,199,204]. Shikimate-3-phosphate uncompetitively inhibits EPSP synthase with a K_i around 10 µM [204].

There is no inhibition of EPSP synthase by nonherbicidal analogues of glyphosate, such as aminomethylphosphinic acid, bis-N-
(phosphonomethyl)glycine, and iminodiacetic acid [183]. No other
PEP-utilizing enzymes other than 3-deoxy-D-arabino-heptulosonate-
7-phosphate (DAHP) synthase have been found to be inhibited by
glyphosate (Table 7). The inhibition of EPSP synthase increases
markedly with pH [183], indicating a possible weak acid at the
glyphosate-binding region.

Glyphosate inhibits several other enzymes of the shikimic acid
pathway to a lesser degree than EPSP synthase (Table 6). These
include DAHP synthase, dehydroquinate synthase, and in *E. coli*,
anthranilate synthase.

Inhibition of EPSP synthase by micromolar levels of glyphosate
causes the accumulation of high levels of shikimate in higher plants

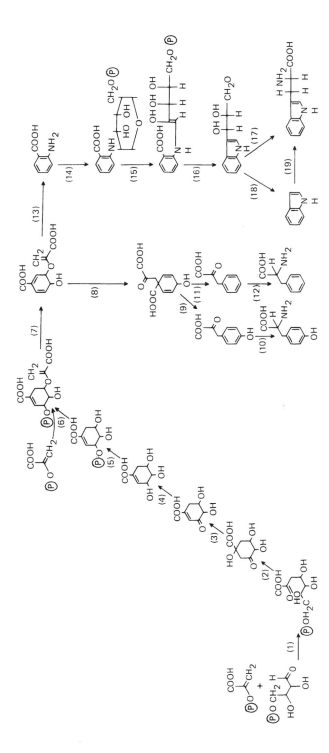

FIG. 6 The shikimic acid pathway. The numbers refer to the enzymes listed in Table 7. Reproduced from *The Herbicide Glyphosate* [3] by permission of the publishers, Butterworths & Co., Ltd.

TABLE 7

Effects of Glyphosate on Enzymes of the Shikimic Acid
Pathway and Aromatic Amino Acid Synthesis[a]

Enzyme	Source
(1.) 3-Deoxy-*D*-arabino-heptulosonate-7-phosphate synthase (EC 4.1.2.15)	*Klebsiella pneumoniae* *E. coli* Mung bean (*Vigna radiata*) (L.) Wilczek *Candida maltosa* 18 species of ascomycetous and basidiomycetous yeasts
(2.) 3-Dehydroquinate synthase (EC 4.6.1.3)	*E. coli* *C. maltosa*
(3.) 3-Dehydroquinate dehydratase (EC 4.2.1.10)	*C. maltosa*
(4.) Shikimate dehydrogenase (EC 1.1.1.25)	Wheat Soybean *C. maltosa*
(5.) Shikimate kinase (EC 2.7.1.71)	*K. pneumoniae* *C. maltosa*
(6.) 5-Enolpyruvylshikimic acid-3-phosphate synthase (EC 2.5.1.19)	*K. pneumoniae* *E. coli* Mung bean Pea Tobacco *Neurospora crassa* *Corydalis sempervirens* (L.) Pers. *C. maltosa*

Effect on extractable activity	Effect on in vitro activity	References
—	None	183
Elevated	Weak inhibition	124, 184
—	Weak inhibition	185
None	Weak inhibition	186, 187, 188, 189
—	Weak inhibition	189
None	Weak inhibition	124, 184
Slight elevation	Weak inhibition	186, 187
Slight elevation	None	186
Elevated	—	190
None	None	191
Slight elevation	None	186
—	None	192
Slight elevation	None	186
—	Strong inhibition	192, 193 194
—	Strong inhibition	192, 195
—	Strong inhibition	192
—	Strong inhibition	196, 197
—	Strong inhibition	198
—	Strong inhibition	199
—	Strong inhibition	200, 201
Slight elevation	Strong inhibition	184

TABLE 7

(continued)

Enzyme	Source
	Salmonella typhimurium
	Bacillis subtilis
	Streptococcus lactis
	Lycopersicon esculentum
	Chlamydomonas reinhardii
	Azotobacter vinelandii
	Pseudomonas fluorescens
	Seratia marcescens
(7.) Chorismate synthase (EC 4.6.1.4)	*K. pneumoniae*
(8.) Chorismate mutase (EC 5.4.99.4)	Wheat
	C. maltosa
	E. coli
(9.) Prephenate dehydrogenase (EC 1.3.1.13)	*C. maltosa*
	E. coli
(10.) Tyrosine aminotransferase (EC 2.6.1.5)	
(11.) Prephenate dehydratase (EC 4.2.1.51)	*C. maltosa*
	E. coli
(12.) Phenylalanine aminotransferase (EC 2.6.1.58)	*C. maltosa*
(13.) Anthranilate synthase (EC 4.1.3.27)	*C. maltosa*
	K. pneumoniae
	E. coli

Effect on extractable activity	Effect on in vitro activity	References
—	Strong inhibition	202
—	Strong inhibition	203
—	Strong inhibition	203
—	Strong inhibition	203
—	Strong inhibition	203
—	Strong inhibition	203
—	Strong inhibition	203
—	Strong inhibition	203
—	None	192
Elevated	—	190
Doubled	None	186
Elevated	None	124, 184
Slight decrease	None	186
Elevated	None	124, 184
—	—	—
None	None	186
Elevated	None	124, 184
Doubled	None	186
Doubled	None	186
—	None	192
Elevated	Weak inhibition	184

TABLE 7

(continued)

Enzyme	Source
(14.) Anthranilate phosphoribosyltransferase (EC 2.4.2.18)	*C. maltosa*
(15.) Phosphoribosyl anthranilate isomerase (EC 4.1.1.48)	*C. maltosa*
(16.) Tryptophan synthase (EC 4.2.1.20)	*C. maltosa*

[a] See Fig. 7 for sites of numbered enzymes in the shikimic acid pathway.

[205,206]. Microorganisms excrete and accumulate both shikimate and shikimate-3-phosphate [203,207]. In higher plants, accumulation of shikimate is primarily in the vacuole [208].

This blockage of the shikimic acid pathway leads to a depletion of the free pool of aromatic acids [181,209–214]. Furthermore, glyphosate greatly lessens increases in phenylalanine levels caused by inhibitors of the enzyme phenylalanine ammonia-lyase (PAL) [215, 216]. Reduction of the free pool level of a substrate does not always signal a concomitant reduction in the rate of flow through that pool. For instance, light causes a reduction in aromatic amino acid pools while increasing the flow rate through the pool [217]. But, when a pool size decrease can be correlated with inhibition of an enzyme responsible for synthesis of the pool, as aromatic amino acid snythesis is dependent on EPSP synthase, one can assume that depletion of the pool is coupled with a much lower flow rate through the pool. Thus, the supply of aromatic amino acids for protein synthesis should be severely limited in glyphosate-treated tissues. This ultimately leads to a slow cessation of growth, just as is observed.

The effects of glyphosate on protein synthesis are poorly documented. Cole et al. [190] found that radiolabeled leucine incorporation into rhizome buds of quackgrass was inhibited by glyphosate; however, incorporation of [14C]phenylalanine was affected to a much lesser extent. The data indicated that inhibition of protein synthesis by glyphosate was due to depletion of the aromatic amino acid pool rather than to direct or indirect interference with or damage to

Effect on extractable activity	Effect on in vitro activity	References
Nearly doubled	None	186
None	None	186
Slight increase	None	186

the protein-synthesizing process.   Others [219] have not found a strong effect of glyphosate on leucine incorporation into protein and no differential effect of glyphosate on incorporation of leucine versus phenylalanine into protein.   Brecke and Duke [154] found that glyphosate inhibited leucine incorporation into protein of isolated leaf cells; however, the effect could be accounted for by inhibition of uptake of labeled leucine.   Duke and Hoagland [219] found no significant effect of glyphosate on incorporation of ^{14}C-labeled tyrosine and phenylalanine into protein of roots of soybean seedlings after 24 hr of treatment.   Significant reduction occurred, however, after 48 hr, a point at which secondary effects may have influenced protein synthesis.   Glyphosate generally has little effect on total soluble protein levels [e.g., Refs. 220, 221].

An anomaly that is very difficult to explain is that glyphosate causes increases in the activity of (PAL) over several days (Fig. 7) [209,210,220,221].   Induction of PAL activity by other inducers has been shown to be due to de novo synthesis [e.g., Refs. 222, 223], and the half-life of PAL in various tissues ranges from 3 to 10 hr [224,225].   Thus, the prolonged and continual increase in PAL activity of glyphosate-treated tissues almost certainly represents increased synthesis of this enzyme.   This conclusion is difficult to reconcile with glyphosate strongly inhibiting protein synthesis.   A possible explanation is that, at herbicidal levels, glyphosate stops protein synthesis in meristematic tissues where it accumulates, but has other, sublethal effects in the majority of tissues where it does not accumulate.

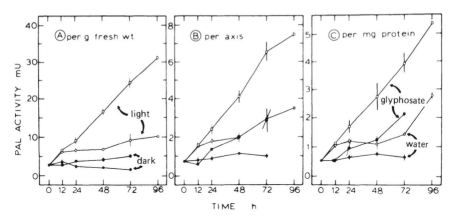

FIG. 7   Effect of glyphosate on extractable PAL activity from
soybean seedlings.   Reproduced from Ref. 210, with permission.

Glyphosate affects amino acid metabolism of nonaromatic amino
acids also; but, no primary effects have been discovered.   Glyphosate
and amino-oxyacetic acid, a potent transaminase inhibitor [225], had
similar effects on nonaromatic amino acid profiles [227]; however, not
all amino acids were affected similarly by both compounds.   A buildup of glutamine and glutamate in glyphosate-treated wheat plants led
Nilsson [211] to suggest that glyphosate is a transaminase inhibitor;
however, no systematic survey of the effects of glyphosate on transaminase activities has been made.   Kilmer et al. [228] suggested that
glutamate buildup in glyphosate-treated cells is the result of un-
regulated flow of substrates for ammonia assimilation into shikimate.
In their studies tricarboxylic acid cycle intermediates, particularly
α-ketoglutarate, succinate, and malate, reversed the inhibitory effects of glyphosate on growth.   Glutamate and aspartate were also
effective as glyphosate antidotes in their system, suggesting that
glyphosate's phytotoxicity may not be due entirely to inhibition of
aromatic amino acid synthesis.   But, at least part of the reversal by
amino acids and TCA cycle intermediates may have been due to inhibition of glyphosate uptake [229].
    Supplying higher plant cell cultures or microorganisms with
exogenous aromatic amino acids has usually alleviated the growth-
retarding effects of glyphosate [124,184,204,205,207,228–232]; however, in a few cases [206,218,233] reversal has not been obtained
or has been quite limited.   In intact higher plants reversal of glyphosate effects with exogenous aromatic amino acids has not been
found in several studies [190,219,220]; however, it has been found
in *Arabidopsis* [231,234] and to a limited extent in Kentucky bluegrass [235].   In glyphosate-treated tissues in which free pools of

aromatic amino acids have been substantially increased, either by feeding them exogenously [219] or by treating with the potent PAL inhibitor, α-amino-oxy-β-phenylpropionic acid (AOPP) [216], glyphosate had about 90% of its growth-inhibiting property. In the feeding studies [219], the aromatic amino acids were readily taken up and incorporated into protein and secondary compounds. Whether the aromatic amino acids accumulated in meristematic areas was not determined, however.

To date, no naturally occurring ecotypes of induced mutants of higher plants that are resistant to glyphosate by virtue of a resistant EPSP synthase have been identified; however, tissue cultures of higher plants [236,237] and bacterial strains [236,238] with elevated EPSP synthase levels are tolerant or resistant to glyphosate. In fact, a strain of cultured cells of the higher plant *Corydalis sempervirens* Pers. is about two orders of magnitude less sensitive to glyphosate than wild-type strains due to a nearly 40-fold increased level of EPSP synthase [204,239]. Similarly, strains of *Petunia hybrida* cell cultures with 15- to 20-fold higher levels of EPSP synthase are about two orders of magnitude less sensitive to glyphosate than a sensitive strain [240]. Strains of *Klebsiella pneumoniae, Pseudomonas* spp., and *Salmonella typhimurium* with glyphosate-insensitive EPSP synthase are insensitive (resistant) to glyphosate [202,203,241–243]. The sensitivity of the EPSP synthase from these resistant strains is two to three orders of magnitude less than in susceptible strains and more than three orders of magnitude less sensitive than higher plant EPSP synthase [203] (Fig. 8). *Escherichia coli* that has had the glyphosate resistance *aro A* allele from a mutagenized *S. typhimurium* strain transferred to it is resistant to glyphosate [202,242]. The same *aro A* gene has been introduced into tobacco plants via *Agrobacterium rhizogenes* T-DNA vectors, producing plants with as much as 50% of the EPSP synthase being resistant [244]. These plants (Fig. 9) are as much as sevenfold less sensitive than untransformed plants, despite having an enzyme that is about 100-fold less sensitive to glyphosate than that of the control [242]. Production of glyphosate-tolerant and resistant plants by these methods has been patented [245]. If EPSP synthase is the only important site of glyphosate action, these results can only be explained in two ways: (a) EPSP synthase is a very limiting enzyme, so that inhibition of half of the enzyme has a pronounced herbicidal effect; or (b) the resistant EPSP synthase has not been introduced into the proper cellular compartment. The latter possibility seems most likely, since EPSP synthase is predominantly, if not entirely, a plastid (e.g., chloroplast) enzyme [197,246,247] in higher plants, and no proof that the gene has been introduced into the plastid has been produced. Movement of proteins from the cytoplasm into the plastid often requires special amino acid sequences [248], which would not be expected on a *S. typhimurium* EPSP synthase.

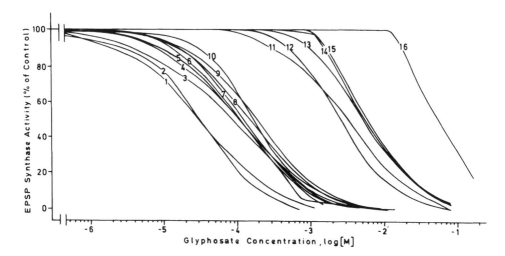

FIG. 8   Activity of EPSP synthases from bacteria and plants as a
function of glyphosate concentration. Data are expressed as %
activity in the absence of glyphosate.   Employed activities ranged
between 8 and 12 pkat per assay.   The numbers indicate the origins
of the enzymes: (1) *Lycopersicon esculentum*; (2) *Corydalis
sempervirens*; (3) *Streptococcus lactis*; (4) *Klebsiella pneumoniae*;
(5) *Bacillus subtilis*; (6) *Chlamydomonas reinhardii*; (7) *Azotobacter
vinelandii*; (8) *Escherichia coli*; (9) *Pseudomonas fluorescens*; (10)
Serratia marcescens; (11) *Pseudomonas maltophilia*; (12) *Pseudomonas
putida*; (13) *Pseudomonas aeruginosa*, W1606; (14) *Pseudomonas
aeruginosa*, H103; (15) *Pseudomonas* sp. ATCC15926; (16) *Klebsiella
pneumoniae* (glyphosate-resistant).   Reproduced from Ref. 203 with
permission.

Production of a higher plant with a glyphosate-resistant EPSP synthase in the plastid should resolve the question of whether this
enzyme is the only important site of action of glyphosate.

C.   Other Physiological and Biochemical Effects

1.   Introduction

The only established direct effects of glyphosate on higher plants
at the molecular level are on the enzymes of the shikimate pathway,
primarily EPSP synthase.   A large number of effects of glyphosate
on other physiological parameters have been reported; however, none
of them have been established as direct effects at the molecular level.

Nevertheless, some of them may be direct effects and many of those that are not primary effects are important in the development of herbicidal symptoms.

2.   Photosynthesis and Cloroplast Development

Glyphosate is not considered to have any direct effect on photosynthesis.   Although it was reported that glyphosate inhibits PS II electron transport in spinach chloroplasts and O_2 evolution in *Scenedesmus* [45], Richard et al. [46] found that these earlier results were probably due to an uncorrected pH effect.   Munoz–Rueda et al. [249] later found glyphosate to inhibit PSII and PSI activities of isolated chloroplasts at a concentration of 250 μM, but, again, there was not proof that pH effects were eliminated.   Glyphosate had no effect on postluminescence decay or chlorophyll fluorescence for

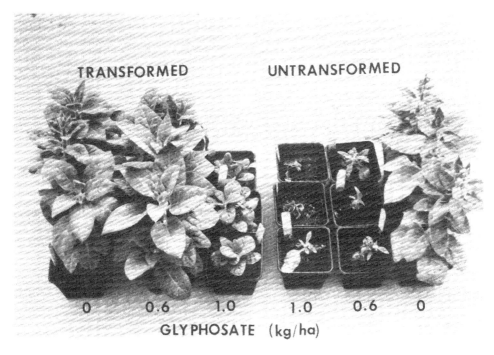

FIG. 9   Glyphosate effects on transformed, glyphosate-tolerant tobacco plants with the EPSP synthase from glyphosate-resistant *S. typhimurium* compared with untransformed tobacco.   The plants were photographed 40 days after being sprayed with 0.6 or 1.0 kg/ha of glyphosate.   Courtesy of Luca Comai of Calgene.

up to 3 days after treatment of wheat or cuckoo flower (*Cardamine pratensis* L.) [250]. Shaner and Lyon [25] and Sprankle et al. [162] found no effects on photosynthesis until 6 and 72 hr, respectively, after treatment with glyphosate. Similarly, Geiger et al. [252] found that reduction of $^{14}CO_2$ fixation by glyphosate was largely due to reduction in stomatal conductance. In the former case, the effect on photosynthesis could be accounted for by effects on stomatal gas exchange. Brecke and Duke [154] found assimilation of $^{14}CO_2$ by bean leaf cells to be more sensitive than protein or RNA synthesis. This is not surprising in that the only known primary site of action is in the plastid, and metabolic disturbances at this intracellular location might be expected to occur at this location before other effects would be observed.

Although there are no pronounced effects of glyphosate on photosynthesis, glyphosate does have relatively strong effects on chlorophyll synthesis [249,253–255]. Glyphosate inhibits the synthesis of the chlorophyll precursor, 5-aminolevulinic acid (ALA) [254]. Incorporation of carbon from glutamate, 2-ketoglutarate, and glycine into ALA in vitro is inhibited; however, incorporation of ALA into chlorophyll is unaffected [256]. The pronounced reduction in catalase content in glyphosate-treated purple nutsedge [257] may also be due to inhibition of ALA synthesis. Both chlorophyll and catalase have porphyrin constituents that are derived from ALA. Other compounds that form especially stable complexes with ferrous iron have been shown to inhibit ALA synthesis [258]. Glyphosate forms an especially stable complex with ferrous iron in vivo [88]. Since glyphosate toxicity can be partially alleviated with 2-ketoglutarate [228], an ALA precursor, inhibition of ALA and porphyrin synthesis may be an important aspect of glyphosate mechanism of action [259]. Glyphosate-inhibited chlorophyll synthesis in tobacco callus was not reversed with aromatic amino acids [260]. Nevertheless, effects on light-induced anthocyanin synthesis, a phenylalanine-dependent process, are about tenfold more sensitive than is the greening process to glyphosate [212].

Although glyphosate has fairly pronounced effects in chloroplast development, its effects on chlorophyll content of green tissues are much less apparent. These effects are likely to be caused by photobleaching as a secondary effect of altered chloroplast function. In both tobacco and soybean discs, Lee [260] found glyphosate to accelerate photodegradation of chlorophyll, particularly in young leaves. Although Ali and Fletcher [255] found chlorophyll loss not to be accompanied by carotenoid loss, Abu–Irmaileh and Jordan [257] found chlorophyll loss to be preceded by carotenoid losses.

3. Nitrate Assimilation

Nitrate reductase, nitrite reductase, and glutamate synthetase are all required for assimilation of nitrogen from nitrate into amino

acids.   Inhibition of nitrite reductase or glutamate synthetase without a concomitant inhibition of nitrate reductase can lead to accumulation of toxic levels of nitrite [261] and(or) ammonia [262].   Glyphosate had little effect on light induction of extractable nitrate
reductase in greening pea explants or in nitrate-induced nitrate
reductase activity in pregreened pea explants [253].   Similarly,
Hoagland [263] found little effect of glyphosate on development of
nitrate reductase in greening soybean cotyledons.   Development of
nitrite reductase activity in greening or pregreened pea tissues; however, was strongly inhibited by glyphosate [253].   Nitrite reductase
is a plastid enzyme and the putative site of glyphosate action, EPSP
synthase, is in the plastid.   Thus, one might expect nitrite reductase
activity to be more sensitive than nitrate reductase activity to the
herbicide.   Furthermore, it is common for herbicides that inhibit
photosynthesis to halt nitrite reductase, while having little effect or
even increasing nitrate reductase activity [264].

An effect of glyphosate on glutamate synthetase has not been
reported in the literature; however, ammonia accumulation, a symptom
of glutamate synthesis inhibition, has been reported [209,221,228].
This ammonia accumulation has been suggested as a factor in glyphosate toxicity [209,228]. Phosphinothricin (DL-homoalanin-4yl(methyl)-
phosphinic acid) is a structural analogue of glyphosate and is a
potent inhibitor of glutamine synthesis, causing buildup of toxic
ammonia levels in plant tissues [18,262].

4.   Respiration

In intact tissues respiration is apparently not an early target
of glyphosate.   Sprankle et al. [162] found respiration to be less
sensitive to glyphosate than photosynthesis in quackgrass plants
and to be unaffected in wheat plants.   Similarly, Cole et al. [253]
found respiration of flax cotyledons to be insensitive to glyphosate,
even at high concentrations, while photosynthesis was markedly affected.   Another study found glyphosate to increase carbon dioxide
production by bean plants to levels tenfold those of control plants
[265].   In root tips of maize [255] and wheat [190] respiration as
measured by tetrazolium reduction was reduced by glyphosate within
6 and 24 hr, respectively.   No effect of glyphosate was measured
on O_2 uptake by isolated bean mesophyll cells; however, there was
some inhibition of liberation of $^{14}CO_2$ from labeled glucose [154].

Glyphosate has been reported to uncouple oxidative phosphorylation in isolated plant [43] and mammalian [44] mitochondria.   These
results, however, were later disputed [47] when they could not be
reproduced.   It was concluded that the results of the earlier papers
were due to impurities of the technical product or to the isopropylamine moiety counterion of the salt.   It was found, however, that
the potassium salt of glyphosate will inhibit protein synthesis, oxygen

uptake (state 3 and state 4 respiration), and passive swelling of maize mitochondria at 10, 100, and 10 µM, respectively [47]. Furthermore, mitochondrial accumulation of calcium in vivo is inhibited by glyphosate [266]. These data suggest that inhibition of respiration could be an important effect of glyphosate, contrary to studies with intact tissues. Whether respiration is rapidly and significantly affected is probably a function of how much glyphosate gets to the mitochondria in vivo. Thus, effects on respiration may be important in apical meristems, but of minor importance in other tissues. The fact that glyphosate is slow acting and has little mammalian toxicity would suggest; however, that respiratory effects play only a minor role in the mechanism of action or that respiratory sites affected are unique to plants.

5. Macromolecule Synthesis

No commercial herbicides are known to directly affect cytoplasmic protein synthesis [182]; however, any effective herbicide will eventually affect this process. Effects of glyphosate on protein synthesis are all indirect and are discussed in detail in the section on amino acid metabolism (Section III. B). Although the effects on protein synthesis are indirect, they are perhaps the second important event in the development of herbicidal damage. This strong effect on protein synthesis may be especially important when glyphosate is used as a harvest aid, in that glyphosate can have profound effects on protein quality and quantity of harvested crops [e.g., Ref. 267].

Similarly, no direct effects of glyphosate on nucleic acid synthesis have been found. Haderlie et al. [218] found glyphosate to slowly inhibit RNA synthesis in carrot cells, but found that inhibition of DNA synthesis was due to inhibition of thymidine uptake. Glyphosate inhibited incorporation of ^{32}P into RNA and DNA of cocklebur root tips within 8 hr [268]. This effect coincided with an increase in ATP. Since there was no reduction in ATP and no evidence of a direct effect of glyphosate on production of nucleic acid precursors, it seems likely that the effects of nucleic acid synthesis are due to a general disruption of cellular function associated with inhibited protein synthesis.

6. Membrane Transport and Ion Movement

It has been known for many years that phosphonic acids, such as glyphosate, are chelators of metal ions [3,269,270]. As discussed earlier, there is evidence that metal ions influence absorption and translocation of glyphosate, as well as activity of glyphosate in the plant. Little effort, however, has gone into determining the effects of glyphosate on movement of metal ions in plant tissues.

Absorption of ^{86}Rb and ^{32}P by isolated bean cells was inhibited before effects on photosynthesis, respiration, or RNA and protein synthesis could be detected [154]. The effect was not due to loss of membrane integrity, decrease in energy supply, or external ion chelation.

Absorption of ^{32}P as orthophosphoric acid was unaffected by glyphosate in cocklebur roots; however, translocation was greatly reduced [271]. Niklicek et al. [272] found glyphosate to greatly inhibit both absorption and translocation of ^{32}P by intact quackgrass plants. The effect on absorption of ^{45}Ca was less; however, there was a marked effect on translocation of ^{45}Ca. The latter results were similar to those of Duke et al. [273], who found glyphosate to greatly retard calcium translocation from roots of soybean seedlings (Fig. 10). Further studies revealed that glyphosate reduced both Ca^{2+} and Mg^{2+} uptake and translocation [266]. Cells produced after glyphosate treatment were particularly deficient in calcium, as revealed by histochemical methods (Fig. 11). Furthermore, the intracellular distribution of Ca^{2+} at the electron microscope level was altered. No effects on K^+ were discerned.

Thus far, there is no clear evidence of whether or not the effect of glyphosate on ion uptake is due to direct interaction of the glyphosate anion with the metal cation. The finding that chlorsulfuron, another amino acid synthesis inhibitor, greatly inhibits ^{45}Ca absorption [274] indicates that at least part of the effect of glyphosate is secondary.

Nilsson [88] found glyphosate to stimulate accumulation of $^{59}Fe^{3+}$ in roots and leaves of wheat plants while inhibiting movement of $^{65}Zn^{2+}$ to the same sites. This finding was supported by the discovery that low levels of glyphosate will stimulate growth of iron-deficient wheat plants, presumably by preventing movement of zinc to produce a more favorable zinc–iron balance.

It is well established that glyphosate has no significant effects on permeability of the plasma membrane [154, 275–278]. Although Watson et al. [275] reported 1 mM glyphosate to slightly stimulate plasmalemma ATPase activity, Cole et al. [253] found no in vitro effects of glyphosate on microsomal adenosine triphosphatase (ATPase) ± K^+ at either pH 6.0 or 9.0, although the extractable activity of pH 9, + K^+ ATPase was increased by treatment of wheat root tips with glyphosate.

7. Secondary Metabolism

Most phenolic secondary compounds are derived from aromatic amino acids, and synthesis of aromatic amino acids is the only confirmed site of action of glyphosate. Thus, the finding that glyphosate has profound effects on secondary compound metabolism is not surprising [214, 279]. The effects of glyphosate on aromatic amino acid-derived phenolic compound synthesis is twofold: (a) accumulation of phenolic compounds that are derivatives of aromatic

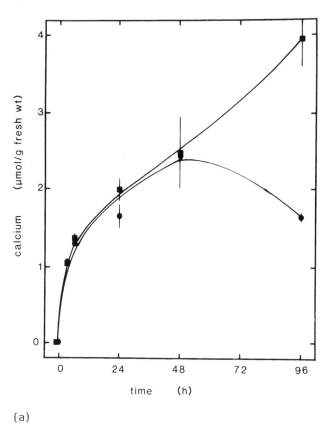

(a)

FIG. 10   Calcium absorption and translocation as measured by
^{45}Ca content of roots (a), hypocotyls (b), and cotyledons (c) of
soybean seedlings exposed to 4 mM Ca^{2+} with (squares) or without
(circles) 0.5 mM glyphosate.   Reproduced from Ref. 273, with
permission.

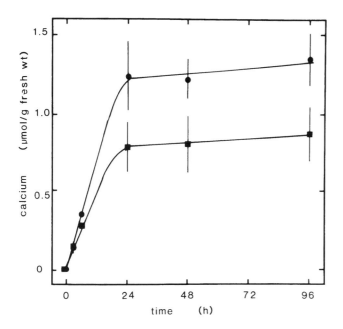

(b)

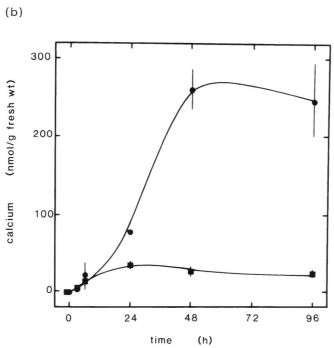

(c)

FIG. 10   (continued)

49

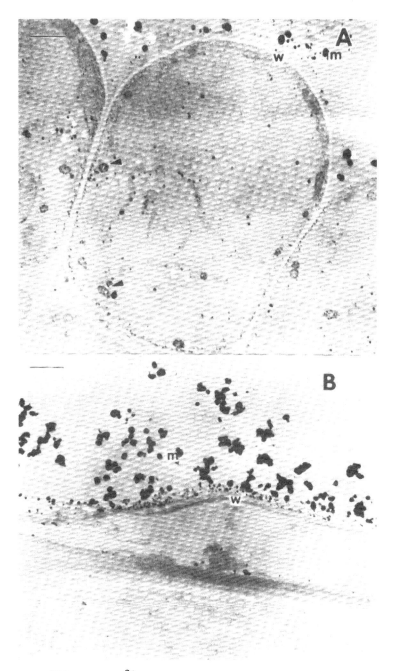

FIG. 11 Ca^{2+} localization in secondary root cells of control (A) and glyphosate-treated (B) soybean seedlings. Samples were taken after 72 hr of the foliar treatment. m = mucilage layer, w = cell wall, arrows indicate mitochondria. Bars = 1.0 μm. (Micrographs courtesy of K. C. Vaughn.)

amino acids is reduced and (b) pools of phenolic compounds derived
from constituents of the shikimate pathway prior to the step catalyzed by EPSP synthase become larger.

The first of these effects is well documented (Table 8). Levels
of all aromatic amino acid-derived secondary compounds that have
been examined are reduced by glyphosate. This reduction is usually
accompanied by an increase in extractable phenylalanine ammonia-
lyase (PAL) activity (Fig. 7, Refs. 190, 210, 214, 216, 220, 291).
In some tissue, however, there is no effect [212] or a decrease [282]
in PAL activity. PAL deaminates phenylalanine to form cinnamic
acid, the first step in production of secondary aromatic compounds
from phenylalanine. The increase in PAL activity is thought to be
the result of alleviation of end-product repression of synthesis, just
as enhancement of extractable PAL activity by the PAL inhibitor
[216,292], α-amino-oxy-β-phenylpropionic acid, is due to this mechanism [293]. There is often a good correlation between induction of
PAL activity, increases in aromatic amino acids, and loss of phenolic
compounds in glyphosate-treated tissues [214,294].

In buckwheat, a species in which PAL/TAL activity is not increased by glyphosate, incorporation of labeled tyrosine into flavonoids is reduced by glyphosate [284]. However, the relative proportion of labeled tyrosine incorporated in flavonoids is not affected,
except at 10 mM, in which case the proportion is increased. In
this study, however, the effect of glyphosate on uptake of tyrosine
was not reported. Duke and Hoagland [219] reported similar effects
of glyphosate on incorporation of labeled phenylalanine and tyrosine
into ethanol-soluble components of soybean seedlings.

Increases in shikimate caused by glyphosate are well documented
[e.g., Refs. 200, 201, 204, 208, 252, 282, 295]; however, increases
in aromatic amino acid precursor-derived benzoic acids are less well
documented (Table 8) [200]. The increased carbon incorporated into shikimate in glyphosate-treated tissues is only a small fraction of
the decrease in carbon incorporated into starch [252].

The implications of glyphosate's effects on secondary metabolism
are not fully apparent. Glyphosate's effect on phenolic metabolism
is unique among herbicides [279,296,297]. Alteration of secondary
compound production could change the interaction of plants with
other organisms. For instance, glyphosate-caused reduction of
glyceollin in soybean [286–289] and medicarpin in lucerne [290]
reduced these species' resistance to fungal pathogens. In greenhouse studies, soil-borne phytopathogenic fungi contribute significantly to the herbicidal activity of glyphosate [298], probably because of glyphosate's lowering resistance to pathogens. Whether
similar effects occur in the field has not been determined. Increases
in shikimate and benzoic acids could also be important under some
circumstances. Shikimate is normally present in only trace amounts
in most plant species, although it can account for as much as 2.5%
of the dry weight in some conifers [299]. The huge increases in

TABLE 8

Effects of Glyphosate on Specific Phenolic Compounds or Phenolic Compound Groups

Compound or compound class	Tissue	Glyphosate concentration (μM)	Duration of treatment (days)	Effect (% difference from control)	References
Anthocyanin	Buckwheat cotyledon	10	1	−50	212
Anthocyanin	Buckwheat seedlings	500	?	−50	280
Anthocyanin	Soybean hypocotyl	500	3	−50	281
Rutin	Buckwheat hypocotyls	1000	1	−93	212
Caffeic acid	Perilla, cell suspension	1000	5	−50	282, 283

Chlorogenic acid	Buckwheat hypocotyls	1000	1	−93	212
Flavonoids	Buckwheat hypocotyls	100	2	−70	284
Glyceollin	Soybean hypocotyl	500	3	−50	285
	Soybean hypocotyl	90	0.5	−15	286, 287
	Soybean leaves	60	3	−80	288, 289
Medicarpin	*Medicago sativa* L. callus cultures	100	2	−70	290
Cinnamoyl putrecines	Tobacco cell cultures	5000	3	−25	206
Gallic acid	*Quercus robur* L. cultures	?	?	+400	200

shikimate caused by glyphosate are potentially important in that shikimate has been reported to be the carcinogenic and mutagenic principle of bracken fern [300,301]. More recently; however, shikimic acid was shown not to be mutagenic in the Ames assay when tested with and without the rat liver microsomal activation system [302]. Nevertheless, shikimate accumulation could be a problem in forage treated with glyphosate [e.g., Ref. 303].

8. Ultrastructural Effects

Despite the tremendous importance of glyphosate as a herbicide, until recently only three ultrastructural studies had been published [304–306]. Unfortunately, in these studies, only effects on mature, photosynthetic tissues were observed. Such tissues are clearly not the initial site of action of glyphosate and any ultrastructural effects observed are certainly far removed spatially and temporally from the site of action.

Effects seen in quackgrass mesophyll cells included disruption of the chloroplast envelope, accumulation of plastoglobuli, swelling of the rough endoplasmic reticulum vesicle formation, wrinkling and detachment of the plasmalemma, and mitochondrial swelling or degeneration [304]. There was no obvious initial site of cellular disruption. In *Sinapis alba*, disruption of chloroplast thylakoids occurred at 2 days, followed by mitochondrial damage at 14 days [306]. Reticulate chloroplast envelopes (peripheral reticula) were caused by glyphosate in the thallus of the liverwort (*Pellia epiphylla*) within 1 day of treatment [305]. Further disruption was characterized by deterioration of grana, detachment of the plasmalemma from the cell wall, and lengthening of mitochondrial cristae.

Recent published studies of ultrastructural effects of glyphosate have focused on root and meristematic tissues [266,307]. Duke et al. [266] found disruption of intercellular calcium localization by histochemical means before ultrastructural abnormalities could be seen in root tissue of soybean seedlings. Meristematic areas of glyphosate-treated soybean roots had symptoms of microtubule loss, cells with heavily lobed nuclei or micronuclei, and cells with arrested stages of division (Fig. 12). Less tubulin was detected by immunochemical methods in glyphosate-treated than in control tissues. Some recovery of microtubule loss during fixation could be obtained by treatment with the strong calcium chelator, EGTA. It was concluded that the abnormal mitosis in glyphosate-treated tissues results from indirect processes, such as inhibition of protein synthesis and effects on calcium.

9. Hormonal Effects

Glyphosate causes a relatively rapid decrease in levels of indole-3-acetic acid (IAA) in plant tissues [308,309]. At least some of the

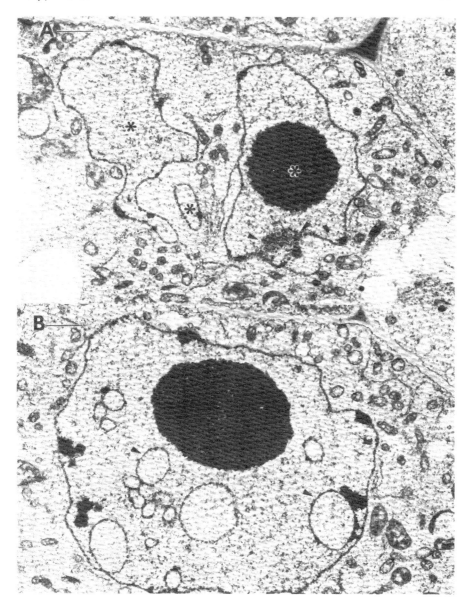

FIG. 12   Abnormal nuclei of glyphosate-treated root tissues. (A)
Lobed nucleus with deep lobes which appear as micronuclei (asterisks).
Arrows mark an area of the nuclear pore.   (B) Abnormal nucleus
with vesicles (arrows) of nuclear membrane.   Bars = 1.0 μM. (Micrographs courtesy of K. C. Vaughn.)

growth effects of glyphosate were due to this reduction in IAA, because exongenous IAA will partially reverse the effects of glyphosate on growth [233]. Loss of IAA in glyphosate-treated tissues is due to both conjugation and oxidation of IAA [308-310]. IAA oxidase, the major enzyme involved in degradation of IAA, is a form of peroxidase. Peroxidases are strongly influenced by phenolic compounds. Thus, the effect of glyphosate on IAA may be an indirect effect of its effects on secondary phenolic compound production.

Plant species with normally high rates of IAA metabolism are more tolerant to glyphosate effects on growth [311], presumably because they are less dependent on IAA. The often observed release of lateral bud dormancy by glyphosate [312-314] is due to drastic alteration of IAA/cytokinin balances [312,314-316].

In addition to altering IAA metabolism, glyphosate reduces IAA transport [316]. This effect was speculated to be due to increased ethylene synthesis. Ethylene production has been reported to be increased [265], little affected [253,317], or decreased [309,317] by glyphosate in different systems. Treatment of plants with gibberellic acid increases the phytotoxicity of glyphosate [318], presumably because acropetal metabolic sinks are activated.

10. Miscellaneous Effects

A number of other obvious secondary and tertiary effects of glyphosate have been described. Glyphosate inhibits transpiration by causing stomatal closure [154,249,252,319], as well as triggering stomatal oscillations [251]. Nafziger and Slife [271] found that glyphosate increased diffusive resistance of treated leaves only. Phenylalanine and tyrosine fed to excised shoots can reverse glyphosate effects on transpiration [181].

Recent evidence indicates that glyphosate disrupts regulation of starch accumulation and carbon allocation in photosynthesizing leaves [253]. Movement of carbon to starch is strongly inhibited, in comparison to glyphosate's effect on export of sugars to metabolic sinks.

Ethane production, an indicator of membrane peroxidation, was greatly enhanced by glyphosate [253]. Such an effect could be the result of cessation of synthesis of radical-scavenging enzymes and (or) disruption of photosynthesis. However, no effect of 20 mM glyphosate was found on glutathione levels in leaves of several species [320]. Glutathione levels were increased in these species by herbicides that increased hydrogen peroxide levels (e.g., paraquat).

Glyphosate was a competitive inhibitor of soybean lactate dehydrogenase at extremely high concentrations [321]. Cellulase-specific activity of abscission zones increased in glyphosate-treated bean plants [265], possibly because of increased ethylene synthesis.

D. Selectivity

Glyphosate is generally considered to be a nonselective herbicide, in that it is generally not sprayed on herbaceous crops except as a harvest aid. There is variation within crop [322], and weed [323, 324] species, indicating that tolerant varieties and biotypes exist; however, the physiological basis for tolerance has not been determined. Most of what is known of the physiological bases of tolerance to glyphosate is the result of attempts to produce resistant crop varieties.

According to Gressel's [325] definition of herbicide resistance, no known glyphosate-resistant biotypes of higher plants have occurred in nature. Resistant plants (not cultured cells) are generally at least two orders of magnitude less susceptible to a herbicide than a nonresistant or susceptible plant. As discussed in the section on aromatic amino acid synthesis (Section III.B), only tolerant or marginally resistant higher plant lines have been produced by selection or genetic transformation. Selected tolerant plants have elevated EPSP synthase levels and amplification of the sensitive gene in susceptible plants results in tolerant plants. Thus far, transformed plants with an enzyme that is resistant are only tolerant.

Various tolerant strains of higher plants have been reported that have not been physiologically characterized [e.g., Refs. 326–328]. In some cases, these glyphosate-tolerant varieties do not grow as well as the glyphosate-susceptible variety when not sprayed with herbicide [229,238]. Curiously, several lines of tobacco have been selected for glyphosate tolerance that have cross tolerance for amitrole and vice versa [327]. Resistance to amitrole is not due to absorption differences. Thus, one might expect some commonality between amitrole and glyphosate mechanisms of action, just as one finds in cross resistance between photosystem II inhibitors [325]. The mechanisms of action of amitrole, however, are thought to be inhibition of histidine and carotenoid synthesis [182]; neither is a process known to be affected by glyphosate.

Species which accumulate shikimate normally may be particularly sensitive to glyphosate due to a particularly inefficient, and thus more susceptible EPSP synthase. In fact, bracken fern, a shikimate-accumulating species, is very easily controlled by glyphosate [329]. This hypothesis has not been researched.

IV. CONCLUSIONS

Glyphosate is perhaps the most important herbicide ever developed. Ironically, its discovery was serendipitous to some extent.

It combines many traits to which it owes it success: high efficacy on most species, low synthesis cost, high stability, virtually no environmental hazard, and rapid and efficient translocation. Its only known site of activity at the molecular levels is EPSP synthase; however, it has not been unequivocally proven that this is the only site of action. The fact that no resistant strains or biotypes have occurred in nature indicates that the site of action is conservatively coded (low genetic plasticity) or that there are multiple sites of action.

ACKNOWLEDGMENTS

I thank the many colleagues who either criticized various versions of this review or provided preprints and other unpublished information for inclusion.

REFERENCES

1. J. E. Franz, in *The Herbicide Glyphosate* (E. Grossbard and D. Atkinson, eds.), Butterworths, London, 1985, pp. 3–17.
2. D. D. Baird, R. P. Upchurch, W. B. Homesley, and J. E. Franz, *Proc. North Cent. Weed Control Conf.*, *26*, 64 (1971).
3. E. Grossbard and D. Atkinson (eds.), *The Herbicide Glyphosate*, Butterworths, London, 1985, pp. 1–490.
4. P. B. Chykaliuk, J. R. Abernathy, and J. R. Gipson, *Bibliography of Glyphosate, Texas Agric. Exp. Stn. Publ.* MP-1443, 1979, pp. 1–87.
5. R. Irani, U.S. Patent 3,455,675 (1969).
6. A. D. F. Toy, P. Forest, and E. H. Uhing, U.S. Patent 3,160,632 (1964).
7. P. C. Hamm, U.S. Patent 3,556,762 (1971).
8. J. E. Franz, U.S. Patent 3,799,758 (1974).
9. P. C. Hamm, U.S. Patent 3,850,608 (1974).
10. J. E. Franz and H. L. Nufer, U.S. Patent 3,971,648 (1976).
11. J. E. Franz, U.S. Patent 3,988,142 (1976).
12. J. E. Franz, U.S. Patent 3,996,040 (1976).
13. K. Moedritzer and R. Irani, *J. Org. Chem.*, *31*, 1603 (1966).
14. G. B. Large, U.S. Patent 4,315,765 (1982).
15. G. B. Large, U.S. Patent 4,341,549 (1982).
16. K. L. Carlson and O. C. Burnside, *Weed Sci.*, *32*, 841 (1984).
17. H. P. Wilson, T. E. Hines, R. R. Bellinder, and J. A. Grande, *Weed Sci.*, *33*, 531 (1985).
18. H. Köcher, *Aspects Appl. Biol.*, *4*, 227 (1983).
19. E. Bakuniak, I. Bakuniak, B. Borucka, and J. Ostrowski, *J. Environ. Sci. Health*, *B18*, 485 (1983).

20. P. J. Diel and L. Maier, *Phosphorus and Sulfur*, *20*, 313 (1984).
21. P. M. Fredericks and L. A. Summers, *Z. Naturforsch.*, *36c*, 242 (1981).
22. R. L. Hilderbrand (ed.), *The Role of Phosphonates in Living Systems*, CRC Press, Boca Raton, FL, 1983, pp. 1–207.
23. P. Knuuttila and H. Knuuttila, in *The Herbicide Glyphosate* (E. Grossbard and D. Atkinson, eds.), Butterworths, London, 1985, pp. 18–22.
24. Herbicide Handbook of the Weed Sci. Soc. Am., 5th Ed., Champaign, IL, 1983, pp. 258–263.
25. S. Shoval and S. Yariv, *Agrochimica*, *25*, 376 (1981).
26. P. Knuuttila and H. Knuuttila, *Acta Chemica Scand.*, *B33*, 623 (1979).
27. L. M. Shkolnikova, M. A. Porai-Koshits, N. M. Dyatlora, G. F. Yaroshenko, M. V. Rudomino, and E. K. Kolova, *J. Struc. Chem.*, *23*, 737 (1982).
28. H. E. L. Madsen, H. H. Christensen, and C. Gottlieb-Peterson, *Acta Chemica. Scand.*, *A32*, 79 (1978).
29. R. D. Wauchope, *J. Agric. Food Chem.*, *24*, 717 (1976).
30. D. Atkinson, in *The Herbicide Glyphosate* (E. Grossbard and D. Atkinson, eds.), Butterworths, London, 1985, pp. 127–133.
31. Environmental Protection Agency, *Fed. Reg.*, *46(70)*, 21631 (1981).
32. Environmental Protection Agency, Code of Federal Regulations: Protection of Environment, 40 CFR, pp. 180–364 (1982).
33. Environmental Protection Agency, *Fed. Reg.*, *47(241)*, 56136 (1982).
34. C. R. Worthing, in *The Pesticide Manual*, Brit. Crop Protect. Council, Croydon, U.K., 1979, p. 292.
35. Monsanto, *Monsanto Material Safety Data Sheet*, 1–4 (1977).
36. T. E. Tooby, in *The Herbicide Glyphosate* (E. Grossbard and D. Atkinson, eds.), Butterworths, London, 1985, pp. 206–217.
37. W. A. Hartman and D. B. Martin, *Bull. Contam. Toxicol.*, *33*, 355 (1984).
38. E. A. Bababunmi, O. O. Olorunsogo, and O. Bassir, *Toxicol. Appl. Pharmacol.*, *45*, 319 (1978).
39. E. C. Spurrier, *PANS*, *19*, 607 (1973).
40. P. D. Harwood, *Science*, *139*, 684 (1963).
41. M. Paru and D. Sendrea, *Clujul Med.*, *58*, 51 (1985).
42. N. V. Vigfusson and E. R. Vyse, *Mutat. Res.*, *79*, 53 (1980).
43. O. O. Olorunsogo, E. A. Bababunmi, and O. Bassir, *Bull. Environ. Contam. Toxicol.*, *22*, 357, (1979).
44. O. O. Olorunsogo and E. A. Bababunmi, *Toxicol. Lett.*, *5 (Sp.1)*, 148 (1980).
45. J. J. S. Van Rensen, in *Proceedings of the Third International Congress on Phytosynthesis* (M. Avron, ed.) Elsevier, Amsterdam, 1974, pp. 684–687.

46. E. P. Richard, J. R. Goss, and C. J. Arntzen, *Weed Sci.*, *27*, 684 (1979).

47. I. Lopez-Brana, A. Delibes, and F. Garcia-Olmedo, *J. Exp. Bot.*, *35*, 905 (1984).

48. E. Hietanen, K. Linnainmaa, and H. Vainio, *Acta Pharmacol. Toxicol.*, *53*, 103 (1983).

49. J. Westerback, K. Rajan, and A. Martell, *J. Am. Chem. Soc.*, *87*, 2567 (1965).

50. J. E. Franz, *Adv. Pestic. Sci.*, *2*, 139 (1979).

51. X. Huang, S. Shan, P. Wu, J. Pu, J. Deng, and X. Yu, *Najing Linxueyuan Xuebao*, *1*, 43 (1984).

52. P. C. Bardalaye, W. B. Wheeler, and H. A. Moye, in *The Herbicide Glyphosate* (E. Grossbard and D. Atkinson, eds.), Butterworths, London, 1985, pp. 263-285.

53. W. G. Richardson, in *The Herbicide Glyphosate* (E. Grossbard and D. Atkinson, eds.), Butterworths, London, 1985, pp. 286-298.

54. G. Pavoni, *Boll. Chim. Union. Ital. Labor. Provinciali*, *9*, 157 (1978).

55. P. Sprankle, C. L. Sandberg, W. F. Meggitt, and D. Penner, *Weed Sci.*, *26*, 673 (1978).

56. M. T. H. Ragab, *Chemosphere*, *2*, 143 (1978).

57. J. C. Young, S. U. Khan, and P. B. Marriage, *J. Agric. Food Chem.*, *25*, 918 (1977).

58. R. L. Glass, *J. Agric. Food Chem.*, *31*, 280 (1983).

59. T. E. Archer and J. D. Stokes, *J. Agric. Food Chem.*, *32*, 586 (1984).

60. A. J. Burns and D. F. Tomkins, *J. Chromatogr. Sci.*, *17*, 333 (1979).

61. A. J. Burns, *J. Assoc. Off. Anal. Chem.*, *66*, 1214 (1983).

62. H. A. Moye and S. J. Scherer, *Anal. Lett.*, *10*, 1049 (1977).

63. H. A. Moye and P. A. St.John, *ACS Symp. Ser.*, *136*, Chapt. 7 (1980).

64. H. A. Moye, C. J. Miles, and S. J. Scherer, *J. Agric. Food Chem.*, *31*, 69 (1983).

65. H. Roseboom and C. J. Berkhoff, *Anal. Chim. Acta.*, *135*, 373 (1982).

66. *Pesticide Analytical Manual*, Monsanto Chem. Co., Vol. 2. Food and Drug Admin., Washington, D. C., Pest. Reg. Sec. 180.364 (1977).

67. R. A. Guinivan, N. P. Thompson, and W. B. Wheeler, *J. Assoc. Off. Anal. Chem.*, *65*, 35 (1982).

68. M. L. Rueppel, L. A. Suba, and J. T. Marvel, *Biomed. Mass Spectrom.*, *3*, 28 (1976).

69. C. L. Deyrup, S.-M. Chang, R. A. Weintraub, and H. A. Moye, *J. Agric. Food Chem.*, *33*, 944 (1985).

70. J. O. Bronstad and H. O. Friestad, *Analyst, 101,* 820 (1976).
71. H. O. Friestad and J. O. Bronstad, *J. Assoc. Off. Anal. Chem., 68,* 76 (1985).
72. G. Ekstrom and S. Johansson, *Bull. Environ. Contam. Toxicol., 14,* 295 (1975).
73. J. N. Seiber, J. M. McChesney, R. Kon, and R. A. Leavitt, *J. Agric. Food Chem., 32,* 681 (1984).
74. H. A. Moye and C. L. Deyrup, *J. Agric. Food Chem., 32,* 192 (1984).
75. D. J. Turner, in *The Herbicide Glyphosate* (D. Grossbard and D. Atkinson, eds.), Butterworths, London, 1985, pp. 221–240.
76. D. L. Shaner, *Weed Sci., 26,* 513 (1978).
77. H. P. Wilson, T. E. Hines, R. R. Bellinder, and J. A. Grande, *Weed Sci., 33,* 531 (1985).
78. P. B. Chykaliuk, J. R. Abernathy, and J. Gipson, *Proc. 32nd Annu. South. Weed Sci. Soc.,* p. 66 (1979).
79. U. Suwannamek and C. Parker, *Weed Res., 15,* 13 (1975).
80. Z. N. Yang, *Biotrop. Newsletter, 23,* 8 (1978).
81. J. L. Hammerton, *PANS, 20,* 425 (1974).
82. D. J. Turner and M. P. C. Loader, *Pestic. Sci., 6,* 1 (1975).
83. J. Zemanek, *Rostl. Vyroba, 25,* 653 (1979).
84. P. J. W. Lutman and W. G. Richardson, *Weed Res., 18,* 65 (1978).
85. H. Schneider, *Mitt. Schweiz. Landwirtschaft, 28,* 37 (1980).
86. A. M. Blair, *Weed Sci., 15,* 101 (1975).
87. G. D. Wills and C. G. McWhorter, *Weed Sci., 33,* 755 (1985).
88. G. Nilsson, in *The Herbicide Glyphosate* (E. Grossbard and D. Atkinson, eds.), Butterworths, London, 1985, pp. 35–47.
89. D. D. Buhler and O. C. Burnside, *Weed Sci., 31,* 163 (1983).
90. D. J. Turner and M. P. C. Loader, *Weed Res., 18,* 199 (1978).
91. P. J. Shea and D. R. Tupy, *Weed Sci., 32,* 802 (1984).
92. P. J. Terry, in *The Herbicide Glyphosate* (E. Grossbard and D. Atkinson, eds.), Butterworths, London, 1985, pp. 375–401.
93. W. H. Ahrens and W. G. Pill, *Hortsci., 20,* 64 (1985).
94. D. Coupland, in *The Herbicide Glyphosate* (E. Grossbard and D. Atkinson, eds.), Butterworths, London, 1985, pp. 25–34.
95. R. H. Shimabukuro, in *Weed Physiology,* Vol. II (S. O. Duke, ed.) CRC Press, Boca Raton, FL, 1985, pp. 215–240.
96. D. Coupland and C. Caseley, *New Phytol., 83,* 17 (1979).
97. B. H. Zandstra and R. K. Nishimoto, *Weed Sci., 25,* 268 (1977).
98. O. Gottrup. P. A. O'Sullivan, R. J. Schraa, and W. H. Vandenborn, *Weed Res., 16,* 197 (1976).

99. D. Coupland, *Pestic. Sci.*, *15*, 226 (1984).
100. J. J. V. Rodrigues, A. D. Worsham, and F. T. Corbin, *Weed Sci.*, *30*, 316 (1982).
101. M. D. Devine and J. D. Bandeen, *Weed Res.*, *23*, 69 (1983).
102. L. Y. Marquis, R. D. Comes, and C. P. Yang, *Weed Res.*, *19*, 335, (1979).
103. J. T. O'Donovan and P. A. O'Sullivan, *Weed Sci.*, *30*, 30, (1982).
104. J. B. Wyrill, III and O. C. Burnside, *Weed Sci.*, *24*, 557 (1976).
105. M. E. Schultz and O. C. Burnside, *Weed Sci.*, *28*, 13 (1980).
106. A. R. Putnam, *Weed Sci.*, *24*, 425 (1976).
107. C. L. Sandberg, W. F. Meggitt, and D. Penner, *Weed Res.*, *20*, 195 (1980).
108. L. Torstensson, in *The Herbicide Glyphosate* (E. Grossbard and D. Atkinson, eds.), Butterworths, London, 1985, pp. 137–150.
109. P. Sprankle, W. F. Meggitt, and D. Penner, *Weed Sci.*, *23*, 224 (1975).
110. J. Zemanek and J. Kubrova, *Sb. Uvti. Ochr. Rost.*, *14*, 297 (1978).
111. P. Sprankle, W. F. Meggitt, and D. Penner, *Weed Sci.*, *23*, 229 (1975).
112. M. L. Rueppel, B. B. Brightwell, J. Schaffer, and J. T. Marvel, *J. Agric. Food Chem.*, *25*, 517 (1977).
113. S. U. Khan and J. C. Young, *J. Agric. Food Chem.*, *25*, 1430 (1977).
114. R. L. Tate and M. Alexander, *Soil Sci.*, *118*, 317 (1974).
115. N. T. L. Torstensson and A. Aamisepp, *Weed Res.*, *17*, 209 (1977).
116. N. S. Nomura and H. W. Hilton, *Weed Res.*, *17*, 113 (1977).
117. F. K. Moore, H. D. Braymer, and A. D. Larson, *Appl. Environ. Microbiol.*, *46*, 316 (1983).
118. H. W. Talbot, L. M. Johnson, and D. M. Munnecke, *Curr. Microbiol.*, *10*, 225 (1984).
119. D. L. Shinabarger, E. K. Schmitt, H. D. Braymer, and A. D. Larson, *Appl. Environ. Microbiol.*, *48*, 1049 (1984).
120. P. L. Eberbach and L. A. Douglas, *Soil Biol. Biochem.*, *15*, 485 (1983).
121. M. M. Mueller, C. Rosenberg, H. Siltanen, and T. Wartiovaara, *Bull. Environ. Contam. Toxicol.*, *27*, 724 (1981).
122. L. J. Moshier and D. Penner, *Weed Sci.*, *26*, 686 (1978).
123. M. Newton, K. M. Howard, B. R. Kelpsas, R. Danhaus, C. M. Lottman, and S. Dubelman, *J. Agric. Food Chem.*, *32*, 1151 (1984).
124. V. Roisch and F. Lingens, *Angew. Chem.*, *13*, 400 (1974).

125. E. Grossbard, in *The Herbicide Glyphosate* (E. Grossbard and D. Atkinson, eds.), Butterworths, London, 1985, pp. 159–185.

126. G. S. Jacob, J. Schaefer, E. O. Stejskal, and R. A. McKay, *J. Biol. Chem.*, *260*, 5899 (1985).

127. E. Cerol and G. Seguin, *C. R. Sean. Acad. Agric.*, *68*, 804 (1982).

128. J. O. Bronstad and H. O. Friestad, in *The Herbicide Glyphosate* (E. Grossbard and D. Atkinson, eds.), Butterworths, London, 1985, p. 200.

129. M. L. Morrison and E. C. Meslow, *For. Sci.*, *30*, 95 (1984).

130. R. D. Comes, V. F. Bruns, and A. D. Kelly, *Weed Sci.*, *24*, 47 (1976).

131. K. H. Bowmer, *Pestic. Sci.*, *13*, 623 (1982).

132. P. R. F. Barrett, in *The Herbicide Glyphosate* (E. Grossbard and D. Atkinson, eds.), Butterworths, London, 1985, p. 365.

133. W. M. Edwards, G. B. Triplett, and R. M. Kramer, *J. Environ. Qual.*, *9*, 661 (1980).

134. J. C. Caseley and D. Coupland, in *The Herbicide Glyphosate* (E. Grossbard and D. Atkinson, eds.), Butterworths, London, 1985, p. 92.

135. M. D. Devine, H. D. Bestman, C. Hall, and W. H. VandenBorn, *Weed Sci.*, *32*, 418 (1984).

136. C. R. Merritt, *Ann. Appl. Biol.*, *101*, 527 (1982).

137. J. B. Wyrill and O. C. Burnside, *Weed Sci.*, *25*, 275 (1977).

138. J. K. Soteres, D. S. Murray, and E. Basler, *Weed Sci.*, *31*, 271 (1983).

139. S. L. Sherrick, H. A. Holt, and F. D. Hess, *Weed Sci.*, in press (1986).

140. D. L. Hensley, D. S. N. Beverman, and P. L. Carpenter, *Weed Res.*, *18*, 287 (1978).

141. P. W. Stahlman and W. M. Phillips, *Weed Sci.*, *27*, 38 (1979).

142. P. A. O'Sullivan, J. T. O'Donovan, and W. H. Hamman, *Can. J. Plant Sci.*, *61*, 391 (1981).

143. G. Nilsson, *Swedish J. Agric. Res.*, *14*, 3 (1984).

144. L. C. Haderlie, F. W. Slife, and H. S. Butler, *Weed Res.*, *18*, 269 (1978).

145. D. J. Penn and J. M. Lynch, *New Phytol.*, *90*, 51 (1982).

146. L. C. Salazar and A. P. Appleby, *Weed Sci.*, *30*, 463 (1982).

147. R. J. Hance, *Pestic. Sci.*, *7*, 363 (1976).

148. C. G. McWhorter, T. N. Jordan, and G. D. Wills, *Weed Sci.*, *28*, 113 (1980).

149. M. S. Ahmadi, L. C. Haderlie, and G. A. Wicks, *Weed Sci.*, *28*, 277 (1980).

150. R. L. Chase and A. P. Appleby, *Weed Res.*, *19*, 241 (1979).

151. T. N. Jordan, *Weed Sci.*, *25*, 448 (1977).
152. F. D. Hess, in *Weed Physiology*, Vol. II, *Herbicide Physiology* (S. O. Duke, ed.), CRC Press, Boca Raton, FL, 1985, p. 191.
153. M. A. Waldecker and D. L. Wyse, *Weed Sci.*, *33*, 299 (1985).
154. B. J. Brecke and W. B. Duke, *Plant Physiol.*, *66*, 656 (1980).
155. E. P. Richard and F. W. Slife, *Weed Sci.*, *27*, 426 (1979).
156. J. A. Gougler and D. R. Geiger, *Plant Physiol.*, *68*, 668 (1981).
157. C. A. Peterson, P. P. Q. DeWildt, and L. V. Edington, *Pestic. Biochem. Physiol.*, *8*, 1 (1978).
158. M. T. Tyree, C. A. Peterson, and L. V. Edgington, *Plant Physiol.*, *63*, 367 (1979).
159. R. S. McAllister and L. C. Haderlie, *Weed Sci.*, *33*, 153 (1985).
160. S. A. Dewey and A. P. Appleby, *Weed Sci.*, *27*, 235 (1979).
161. J. J. Kells and C. E. Rieck, *Weed Sci.*, *27*, 235 (1979).
162. P. Sprankle, W. F. Meggitt, and D. Penner, *Weed Sci.*, *23*, 235 (1975).
163. J. A. Gougler and D. R. Geiger, *Weed Sci.*, *32*, 546 (1984).
164. B. H. Zandstra and R. K. Nishimoto, *Weed Sci.*, *25*, 268 (1977).
165. J. S. Claus and R. Behrens, *Weed Sci.*, *24*, 149 (1976).
166. G. R. Leather and J. R. Frank, *HortSci.*, *20*, 70 (1985).
167. T. B. Klevorn and D. L. Wyse, *Weed Sci.*, *32*, 744 (1985).
168. P. E. Keeley, C. H. Carter, and R. J. Thullen, *Weed Sci.*, *34*, 25 (1986).
169. D. S. N. Beverman, D. L. Hensley, and P. L. Carpenter, *HortSci.*, *19*, 296 (1984).
170. M. D. Devine and J. D. Bandeen, *Weed Res.*, *23*, 69 (1983).
171. P. C. Lolas and H. D. Coble, *Weed Res.*, *20*, 267 (1980).
172. T. Whitwell, P. Banks, E. Basler, and P. W. Santelmann, *Weed Sci.*, *28*, 93 (1980).
173. G. D. Wills, *Weed Sci.*, *26*, 509 (1978).
174. T. B. Klevorn and D. L. Wyse, *Weed Sci.*, *32*, 402 (1984).
175. D. Coupland, *Weed Res.*, *23*, 347 (1983).
176. M. D. Devine, J. D. Bandeen, and B. D. McKersie, *Weed Sci.*, *31*, 461 (1983).
177. D. R. Shaw, T. F. Peeper, and E. Basler, *Plant Growth Regul.*, *3*, 79 (1985).
178. P. B. Chykaliuk, T. F. Peeper, and E. Basler, *Weed Sci.*, *30*, 6 (1982).
179. M. E. Foley and L. M. Wax, *Physiol. Plant.*, *56*, 482 (1982).
180. R. A. Martin and L. V. Edgington, *Pestic. Biochem. Physiol.*, *16*, 87 (1981).
181. D. L. Shaner and J. L. Lyon, *Weed Sci.*, *28*, 31 (1980).

182. S. O. Duke, in *Weed Physiology*, Vol. II, *Herbicide Physiology* (S. O. Duke, ed.), CRC Press, Boca Raton, FL, 1985, p. 91.

183. H. C. Steinrücken and N. Amrhein, *Eur. J. Biochem.*, *143*, 351 (1984).

184. V. Roisch and F. Lingens, *Hoppe—Seyler's Z. Physiol. Chem.*, *361*, 1049 (1980).

185. J. L. Rubin, C. G. Gaines, and R. A. Jensen, *Plant Physiol.*, *70*, 833 (1982).

186. R. Bode, C. Melo, and D. Birnbaum, *Biochem. Physiol. Pflanz.*, *179*, 775 (1984).

187. R. Bode, C. Melo, and D. Birnbaum, *Arch. Microbiol.*, *140*, 83 (1984).

188. R. Bode, C. M. Ramos, and D. Birnbaum, *FEMS Microbiol. Lett.*, *23*, 7 (1984).

189. R. Bode, F. Schauer, and D. Birnbaum, *Biochem. Physiol. Pflanz.*, *181*, 39 (1986).

190. D. J. Cole, A. D. Dodge, and J. C. Caseley, *J. Exp. Bot.*, *31*, 1665 (1980).

191. S. O. Duke and R. E. Hoagland, *Weed Sci.*, *29*, 297 (1981).

192. H. C. Steinrücken and N. Amrhein, *Biochem. Biophys. Res. Commun.*, *94*, 1207 (1980).

193. N. Amrhein, J. Shab, and H. C. Steinrücken, *Naturwissenschaften*, *67*, 356 (1980).

194. D. L. Anton. L. Hedstrom, S. M. Fish, and R. H. Abeles, *Biochemistry*, *22*, 5903 (1983).

195. K. Duncan, A. Lewendon, and J. R. Coggins, *FEBS Lett.*, *65*, 121 (1984).

196. D. M. Mousdale and J. R. Coggins, *Planta*, *160*, 78 (1984).

197. D. M. Mousdale and J. R. Coggins, *Planta*, *163*, 241 (1985).

198. J. L. Rubin, C. G. Gaines, and R. A. Jensen, *Plant Physiol.*, *75*, 839 (1984).

199. M. R. Boocock and J. R. Coggins, *FEBS Lett.*, *154*, 127 (1983).

200. N. Amrhein, H. Holländer-Czytko, J. Leifeld, A. Schulz, H. C. Steinrücken, and H. Topp, in *Journee internationales d'etudes due Groupe Polyphenols*, Bulletin de Liaison, Vol. II (A. M. Boudet and R. Ranjeva, eds.), Toulouse, 1982, p. 21.

201. N. Amrhein, D. Johänning, J. Schab, and A. Schulz, *FEBS Lett.*, *157*, 191 (1983).

202. D. M. Comai, L. Sen, and D. Stalker, *Science*, *221*, 370 (1983).

203. A. Schulz, A. Krüper, and N. Amrhein, *FEMS Microbiol. Lett.*, *28*, 297 (1985).

204. C. C. Smart, D. Johänning, G. Müller, and N. Amrhein, *J. Biol. Chem.*, *260*, 16338 (1985).

205. N. Amrhein, B. Deus, P. Gehrke, and H. C. Steinrücken, *Plant Physiol.*, *66*, 830 (1980).

206. J. Berlin and L. Witte, *Z. Naturforsch.*, *36c*, 210 (1981).
207. R. Bode, G. Kunze, and D. Birnbaum, *Biochem. Physiol. Pflanz.*, *180*, 613 (1985).
208. H. Holländer–Czytko and N. Amrhein, *Plant Sci. Lett.*, *29*, 89 (1983).
209. R. E. Hoagland, S. O. Duke, and C. D. Elmore, *Plant Sci. Lett.*, *13*, 291 (1978).
210. S. O. Duke, R. E. Hoagland, and C. D. Elmore, *Physiol. Plant.*, *46*, 307 (1979).
211. G. Nilsson, *Swed. J. Agric. Res.*, 7, 153 (1977).
212. H. Holländer and N. Amrhein, *Plant Physiol.*, *66*, 823 (1980).
213. A. Ekanayake, R. L. Wichremasinghe, and H. D. S. Liyanage, *Weed Res.*, *19*, 39 (1979).
214. S. O. Duke and R. E. Hoagland, in *The Herbicide Glyphosate* (E. Grossbard and D. Atkinson, eds.), Butterworths, London, 1985, p. 75.
215. B. Laber, H.-H. Kiltz, and N. Amrhein, *Z. Naturforsch.*, *41c*, 49 (1986).
216. S. O. Duke, R. E. Hoagland, and C. D. Elmore, *Plant Physiol.*, *65*, 17 (1980).
217. S. O. Duke and A. W. Naylor, *Plant Sci. Lett.*, *6*, 361 (1976).
218. L. C. Haderlie, J. M. Widholm, and F. W. Slife, *Plant Physiol.*, *60*, 40 (1977).
219. S. O. Duke and R. E. Hoagland, *Weed Sci.*, *29*, 297 (1981).
220. S. O. Duke and R. E. Hoagland, *Plant Science Lett.*, *11*, 185 (1978).
221. R. E. Hoagland, S. O. Duke, and C. D. Elmore, *Physiol. Plant*, *46*, 357 (1979).
222. G. P. Bolwell, J. N. Bell, C. L. Cramer, W. Schuch, C. J. Lamb, and R. A. Dixon, *Eur. J. Biochem.*, *149*, 411 (1985).
223. K. Hennessy and E. Kieff, *Science*, *227*, 1240 (1984).
224. G. J. Acton and G. Gupta, *Biochem. J.*, *184*, 367 (1979).
225. B. Betz, E. Schafer, and K. Hahlbrock, *Arch. Biochem. Biophys.*, *190*, 126 (1978).
226. R. A. John, A. Charteris, and L. J. Fowler, *Biochem. J.*, *171*, 771 (1978).
227. R. E. Hoagland and S. O. Duke, *Plant Cell Physiol.*, *23*, 1081 (1982).
228. J. L. Kilmer, J. M. Widholm, and F. W. Slife, *Plant Physiol.*, *68*, 1299 (1981).
229. E. D. Nafziger, J. M. Widholm, and F. W. Slife, *Plant Physiol.*, *71*, 623 (1983).
230. E. G. Jaworski, *J. Agric. Food Chem.*, *20*, 1195 (1972).
231. P. M. Gresshoff, *Aust. J. Plant Physiol.*, *6*, 177 (1979).
232. G. S. Byng, R. J. Whitaker, and R. A. Jenson, *Can. J. Bot.*, *63*, 1021 (1985).

233. T. T. Lee, *Weed Res.*, *20*, 365 (1980).
234. P. M. Gresshoff, *Arabidopsis Inf. Serv. Oct.*, 3 (1979).
235. E. V. Parups and W. E. Cordukes, *Phyton*, *43*, 57 (1983).
236. N. Amrhein, D. Johänning, J. Schab, and A. Schulz, *FEBS Lett.*, *157*, 191 (1983).
237. E. D. Nafziger, J. M. Widholm, H. C. Steinrücken, and J. L. Kilmer, *Plant Physiol.*, *76*, 571 (1984).
238. S. G. Rogers, L. A. Brand, S. B. Holder, E. S. Sharps, and M. J. Brackin, *Appl. Environ. Microbiol.*, *46*, 37 (1983).
239. N. Amrhein, D. Johänning, and C. C. Smart, in *Primary and Secondary Metabolism of Plant Cell Cutures* (K. H. Neumann, ed.), Springer-Verlag, Heidelberg, 1985, p. 356.
240. H. C. Steinrücken, A. Schulz, N. Amrhein, C. A. Porter, and R. T. Fraley, *Arch. Biochem. Biophys.*, *244*, 169 (1986).
241. A. Schulz, D. Sost, and N. Amrhein, *Arch. Microbiol.*, *137*, 121 (1984).
242. D. M. Stalker, W. R. Hiatt, and L. Comai, *J. Biol. Chem.*, 260 4724 (1985).
243. D. Sost, A. Schulz, and N. Amrhein, *FEBS Lett.*, *173*, 238 (1984).
244. L. Comai, D. Facciotti, W. R. Hiatt, G. Thompson, R. E. Rose, and D. M. Stalker, *Nature*, *317*, 741 (1985).
245. L. Comai, U.S. Patent 4,535,060 (1985).
246. A. Bitsch, R. Trihbes, and G. Schultz, *Physiol. Plant.*, *61*, 617 (1984).
247. R. A. Jensen, *Physiol. Plant.*, *66*, 164 (1986).
248. K. C. Vaughn and S. O. Duke, in *Models in Plant Physiology* (D. W. Newman and K. G. Wilson, eds.), CRC Press, Inc., Boca Raton, FL, in press, 1987.
249. A. Munoz-Rueda, C. Gonzalex-Murua, J. M. Becerril, and M. F. Sanchez-Diaz, *Physiol. Plant.*, *66*, 63 (1986).
250. R. M. Devlin, I. I. Zbiec, A. J. Murkowski, and S. J. Karczmarczyk, *Weed Res.*, *21*, 133 (1980).
251. D. L. Shaner and J. L. Lyon, *Plant Sci. Lett.*, *15*, 83 (1979).
252. D. R. Geiger, S. W. Kapitan, and M. A. Tucci, *Plant Physiol.*, *82*, 468 (1986).
253. D. J. Cole, J. C. Caseley, and A. D. Dodge, *Weed Res.*, *23*, 173 (1983).
254. L. M. Kitchen, W. W. Witt, and C. E. Rieck, *Weed Sci.*, *29*, 513 (1981).
255. A. Ali and R. A. Fletcher, *Can. J. Bot.*, *56*, 2196 (1978).
256. L. M. Kitchen, W. W. Witt, and C. E. Rieck, *Weed Sci.*, *29*, 571 (1981).
257. B. E. Abu-Irmaileh and L. S. Jordan, *Weed Sci.*, *26*, 700 (1978).

258. G. W. Miller, A. Denney, J. Pushnik, and M.-H. Yu, *J. Plant Nutr.*, *5*, 289 (1982).

259. D. J. Cole, in *The Herbicide Glyphosate* (E. Grossbard and D. Atkinson, eds.), Butterworths, London, 1985, p. 48.

260. T. T. Lee, *Weed Res.*, *21*, 161 (1981).

261. L. A. Klepper, *Plant Physiol.*, *64*, 273 (1979).

262. A. Wild and R. Mandersheid, *Z. Naturforsch.*, *39c*, 500 (1984).

263. R. E. Hoagland, *Plant Cell Physiol.*, *26*, 565 (1985).

264. S. H. Duke and S. O. Duke, *Physiol. Plant.*, *62*, 485 (1984).

265. B. E. Abu-Irmaileh, L. S. Jordan, and J. Kumamoto, *Weed Sci.*, *27*, 103 (1979).

266. S. O. Duke, K. C. Vaughn, and R. D. Wauchope, *Pestic. Biochem. Physiol.*, *24*, 384 (1985).

267. A. L. Cereira, A. W. Cole, and D. S. Luthe, *Weed Sci.*, *33*, 1 (1985).

268. M. E. Foley, E. D. Nafziger, F. W. Slife, and L. M. Wax, *Weed Sci.*, *31*, 76 (1983).

269. R. P. Carter, R. L. Carroll, and R. R. Irani, *Inorg. Chem.*, *6*, 939 (1967).

270. M. I. Kabachnik, T. Ya Medred, N. M. Dyatlova, and M. V. Rudomino, *Russ. Chem. Rev.* (Engl. Transl.), *43*, 733 (1974).

271. E. D. Nafziger and F. W. Slife, *Weed Sci.*, *31*, 874 (1983).

272. L. Niklicek, E. Bergmannova, and L. Taimv, *Sb. Uvti. Ochr. Rostl.*, *19*, 147 (1983).

273. S. O. Duke, R. D. Wauchope, R. E. Hoagland, and G. D. Wills, *Weed Res.*, *23*, 133 (1983).

274. J. Crowley and G. N. Prendeville, *Weed Res.*, *25*, 341 (1985).

275. M. C. Watson, P. G. Bartels, and K. C. Hamilton, *Weed Sci.*, *28*, 122 (1980).

276. R. A. Fletcher, P. Hildebrand, and W. Akey, *Weed Sci.*, *28*, 671 (1980).

277. M. C. O'Brien and G. N. Prendeville, *Weed Res.*, *19*, *331* (1979).

278. G. N. Prendeville, and G. F. Warren, *Weed Res.*, *30*, 251 (1977).

279. S. O. Duke, *ACS Symp. Ser.*, *268*, 113 (1985).

280. A. Tohver and U. Onnepalu, *Eesti. Nsv. Tead. Akad. Toim. Biol.*, *31*, 213 (1982).

281. R. E. Hoagland, *Weed Sci.*, *28*, 393 (1980).

282. N. Ishikura and Y. Takeshima, *Plant Cell Physiol.*, *25*, 185 (1984).

283. N. Ishikura, M. Iwata, and S. Mitsue, *Bot. Mag. Tokyo*, *96*, 111 (1983).

284. U. Margna, L. Laanest, E. Margna, and T. Vainjarv, *Z. Naturforsch.*, *40c*, 154 (1985).

285. E. W. B. Ward, *Physiol. Plant Pathol.*, *25*, 381 (1984).
286. M. T. Keen, M. J. Holliday, and M. Yoshikawa, *Phytopathology*, *72*, 1467 (1982).
287. M. J. Holliday and N. T. Keen, *Phytopathology*, *72*, 1470 (1982).
288. P. Moesta and H. Grisebach, *Physiol. Plant Pathol.*, *21*, 65 (1982).
289. H. Grisebach, H. Borner, and P. Moesta, *Ber. Dtschs. Bot. Ges.*, *95*, 619 (1982).
290. A. O. Latunde-Dada and J. A. Lucas, *Physiol. Plant Pathol.*, *26*, 31 (1985).
291. S. Margit, C. Jolan, and K. Erzebet, *Bot. Kozl.*, *70*, 151 (1983).
292. N. Amrhein and J. Gerhardt, *Biochem. Biophys. Acta*, *583*, 434 (1979).
293. N. Noe and H. U. Seitz, *Planta*, *154*, 454 (1982).
294. R. E. Hoagland and S. O. Duke, *ACS Symp. Ser.*, *181*, 175 (1982).
295. N. Amrhein, B. Beus, P. Gehrke, H. Holländer, J. Schab, A. Schulz, and H. C. Steinrücken, *Proc. PGRWG*, *8*, 99 (1981).
296. R. E. Hoagland and S. O. Duke, *Weed Sci.*, *29*, 433 (1981).
297. R. E. Hoagland and S. O. Duke, *Weed Sci.*, *31*, 845 (1983).
298. G. S. Johal and J. E. Rahe, *Phytopathology*, *74*, 950 (1984).
299. T. Popoff and O. Theander, *Appl. Polym. Symp.*, *28*, 1341 (1976).
300. I. A. Evans and J. Mason, *Nature*, *208*, 913 (1965).
301. I. A. Evans, *Bot. J. Linnean Soc.*, *73*, 105 (1976).
302. L. B. Jacobsen, C. L. Richardson, and H. G. Floss, *Lloydia*, *41*, 450 (1978).
303. R. Jones and J. M. Forbes, *Anim. Prod.*, *38*, 301 (1984).
304. W. F. Campbell, J. O. Evans, and F. C. Reed, *Weed Sci.*, *24*, 22 (1976).
305. S. Pihakaski and K. Pihakaski, *Ann. Bot.*, *46*, 133 (1980).
306. M. Uotila, K. Evjen, and T.-H. Iverson, *Weed Res.*, *20*, 153 (1980).
307. K. C. Vaughn and S. O. Duke, *Pestic. Biochem. Physiol.*, *26*, 56 (1986).
308. T. T. Lee, *Physiol. Plant.*, *54*, 289 (1982).
309. T. T. Lee, T. Dumas, and J. J. Jevnikar, *Pestic. Biochem. Physiol.*, *20*, 354 (1983).
310. T. T. Lee, *J. Plant Growth Regul.*, *1*, 37 (1982).
311. T. T. Lee, and T. Dumas, *J. Plant Growth Regul.*, *4*, 29 (1985).
312. T. T. Lee, *J. Plant Growth Regul.*, *3*, 227 (1984).
313. J. R. Baur, R. W. Bovey, and J. A. Veech, *Weed Sci.*, *25*, 238 (1977).

314. R. Scorza, W. V. Welker, and C. J. Duhn, *HortSci.*, *19*, 66
      (1984).
315. J. R. Baur, *Weed Sci.*, *27*, 69 (1979).
316. J. R. Baur, *Plant Physiol.*, *63*, 882 (1979).
317. T. T. Lee and T. Dumas, *Plant Physiol.*, *72*, 855 (1983).
318. J. P. Sterrett and R. H. Hodgson, *Weed Sci.*, *31*, 396 (1983).
319. D. L. Shaner, *Weed Sci.*, *26*, 513 (1978).
320. I. K. Smith, *Plant Physiol.*, *79*, 1044 (1985).
321. M. Press, J. Barthova, and S. Leblova, *Biol. Plant.*, *25*, 274
      (1983).
322. T. N. Jordan and R. R. Bridge, *Agron. J.*, *71*, 927 (1979).
323. F. P. DeGennaro and S. C. Weller, *Weed Sci.*, *32*, 472
      (1984).
324. C. T. Bryson and G. D. Wills, *Weed Sci.*, *33*, 848 (1985).
325. J. Gressel, in *Weed Physiology*, Vol. II, *Herbicide Physiology*
      (S. O. Duke, ed.), CRC Press, Boca Raton, FL, 1985, p.
      159.
326. J. King and A. Maretzki, *Physiol. Plant.*, *58*, 457 (1983).
327. S. R. Singer and C. N. McDaniel, *Plant Physiol.*, *78*, 411
      (1985).
328. L. G. Hickok, *ACS Symp. Ser.*, *334*, 53 (1987).
329. R. C. Kirkwood and A. Hinshalwood, *Proc. Royal Soc.*
      (Edinburgh), *86B*, 179 (1985).

Chapter 2

POLYCYCLIC ALKANOIC ACIDS

STEPHEN O. DUKE

United States Department of Agriculture
Agricultural Research Service
Southern Weed Science Laboratory
Stoneville, Mississippi

WILLIAM H. KENYON*

MSU-DOE Plant Research Laboratory
Michigan State University
East Lansing, Michigan

*Current affiliation: Agricultural Products Department, E. I. du Pont
de Nemours & Company, Inc., Wilmington, Delaware.

I. INTRODUCTION

A. History

During the past 15 years, thousands of herbicides with varying chemistries have been patented for control of gramineous weeds. Many of these herbicides have almost identical selectivities and produce similar phytotoxic symptoms in the field despite their different chemistries. Also, there is evidence that many of these compounds have similar, if not identical, mechanisms of action [1]. The largest group of these new herbicides with a common structural identity are the polycyclic alkanoic acids (PCAs), that is compounds comprised of an alkanoic acid with a group containing more than one ring structure (one is usually a phenyl ring) attached to an asymmetric, noncarbonyl carbon of the alkanoic acid. Several hundred patent applications of more than twenty companies have been filed for herbicides or synthesis of compounds in this category during the last 15 years. There are other chemistries (e.g., sethoxydim; 2-[1 (ethoxyimino)butyl]-5-[2-(ethylthio)propyl]-3-hydroxy-2-cyclohexen-1-one and chlorfenprop-methyl) which do not fit this chemical definition and have similar selectivities and cause comparable herbicidal effects in the field. This chapter, however, is limited to one structurally related group—the PCAs.

Napropamide [N,N-diethyl-2(1-naphthalenyloxy)propanamide] is apparently bioactivated in susceptible plants by metabolism to 1-(2-naphthoxy)propionic acid and methyl-1-(2-naphthoxy)propionate [2,3]. Technically, these compounds fit out structural definition of PCA herbicides, however, their activity and probable mechanism of action are different from those of the PCAs that are discussed in this chapter. The obvious structural difference between these napropamide derivatives and the grass-killing PCAs is that the two-ring structures are separated by a minimum distance of an ester linkage in the PCAs. Thus, there is apparently a structure/activity requirement in the PCAs for an appropriate distance and/or potential geometric configuration between the ring structures.

Unfortunately, many PCA compounds have been termed "diphenyl ethers" or "phenoxyphenoxy" herbicides, even though few chemically fit these chemical categories. Also, such common nomenclature usage can confuse this chemical group with photobleaching p-nitro-substituted diphenyl ethers, an entirely different herbicide group [4]. In some older reviews of diphenyl ethers, phenoxyphenoxy members of the PCAs have been grouped with p-nitro-substituted diphenyl ethers [5].

The PCAs can be further divided into the oxyphenoxy alkanoic acids (generally termed the fops, e.g., diclofop) and the benzoyl-N-phenyl phenoxy propanoic acids (often termed props, e.g., flamprop) (Fig. 1). The oxyphenoxys contain phenoxyphenoxy, pyridinyloxyphenoxy, quinoxalinyloxphenoxy, and benzoxazolyloxy-phenoxy propanoic acids.

The first group of PCAs to be developed as herbicides were the benzoyl-N-phenyl propanoic acids [6], which were developed in re-sponse to a particular weed problem—wild oats (Avena fatua L.). After long-term spraying of phenoxyalkanoic acid herbicides, which do not control gramineous weed species, wild oats became a serious problem in small grain crops. Benzoylprop-ethyl has been used since the early 1970s, but has since been replaced by flamprop-methyl. The phenoxyphenoxy propanoic acid herbicide diclofop-methyl [methyl ester of (±)-2-[4-(2,4-dichlorophenoxy)phenoxy] propanoic acid] was found to have excellent herbicidal selectivity for wild oats in several monocot crops and began to be used in 1976. The phenoxyphenoxy propanoic acids are also effective against other gramineous weed species and are used in barley, wheat, rye, and many dicot crops.

This chapter will deal primarily with those PCA herbicides which are currently in commercial production.

B. Physical and Chemical Properties

The physical and chemical properties of representative examples of several commercialized PCA ester herbicide are provided in Table 1. The structures of representative examples of various chemical types of PCA esters are shown in Figure 1. The structures of these compounds are so diverse that there are no general physical or chemical properties. All commercialized versions of the PCAs are propanoic acid esters. Thus, virtually all research on PCA herbi-cides has been done with PCA esters. Because hydrolysis of the PCA ester to the PCA is generally rapid (to be discussed later), distinguishing between the ester and acid effects is often difficult. From a biological and environmental standpoint, the most important physical difference between the acid and ester of a PCA herbicide is probably solubility in water. The acid is more soluble than the ester, but not necessarily drastically so. For instance, the solubili-ties in water at 20°C of quizalofop and its ethyl ester are 10.2 and 3 ppm, respectively [9]. All PCAs with grass killing activity have an asymmetric carbon adjacent to the carboxyl groups of the alkanoic acid. One constituent of this asymetric carbon is polycyclic. Al-though these herbicides are usually sold as racemic mixtures, there is evidence (see Section I.D) that only one enantiomer is herbicidal-ly active.

OXYPHENOXYS

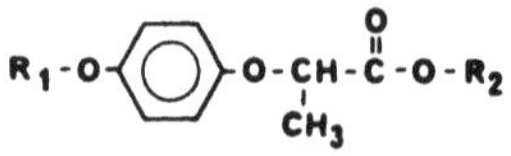

$$R_1-O-\text{⟨⟩}-O-CH-\overset{O}{\overset{\|}{C}}-O-R_2$$
$$CH_3$$

HERBICIDE	R_1	R_2
fluazifop-butyl	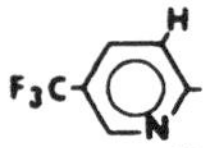 F_3C- (pyridine, H)	C_4H_9
haloxyfop-methyl	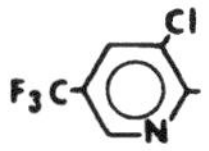 F_3C- (pyridine, Cl)	CH_3
fenoxaprop-methyl	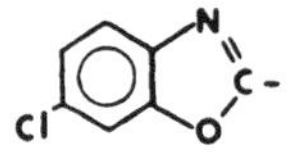 Cl- (benzoxazole) C-	C_2H_5
fenthiaprop-ethyl	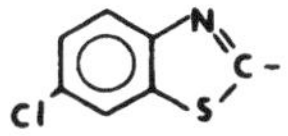 Cl- (benzothiazole) C-	C_2H_5
quizalofop-ethyl	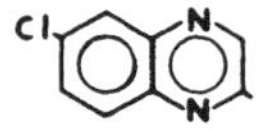 Cl- (quinoxaline)	C_2H_5
diclofop-methyl	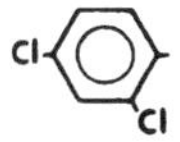 Cl-⟨⟩-Cl	CH_3
clofop-isobutyl	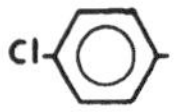 Cl-⟨⟩-	$CH_2-CH\overset{CH_3}{\underset{CH_3}{}}$
trifop-methyl	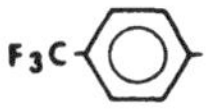 F_3C-⟨⟩-	CH_3
metriflufen	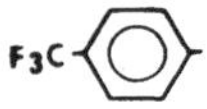 F_3C-⟨⟩-	H

BENZOYL-N- PHENYLS

$$R_1-\text{⟨⟩}-\underset{R_2}{}N-CH-\overset{O}{\overset{\|}{C}}O-R_3$$
$$\overset{CO-\text{⟨⟩}}{} \quad CH_3$$

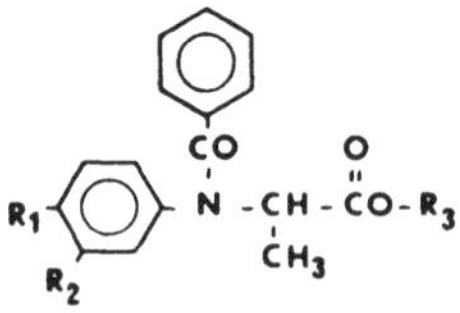

HERBICIDE	R_1	R_2	R_3
benzoylprop-ethyl	Cl	Cl	C_2H_5
flamprop-methyl	Cl	F	CH_3

FIG. 1 Structures of various polycyclic alkanoic acids.

TABLE 1

Physical and Chemical Properties of Several Commercially Available PCA Ester Herbicides

Chemical group	Phenoxyphenoxy	Pyridinyloxyphenoxy	Quinoxalinyloxphenoxy	Benzoyl-N-phenyl
Herbicide	Diclofop-methyl	Fluazifop-butyl	Quizalofop-ethyl	Flamprop-methyl
Trade name of formulated product	Hoelon	Fusilade	Assure	Mataven
Molecular wt.	341	383	372.8	335.5
Water (g/l) solubility	3 at 22°	0.2×10^{-3} at ambient	3.0×10^{-3} at 20°	3.5×10^{-2} at 20°
MP, BP, decomposition point (°C)	39–41, 175–175, 288.4	−20, None at atmospheric pressure, 210	91.7–92.1, 220, ?	81–82, ?, 300
Vapor pressure (mmHg)	0.258×10^{-6} at 20° 0.113×10^{-5} at 30°	4.1×10^{-7} at 20°	3×10^{-7} at 20°	3.5×10^{-7} at 20°
Color	None	Light straw	White	Light tan
Odor	None	None	?	Mild
Physical state	Solid	Liquid	Crystalline solid	Crystalline solid
Specific gravity (g/cm^3)	1.3 at 40°	1.21 at 20°	1.35 at 20°	0.44

Source: From Refs. 7–11.

C. Toxicological Properties

Short-term and acute toxicological properties of several of the commercialized PCA esters are listed in Table 2. None of the commercialized PCAs are very acutely toxic, except to aquatic organisms. However, in the acid form, the acute toxicity of PCAs to aquatic life may be considerably reduced. For instance, haloxyfop is more than 3000-fold less toxic to minnows in the acid form than in the methyl ester form [10]. This may be a result of better absorption of the more lipophilic ester. No information on the mechanism of animal toxicity of the PCAs has been published.

Several effects of PCAs, however, have been found on mammalian physiological and biochemical systems. For instance diclofop-methyl stimulates calmodulin-dependent cyclic nucleotide phosphodiesterase from bovine brain twofold at 100 μM [12]. In vitro assembly of brain tubulin into microtubules was inhibited by 25% at 100 μM diclofop-methyl [13], however, such a small effect at such a high dose is not likely to be important. Benzoylprop-ethyl and flamprop-isopropyl interfered with energy transfer and inhibited electron transfer in rat liver mitochondria [14]. The effects were thought to be secondary effects of decreasing membrane fluidity. The free acids were far less effective than the esters. Since animals readily metabolize PCA esters to acids, little toxicity could be expected from these mechanisms.

The results of mutagenicity tests and tests of long-term sublethal doses of several PCAs are shown in Table 3. No significant mutagenic or sublethal effects of commercialized PCAs have been reported. Consequently, tolerance levels are in the 1 ppmw range. The tolerance limit for diclofop-methyl and/or its acid under the Federal Food, Drug, and Cosmetic Act is 0.1 ppm in flax (*Linium usitatissinum* L.)[15], soybean [*Glycine max* (L.) Merr.] [7], wheat (*Triticum aestivum* L.) [7], and barley (*Hordeum vulgare* L.) [7] seed. The tolerances for fluazifop in soybean and cotton (*Gossypium hirsutum* L.) seed oil are 2.0 and 0.2 ppmw, respectively [16].

D. Synthesis and Analytical Methods

This review will only briefly mention synthesis of PCAs because of the lack of significant literature on this subject, especially for the commercialized PCAs. Papers have been published, however, which give some information on synthesis of phenoxyphenoxy-propionic acids and their derivatives [17] and quinoxalinyloxyphenoxys [18]. Virtually all published information on PCA synthesis is contained in the patent literature. As mentioned earlier, this literature is extensive and complex.

There is considerable evidence that one enantiomer of a PCA is much more herbicidally active than the other. The laevorotatory

TABLE 2

Toxicity of Several Important PCAs in Short-Term Studies

Herbicide	Test	Species	LD_{50} (mg/kg)	LC_{50} (ppm)
Haloxyfop-methyl	Acute oral	Rat (male)	393	—
		Rat (female)	599	—
		Mallard duck	>2150	—
	Acute dermal	Rabbit	>5000	—
	Dietary (8 days)	Bobwhite quail	—	>5620
	Fresh water	Fathead minnow	—	0.3
	(96 hr exposure)	Bluegill	—	0.2
		Rainbow trout	—	0.4
	(48 hr exposure)	Daphnia spp.	—	6.2
Diclofop-methyl	Acute oral	Rat	557–580	—
		Dog	>1600	—
		Bobwhite quail	4400	—
	Acute dermal	Rat	>5000	—
		Rabbit	180	—

TABLE 2

(continued)

Herbicide	Test	Species	LD_{50} (mg/kg)	LC_{50} (ppm)
Flamprop-methyl	Dietary (8 days)	Bobwhite quail	—	13,000
		Mallard ducks	—	20,000
	Acute oral	Pheasants, domestic fowl, mallards	>1000	—
		Pigeons, bobwhite quail	4640	—
		Rat	1210 in DMSO	—
		Mouse	720 in DMSO	—
		Dog	>2000	—
	Acute IP	Rat	—	350–500 in DMSO
Fluazifop-butyl	Acute oral	Rat (male)	4096	—
	Acute oral	Mallard duck	>17,000	—
	Acute dermal	Rabbit	>2420	—
	Fresh water (4 days)	Rainbow trout	—	1.6

Quizalofop-ethyl		
Daphnia magna (2 days)	—	10
Acute oral		
Rat (male)	1670	—
Mallard duck	>2000	—
Bobwhite quail (8-day dietary)	—	>5620
Fresh water		
Rainbow trout (96 hr exposure)	—	10.7
Bluegill sunfish	—	0.46–2.8
Daphnia (48 hr exposure)	—	6.4

Source: From Refs. 7–11.

TABLE 3

Sublethal and Mutagenic Effects of Several PCAs

Chemical	Test	Test level	Result
Diclofop-methyl	Ames	Up to 5 mg	Negative in two studies with 4 bacterial strains with and without liver preparation
	Rat reproduction	>30 ppm	Negative
	Micronucleus test (mice)	?	Negative
	Mutagenicity test with dominant lethal mice	>100 mg/kg	Negative
	skin irritation (rabbit)	?	Slightly irritating
Flamprop-methyl	3-Month feeding studies		
	Rats	5000 ppm	Minor effects
		10 ppm	No effects

Quizalofop-ethyl			
	Dogs	1000 ppm	Minor effects
		10 ppm	No effects
	3-Month feeding		
	Rats	40 ppm in diet	No effect
	Mutagenicity	?	Negative results in Ames bacterial assay, mouse micronucleus assay, in vitro cytogenic assay, in vitro unscheduled DNA synthesis assay, and in vitro Chinese hamster ovary assay
	Teratogenicity	300 mg/kg/day	No effect
		(rat)	
		60 mg/kg/day	No effect
		(rabbit)	

Source: From Refs. 7 and 8.

(R) enantiomer of flamprop-isopropyl was shown to have twice the
activity of the racemates against wild oats [19]. Resolved isomers
of both flamprop-methyl and flamprop-isopropyl are marketed [20].
The dextrorotatory (R) enantiomer of diclofop-methyl was found to
be the herbicidal principle of the racemate [21]. Optically pure
(R)(+)-2-[4-(6-chloro-2-quinoxalinyloxy)phenoxy] propanoate was
430 and 65 times more herbicidally effective on rice (*Oryza sativa*
L.) and barnyardgrass (*Echinochloa crusgalli*), respectively, as the
(S)(−)-enantiomer [22]. The S enantiomer of fluazifop-butyl had no
effects on fresh weight of wild oat and maize (*Zea mays* L.) seedlings at rates (160 g ha^{-1}) with which the R enantiomer reduced
fresh weight by about 90 and 95%, respectively [23]. Grass weed
control in winter oil seed rape was much better with the R than the
S enantiomer of fluazifop-butyl [24]. In pre-emergence treatments
of maize with fluazifop-butyl there were, however, no significant
differences between the R and S enantiomers [23]. This was explained by possible stereochemical transformations in soil (see
Section II.B). Published methods are available for commercial synthesis of excesses of the herbicidally active R enantiomers of PCAs
[25,26]. In fact, fluazifop-butyl is now sold in the resolved R
enantiomer form (Fusilade 2000).

     Methods for analytical determination of various PCAs are well
documented (Table 4). Virtually all published methods involve gas
chromatography (GC) or gas-liquid chromatography (GLC) separation
of extracted PCAs, followed by electron capture detection. Although
bioassays for PCAs exist [e.g., Refs. 37, 38], bioassays are generally poor substitutes for chemical analysis.

     An immunoassay for diclofop-methyl has been developed which
has a capability of detecting 0.23 ppm of the herbicide in relatively
crude biological samples [39]. Both an enzyme-linked and a fluoro-
immunoassay were successfully tested. The de-esterified acid form,
diclofop, had only 34% cross reactivity with the antibody, while the
2-methoxy-1-methyl-2-oxoethyl ester of diclofop had 59% cross reactivity. Compounds, including other PCA herbicides, which did
not posses the 4-(2,4-dichlorophenoxy) phenoxy moiety of diclofop
had no cross reactivity with the antibody.

E. Formulations

     Because of the large number and variety of PCAs that are avail-
able, there are no generalities that can be made regarding formulations other than that commercial versions of PCAs are all esters.
However, because of their relatively low water solubility, they are
generally formulated as emulsifiable concentrates. In some cases
(e.g., haloxyfop-methyl), addition of nonphytotoxic oil concentrates
containing a surfactant to spray solutions is recommended. For
instance, addition of crop oil concentrates increased efficacy of

TABLE 4

Analytical Methods for Determination of Various PCAs

Herbicide	Derivatization	Separation	Detection	Sample type	References
Diclofop and its methyl ester	Pentafluoro-benzylation	GLC	Electron capture	Soil	27, 28
	None	GC	Electron capture	Soil	29–32
	None	GLC	Electron capture	Water	33
	None	HPLC	UV spectrophotometry	Formulations	34
Benzolyprop-ethyl and methyl	None	GC	Electron capture	Soil	29
	None	GC	Electron capture	Water	33
	None	GC	Electron capture	Sediment	32
Flamprop-methyl	None	GC	Electron capture	Soil	29
Fluazifop-butyl	None	GC	Flame ionization	Formulations	35
Haloxyfop-methyl	None	TLC	UV	Soil	36

TABLE 5

Interactions of PCA Herbicides with Other Herbicides, Adjuvants, Additives, and Safeners

PCA	Interactant	Type of Interaction	References
Herbicides			
Diclofop	Chlorsulfuron	Antagonism of diclofop	45
Diclofop-methyl	MCPA	Antagonism of diclofop-methyl	46—48
	Bentazon	Antagonism of diclofop-methyl	49
	2,4-D	Antagonism of diclofop-methyl	50—52
	2,4-DB	Antagonism of diclofop-methyl	48
	Desmedipham	Antagonism of diclofop-methyl	53
	Metribuzin	Antagonism of diclofop-methyl	54
	Metamitron	Synergism	55
Flamprop	Metribuzin	Antagonism of flamprop	56
	Chlorsulfuron	Antagonism of flamprop	45
Flamprop-methyl	MCPA	Antagonism of flamprop-methyl	56
	2,4-D	Antagonism of flamprop-methyl	56, 57
	Dicamba	Antagonism of flamprop-methyl	57
	Bromoxynil	Antagonism of flamprop-methyl	57

	2,4-D	Antagonism of flamprop-isopropyl	57
	Dicamba	Antagonism of flamprop-isopropyl	57
Benzoylprop-ethyl	MCPA	Antagonism of benzoylprop-ethyl	57
	Bromoxynil	Antagonism of benzoylprop-ethyl	57
	Dicamba	Antagonism of benzoylprop-ethyl	57
	2,4-D	Antagonism of benzoylprop-ethyl	56, 57
	MCPA	Antagonism of benzoylprop-ethyl	57

Adjuvants and Additives

Diclofop-methyl	Renex 36	Synergism	58
Fluazifop-methyl	Crop oil	Synergism	40
Flamprop-methyl	Ammonium nitrate/urea	Antagonism	59

Safeners

Fluazifop-butyl	1,8-Naphthalic Anhydride	Antagonism	60
	Cyometrinil	Antagonism	60
	N-(1,3-dioxolan-2-yl-methoxy)imino benzeneaceto-nitrile	Antagonism	60

fluazifop-butyl and haloxyfop-methyl [40]. Nonionic wetting agents added to PCA tank mixes of diclofop generally had no effect on efficacy of the herbicide [41]. New methods of formulation such as low volume spraying in various oils [42] and granulated formulations [43,44] show excellent potential with the PCAs.

Considerable information exists on the compatibility of other herbicides, adjuvants, and safeners with PCAs. This information is summarized in Table 5.

II. DEGRADATION PATHWAYS

A. Degradation in Plants

The first chemical reaction of PCA esters and related esters in most species is hydrolysis to the parent acid [61–70]. This hydrolysis is not a degradative step in terms of herbicidal activity, but it is a bioactivation step. It is assumed that, in plants, the hydrolysis of the ester is accomplished by an esterase, specifically a carboxylesterase. This enzyme has been partially purified and characterized from oat (*Avena sativa* L.) [65] and wild oat [62].

After hydrolysis, a number of chemical reactions can occur. These include conjugation to another ester conjugate in susceptible species, and aryl-hydroxylation [67] and phenolic conjugations in resistant species [71]. Although some differences in metabolism of PCA herbicides in susceptible and resistant species have been found, these differences do not entirely explain differential sensitivity to the herbicide (see Section III.C). A few specific examples of PCA metabolism in plants will be considered.

1. Benzoylprop-ethyl

Benzoylprop-ethyl is rapidly esterified to benzoylprop in sensitive species such as wild oats [62,63]. The carboxylesterase responsible for this de-esterification has been partially purified from wild oat [62] and oats [64]. Wheat detoxifies benzoylprop acid, preventing large accumulations of the toxic form of the herbicide [63]. Detoxication is slower in susceptible oats.

Environment can strongly affect metabolism of herbicides by plants. For instance, under low light wild oats de-esterified more benzoylprop-ethyl than under high light, resulting in a buildup of benzoylprop acid [62]. Conjugation of the acid form was not affected by low light until 12 days after treatment, presumably due to limited carbohydrate. Under low nutrient level conditions a higher percentage of benzoylprop-ethyl accumulated in wild oat than in high nutrient-treated plants, however, the high nutrient-treated plants absorbed more ^{14}C-labeled herbicide [72].

2.   Diclofop-methyl

Metabolism of diclofop-methyl in plants has been well studied, compared to other PCAs.   The subject of plant metabolism of diclofop-methyl has been reviewed by Shimabukuro [71] as part of a more extensive review.   A schematic of the metabolic processes known to occur in tolerant and susceptible species is shown in Figure 2.

Hydrolysis of the ester occurs in both susceptible and tolerant species, however, in susceptible species, conjugation to an ester conjugate is more rapid than in tolerant species [61].   In tolerant species, the free acid is aryl-hydroxylated, followed by conjugation to a phenolic conjugate [73].   Since aryl-hydroxylation is irreversible and ester conjugation is not, the basis for tolerance in some species may be inactivation through aryl-hydroxylation and/or phenolic conjugation.

In wild oat, moisture stress did not significantly affect the pattern of metabolism of diclofop methyl in wild oats [74].

3.   Flamprop

Both the methyl and isopropyl esters of flamprop have been studied.   Both are hydrolyzed to flamprop, but the methyl ester is hydrolyzed somewhat faster [75].   Flamprop-methyl is also known to

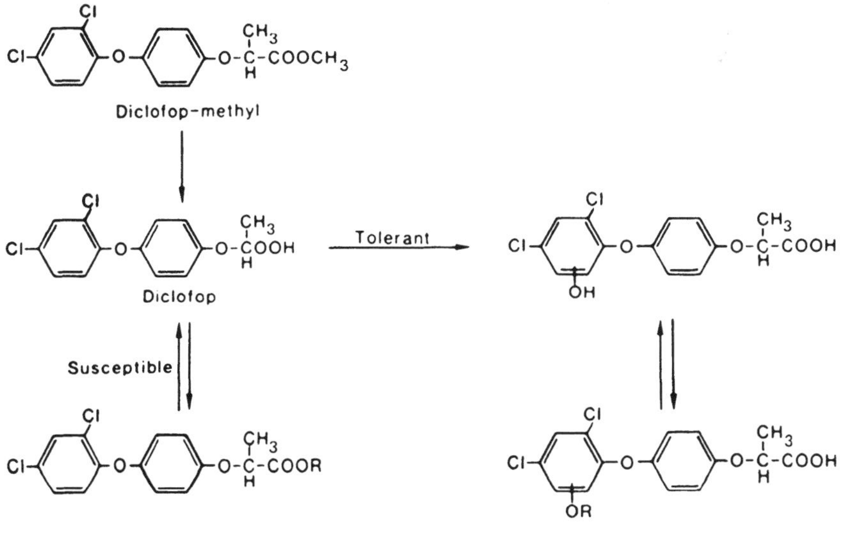

FIG. 2   Metabolism of diclofop-methyl in tolerant and susceptible species of higher plants.   Reproduced from Ref. 71.

be conjugated [76] to the malonylglucose ester [77], apparently in a stepwise fashion with esterification to glucose first, followed by acetylation of the hexosyl derivative with malonic acid. Pillmoor et al. [78] found flamprop to be rapidly metabolized in oats, with up to 85% degradation with 24 hr. The major metabolic fraction appeared to be sugar conjugates of flamprop.

4. Fluazifop-Butyl

De-esterification is the first metabolic process for fluazifop-butyl as with other PCA esters [79]. In *Elymus repens* (L.), Coupland [80] found metabolism of fluazifop-butyl to be characterized by: (a) relatively high amounts of radioactivity remaining in tissue residues after extraction; (b) small amounts of polar and nonpolar conjugates; and (c) relatively high levels of the PCA acid. Warmer temperatures favored de-esterification of fluazifop-butyl in rhizomes.

5. Haloxyfop-Methyl

Haloxyfop-methyl is rapidly hydrolyzed to haloxyfop in plant tissues and then is more slowly metabolized to more polar metabolites [81]

B. Degradation in Soil

In general, most of the loss of PCAs from soil is through biodegradation, rather than leaching or volatilization. For example, neither diclofop-methyl nor diclofop leaches significantly and no loss of diclofop-methyl from a wheat field could be accounted for by volatilization over a 7-day period [82]. An exception may be haloxyfop, which is more volatile than other commercially available PCAs. As in plants, degradation in soil usually begins with rapid hydrolysis of the ester to the PCA acid, followed by slower degradative steps.

In Canadian prairie soils at 20°C, diclofop-methyl is rapidly hydrolyzed and then slowly degrades with a half life of about 28 days [82]. By October, after a spring application of diclofop-methyl in the field in Saskatchawan, only 8% of the diclofop remained in the soil as the free acid and no ester was detected [29]. The metabolite, 4-(2,4-dichlorophenoxy)phenol, was detected in soil treated with diclofop-methyl [84].

In similar studies, flamprop-methyl was shown to be rapidly hydrolyzed to the free acid [29]. In the field, 20% of the herbicide was still present in the ester form in October after a May application. Six to 16% was present as the acid. No residues or herbicide were found below a soil depth of 5 cm in studies with diclofop-methyl or flamprop-methyl. Hitchings and Roberts [85,86] found the free acid and 3-chloro-4-fluoroaniline to be the principal degradation points of flamprop-methyl. They occurred mainly in bound forms, with only limited metabolism to CO_2.

Similarly, in studies with HOE-35609 (fenthiaprop-ethyl) and HOE 33171 (fenoxaprop-ethyl), less than 20% of recovered radiolabel was found at soil depths greater than 5 cm after one month [87]. The esters of both compounds rapidly hydrolyzed in the soil. Less than 50% of the ester plus the free acid of HOE 33171 and HOE 35609 were present in the soil after less than a week or a month, respectively.

Several environmental factors influence degradation of PCAs in soil. For instance, hydrolysis of the ester of diclopfop-methyl is more rapid at soil moisture levels above the wilting point for plants [88]. Similarly, hydrolysis of the ester bond of haloxyfop-methyl was much slower in air-dried soil than in moist soil [36]. Hydrolysis of diclofop-methyl is slower at pH 5.5 than at higher pHs [83]. There was no effect of pH in the 5.5 to 7.5 range on subsequent degradation rate or metabolites formed. Anaerobic conditions greatly reduce the rate of diclofop-methyl degradation [70,89].

Soil type can strongly influence the degradation rates of PCAs. For example, the respective half life of extractable radioactivity from soils treated with [^{14}C]haloxyfop was 92, 38, and 27 days for clay loam, heavy clay, and sandy loam soils [36]. The half lives of benzoylprop-ethyl and flamprop-isopropyl were 4 to 16 weeks and 4 to 20 weeks, respectively, depending on location [90,91].

There is evidence that the relatively inactive S enantiomer of PCAs may be activated by racemization reactions in the soil. After hydrolysis ($t_{1/2} < 2$ hr), the S enantiomer fluazifop-butyl was found to undergo inversion with a half life of 1 to 2 days [23]. After 2 days of application of racemic mixture (R:S, 1:1), the R:S ratio was 3:1, indicating that racemization reactions in soil favor the R form. A schematic of these findings are illustrated in Figure 3.

C. Degradation in Animals

Little work has been published on metabolism of PCA herbicides in animals. Early work with a series of PCAs (benzoylprop-ethyl, flamprop-methyl, flamprop-isopropyl) showed that these herbicides were rapidly excreted by cattle, pigs, and poultry after metabolic de-esterification [92]. They concluded that the parent acids have ideal physical properties for excretion via urine and feces and were chemically unsuited for transport to milk and eggs. De-esterification in mammals is mediated via a mono-oxygenase [93].

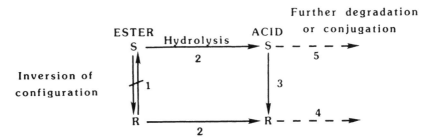

FIG. 3   Transformation of enantiomers of fluazifop in different
environments. (Adapted from Ref. 23.) (1) Absent in soil and
plant; (2) Rapid in soil and plant; (3) Rapid in soil; absent or
very slow in plant; (4) Predominant in soil; significant in plant;
(5) Limited in soil; significant in plant.

D.   Degradation in Microorganisms

Nothing has been published on metabolic degradation of PCAs
by defined microbes, however, numerous laboratory studies of the
microbial degradation mechanisms of PCAs in soils have been made.
The metabolites that have been identified in these studies are listed
in Table 6. Only the phenol and phenetole derivatives of PCAs
have been identified as breakdown products. Since the PCAs are
known to breakdown eventually into CO_2 and other small molecules,
obviously further study is needed to elucidate mechanisms and pathways of microbial degradation of this herbicide class.

E.   Degradation by Photodecomposition

Little information is available on this subject, and that available
is sketchy. Table 7 lists the relative susceptibility to photodecomposition of three PCAs by ultraviolet (UV) irradiation. Photodecomposition of PCAs is not likely to be an important mechanism of dissipation in field situations when these herbicides are soil incorporated,
however, photodecomposition may be a problem with foliar applications.
For instance, haloxyfop was found to be more photolabile to laboratory
UV lights (300–400 nm) than 2,4-D or bentazon [3-(1-methyl ethyl)-
(1H)-2,1,3-benzothiadiazin-4(3H)-one 2,2-dioxide] [95] and the photo-
lability was enhanced by a wide range of adjuvants [95]. Haloxyfop
photodegradation was particularly rapid in mineral oil, with a 92%
loss after 6 hr.

TABLE 6

Microbially Produced Metabolites of Various PCAs Identified in Soil
Degradation Studies

PCA	Metabolites	Reference
Diclofop	4-(2,4-dichlorophenoxy)phenol	29, 84, 89
	4-(2,4-dichlorophenoxy)phenetole	29, 83
Fenoxyprop (HOE-53022)	chlorobenzoxazolone	87
Fenthioprop (HOE-4336)	chlorobenathiazolone	87
Haloxyfop	2-[4-(3-chloro-5-(trifluoromethyl)-2-pyridinyl)oxy]phenol[a]	94

[a]Tentative identification.

F. Movement and Persistence in Water

    In general, PCAs are strongly absorbed to soil and have very low
solubility in water. As a result, in agricultural situations very little
PCA herbicides enter ground water, streams, or other aquatic environments through leaching or runoff of soil-applied PCAs or PCAs washed
off of foilage. Diclofop is relatively soluble in water, making it an exception. Nevertheless, even with diclofop, less than 10% of soil-applied

TABLE 7

Resistance to Decomposition of Several PCA
Herbicides by Ultraviolet Light

	Resistance	Reference
Diclofop-methyl	Low	7
Flamprop-methyl	High	7
Fluazifop-butyl	High	7

herbicide was found to be leached to greater than 10-cm depths in a
variety of soils by 10 cm of water [96]. In a study in which 2.3
μg/L of diclofop was found in tailwaters (runoff) after irrigation,
6.1 μg/L was found in drain canal water [97]. In the same study
drain canal levels of two non-PCA herbicides, triallate [S-(2,3,3-
trichloro-2-prophenyl)bis(1-methylethyl) carbamothioate] and tri-
fluoralin (α,α,α-trifluoro-2,6-dinitro-N-N-dipropyl-p-toluidine), were
much lower than tailwater concentrations.

III. MODE OF ACTION

A. Uptake and Translocation

    PCA herbicides are readily absorbed by root and leaf tissues.
A summary of several foliar absorption studies is given in Table 8.
Although absorption varies with species and with the PCA herbicide
studied, it can be generalized that PCA herbicides are relatively well
absorbed. Nevertheless, adjuvants and environmental parameters
can strongly alter the rates and, to a lesser extent, the final level
of absorption.

    The absorption values in Table 8 may not reflect accurately the
amount of herbicide to have actually penetrated the cuticle, due to
incomplete removal of the herbicide from the cuticle. In a very care-
ful study, 42% of applied flamprop-methyl was found to be retained
by the cuticle of wheat 10 days after application [107]. This study
indicated that lipophilic compounds like PCA esters can be strongly
adsorbed to the cuticle epicuticular wax, slowing desorption into the
apoplast and making removal by leaf washes difficult. Various adjuvants which temporarily dissolve epicuticular waxes may enhance
adsorption to epicuticular waxes, increasing rainfastness, while slowing movement to the apoplast.

    In field situations, rainfall after foliar application can greatly re-
duce the absorption of applied herbicide by reducing the amount of
herbicide in contact with the plant. Various adjuvants can increase
the "rainfastness" of a herbicide and/or increase the rate of absorption. Little work has been done on the rainfastness of PCA herbicides. About two-thirds of the fluazifop-butyl applied in a commercial
formulation with an additional surfactant was washed off of *Elymus
repens* foliage by 1 cm of simulated precipitation given 10 min after
spraying the plants with the herbicide [108].

    Absorption is generally positively correlated with relative humid-
ity (RH). Harrison and Wax [104] found foliar absorption of
haloxyfop-methyl in maize was higher three days after application
at 70% than at 30% RH. In soybeans and johnsongrass [*Sorghum
halepense* (L.) Pers.], however, McWhorter and Wills [105] found
no difference in absorption of haloxyfop-methyl at 45 and 95% RH

TABLE 8

Absorption of PCA Herbicides After Foliar Application

Herbicide	Species	Absorption (% of applied)	Time after application (hours)	Reference
Diclofop-methyl	Barnyardgrass	65	72	98
	Proso milet	69	72	98
	Cucumber	77	72	98
	Sunflower	60	96	99
	Wheat	90	96	100
	Wild oat	70	96	100
Fluazifop	Bermudagrass	87	48	101
Fluazifop-butyl	Setaria viridis	95	24	102
	Quackgrass	>80	48	103
	Soybean	>90	48	103
Haloxyfop-methyl	Corn	65	5	104
	Soybean	>75	6	81
		>80	48	105
	Shattercane	>75	6	81
	Johnsongrass	>60	48	105
	Yellow foxtail	>80	6	81
Metriflufen (HOE-29152)	Johnsongrass	>60	48	106
	Soybean	>60	48	106

two days after treatment. Wills and McWhorter [101] found absorption of fluazifop to be greater at 100 than 40% RH in bermudagrass.
McWhorter [106], however, found relative humidity did not influence
the foliar absorption of metriflufen by soybean or johnsongrass.
Moisture stress has been reported to have little effect on absorption
of diclofop-methyl by wild oat leaves [74,109].

Within metabolic ranges, increased temperatures generally increase
absorption of PCAs. Foliar absorption of metriflufen in soybeans was
significantly higher at 27°C than at 18°C and higher at 35°C than at
27°C [106]. No effect of temperature, however, was detected in
johnsongrass. Fluazifop absorption by bermudagrass [Cynodon
dactylon (L.) Pers.] was greater at 35 than at 27°C and greater at
27 than at 18°C [101]. Haloxyfop-methyl absorption was greater in
soybeans at 35 than at 18°C, but not in johnsongrass [105]. Thus,
humid, warm conditions generally favor PCA uptake. For instance,
sunflowers (Helianthus annuus L.) absorbed about 30% more foliarly
applied diclofop-methyl at 30°C and 90% RH than at 10°C and 40%
RH [99].

The portion of the plant in contact with the herbicide can strongly affect absorption. Diclofop-methyl was absorbed 64% and 95%
better by the base of the first and second leaves, respectively, of
wild oats than by the leaf tips [110]. This difference may have been
due to less developed cuticle in the younger, basal portions of the
leaves, however, Whitehouse et al. [111] found that epicuticular wax
was probably not a barrier to diclofop-methyl absorption of wild oats.
Absorption of diclofop-methyl by roots of oat and wheat was more
than three times greater in the apical than basal root regions [112].

Surfactants and/or adjuvants can also greatly increase absorption
of PCA herbicides. For instance, a mineral oil surfactant increased
absorption of fluazifop in bermudagrass fourfold at 18°C and 40% RH
and 30% at 35°C and 40% RH [101]. Nalewaja et al. [113] found that
stearic, oleic, and linoleic acids and their methyl esters significantly
increased absorption of fluaxifop in oats. Petroleum oil and methylated sunflower oil increased absorption of fluaxifop, diclofop, and
CGA-82725 {(±)-2-[4-(3,5-dichloro-2-pyridyloxy]propanoic acid)-2-
propynyl ester} in oats. Refined sunflower oil increased absorption
of CGA-82725 and diclofop, but had no effect on fluazifop absorption.
Fluazifop absorption by oats was not affected by palm, safflower,
soybean, or linseed oil additives, however, several chemically un-
defined emulsifiers increased absorption [114]. A phytobland oil
adjuvant increased absorption of haloxyfop-methyl in soybeans and
johnsongrass [105].

Other compounds can reduce absorption of PCAs. For example,
the herbicide MCPA significantly reduces uptake of diclofop-methyl by
wild oat leaves [115]. The inhibition is apparently due to chemical interactions outside the plant. In MCPA-pretreated plants, no effect was

found on diclofop uptake [116]. Although bentazon and 2,4-D antagonize diclofop, neither affects absorption of diclofop or diclofop-methyl [117]. Hall et al. [117] found the methyl ester of diclofop to be absorbed significantly better than diclofop, although both were absorbed more than 80% three days after treatment in both soybean and oat.

Although absorption of PCA herbicides is generally good, translocation is considerably reduced and more variable. For instance, in five species, less than 2% of the ^{14}C from [^{14}C]diclofop in a treated leaf was translocated over a five-day period. Most of the translocated ^{14}C was moved acropetally [98]. Ten days after foliar treatment of sunflowers with diclofop-methyl, about 2% of the herbicide absorbed was translocated from the treated leaf [99]. Approximately equal amounts of herbicide were translocated to roots, to the shoot above the treated leaf, and to the shoot below the treated leaf. Less than 5% of root-absorbed diclofop-methyl was translocated in wheat and oat roots [112]. Hall et al. [117] found about 5% of absorbed diclofop-methyl to be translocated in oat and soybean. Radioactivity accumulated in the intercalary meristem of oat. In wheat, 10 days after treatment, about 5% of applied flamprop-methyl was translocated from the treated leaf [107]. Within the treated leaf, movement was primarily acropetal. Nalewaja et al. [113] increased translocation of diclofop in oats during three days after application from about 10% of absorbed to about 30 and 20% of absorbed with petroleum oil and methylated sunflower oils, respectively, as adjuvants. Less than 1% of ^{14}C from applied quizalofop-methyl was translocated from the treated leaf of *Echinochloa crus-galli* 6 days after treatment [118]. Approximately one-third of absorbed CGA-82775 was translocated in oats within two days after application and this percent of absorbed herbicide was not greatly changed by petroleum or sunflower oil adjuvants, even though the total absorbed herbicide was increased 5- to 7-fold by oil adjuvants [113].

Even though haloxyfop-methyl is rapidly absorbed, it is also poorly translocated. In corn, 5 hours after foliar treatment, 3% of recovered ^{14}C was translocated from the treated leaf area which contained 43% of recovered radioactivity [104]. Within the treated leaf, translocation was primarily acropetal. Wilhm et al. [120], however, found almost 50% of foliar-applied haloxyfop to be translocated within 96 hr after treatment. McWhorter and Wills [105] also found metriflufen to be mostly acropetally translocated in treated johnsongrass leaves.

Fluazifop-butyl appears to be translocated better than other PCAs. In a rare published structure-activity analysis of PCAs, Kimura et al. [119] found haloxyfop-methyl to translocate in bermudagrass much better than its phenoxyphenoxy analogue. Generally, phenoxyphenoxy analogues of pyridinyloxyphenoxy compounds did not translocate as well as the pyridinyloxyphenoxy compound.

Kells et al. [103] found up to 25% of absorbed fluazifop-butyl to be translocated, 144 hr after treatment of quackgrass. There was little difference in translocation in susceptible quackgrass and resistant soybeans. Translocation was both basipetal and acropetal in both species, with ^{14}C accumulation in meristems. Carr et al. [102] also found that ^{14}C from radiolabeled fluazifop-butyl accumulates in meristems, however, they found only 2% of applied ^{14}C was translocated. They found most of the translocated ^{14}C to be in the form of fluazifop acid and none as fluazifop-butyl. Wills and McWhorter [101], however, found up to 60% of fluazifop applied to bermudagraws to be translocated from the treated area. Similarly, Nalewaja et al. [113] found approximately that 60% of fluazifop absorbed by oats was translocated. Although several fatty acids and their methyl esters, petroleum oil, and methylated sunflower oil increased absorption of the herbicide, they did not increase the percent of absorbed herbicide that was translocated. The pattern of translocation of fluazifop in oats was not changed by a petroleum oil additive [113]. Translocation was both acropetal and basipetal.

As with absorption, higher temperatures and relative humidities generally increase translocation of PCA herbicides. For instance, increased RH and temperature increased translocation of metriflufen in johnsongrass and soybeans [106]. At 35° and 100% RH, almost 75% of ^{14}C from fluazifop applied to bermudagrass was translocated, whereas at 18° and 40% RH less than half of absorbed ^{14}C from fluazifop was translocated [101]. Water stress greatly reduced translocation of diclofop-methyl in wild oat [74]. Fluazifop-butyl translocation was significantly greater in quackgrass under bright light than in shade [103], probably due to increased phloem transport activity. Adjuvants can also influence haloxyfop translocation. For example, petroleum oil concentrate increased translocation of foliar-applied haloxyfop-methyl in corn, while polyoxyethylene sorbitan monolaurate reduced its translocation [104]. Although a mineral oil surfactant increased absorption of fluazifop, it reduced the proportion of ^{14}C from [^{14}C]fluazifop translocated [120].

The relatively poor translocation of most PCA herbicides may be a result of rapid cellular uptake by the cells most proximal to the site of application, reducing the level of herbicide available for translocation. Generally, lipophilic compounds, such as many PCA esters, are not phloem-mobile and cross the plasmalemma easily. Cultured soybean and yellow foxtail (*Setaria glauca* (L.) Beauv.) both readily absorbed [^{14}C]haloxyfop-methyl from culture medium [81]. Once in the symplast, conjugation or other alteration of the herbicide, depending on the plant species and PCA, can further limit its potential for translocation. Pillmoor et al. [78] found conjugation of flamprop to cause retention of the herbicide by the symplast, limiting its availability for transport. The de-esterified,

less lipophilic form of the herbicide is probably the principal translocated form of most PCA herbicides.

B. Physiological Effects

The physical and biochemical effects of the structurally related
PCA herbicides appear to be similar. Herbicidal injury of the PCAs
usually follows the pattern of: (a) arrested growth, (b) chlorosis,
and (c) ultimately meristematic necrosis. Investigations on the
mechanism of action of the PCAs have focused on four observations.
Low concentrations of the herbicides cause (a) membrane disruption,
(b) growth inhibition of apical, internodal, and root meristems, (c)
disruption of auxin action, and (d) inhibition of fatty acid biosynthesis. Evidence in support of each of these observations and other,
more general effects will be presented and general conclusions on
the mechanism of action of the PCAs will be offered.

1. Membrane Disruption

One of the earliest visible signs of herbicidal injury at the point
of contact in leaves is a water-soaked appearance followed by chlorosis
and necrosis [121,122]. Ultrastructural studies revealed a disruption
of internal cellular membranes in both sensitive oat and insensitive
wheat exposed to diclofop-methyl, with chloroplasts exhibiting the
most sensitivity, although plasmalemma damage and membrane reticulation and extraplastidic myelin figures were common [122]. In *Triticum
monococcum* L. cell suspension cultures, Davis and Brezeanu [123]
found diclofop-methyl to cause extensive myelin figures, plastid
disorganization, and reticulation. No ultrastructural studies of effects of PCAs on meristems have been published. In unpublished
work, however, Vaughn [124] has found haloxyfop-methyl to cause
extensive vacuolization and vacuolar engulfment of cytoplasm in root
meristems of grain sorghum [*Sorghum bicolor* (L.) Moench] seedlings
(Fig. 4).

Using direct measurements of membrane permeability, Crowley
and Prendeville [125] and O'Leary et al. [126] found that diclofop-
methyl caused an increase in leaf cell membrane leakage. Hoppe
[127] found that diclofop-methyl causes leakage of amino compounds
from maize root tips. These results suggest that one of the early
effects of this herbicide is a disruption of cell membrane integrity.
Generally the cause of cell leakage may be direct, a result of membrane disruption, or indirect, through a disruption of energy trans-
duction in chloroplasts or mitochondria [128]. Cohen and Morrison
[129] measured the effect of diclofop-methyl on isolated wheat and
oat mitochondria. They found that the herbicide affected the
permeability properties of the inner mitochondrial membrane from
both plants, similar to an uncoupler of electron transport. They

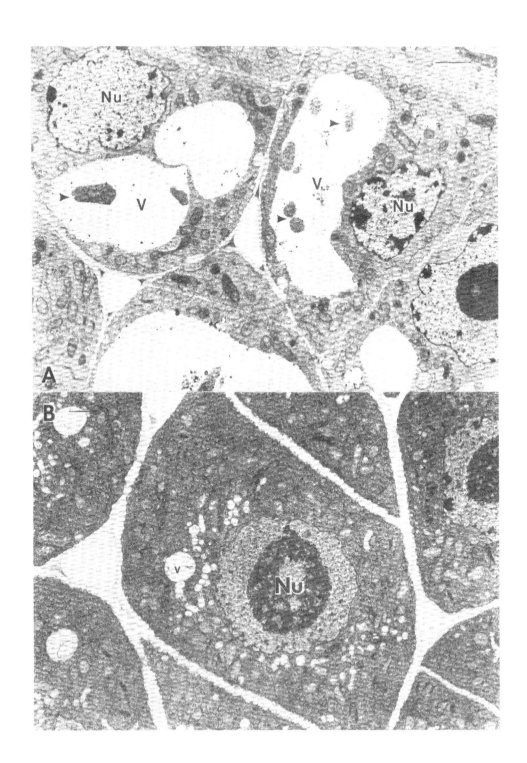

98

concluded that diclofop-methyl does not have a specific effect on mitochondria but rather affects membrane permeability in a more general
way.

Wright and Shimabukuro [130], using oat and wheat coleoptiles
and Lucas et al. [131] using *Chara* cells demonstrated that diclofop
can cause a partial collapse of the plasma membrane potential. This
effect may be caused by diclofop acting as a proton ionophore or
directly interfering with the plasmalemma ATPase. Rather high concentrations (100 µm) were required for the effect, however, which
suggests that alteration of the membrane potential may not be a
primary effect of the herbicide. Ikai et al. [118] found 10 µm
quizalofop-ethyl to cause cellular leakage of oat coleoptile tissue by
14 hr after exposure.

2. Inhibition of Growth and Development

Growth inhibition resulting in a stunted growth habit is a common
feature of sensitive plants treated with PCAs [75,132–135]. Shoots
as well as roots are affected by diclofop-methyl and these effects are
associated with decreases in photosynthetic activity, photosynthate
translocation to roots, and ATP levels and accumulation of sugars in
the shoots [136]. Both diclofop-methyl and its parent acid, diclofop
inhibited root growth in oat but not wheat [137]. Furthermore, root
growth was more sensitive to diclofop than to diclofop-methyl, suggesting that diclofop may be the molecule which causes root inhibition
in vivo.

Morrison et al. [56], in a detailed histological study of the effects
of benzoylprop-ethyl and flamprop-methyl on growth of wild oats, observed that both cell division and cell elongation were inhibited.
These effects resulted in a strong reduction in the growth of leaves
and internodes of herbicide-treated plants. Both herbicides also inhibited vascular development and this caused retarded development
of the apical meristem.

The effect of diclofop-methyl on maize is similar to that on wild
oats, causing inhibited radicle growth [138]. It was found that a
known cell division inhibitor, hydroxylurea, reduced the effect of
diclofop-methyl, suggesting that the herbicide also affected cell
division in maize.

FIG. 4   Cells from the apical meristem of grain sorghum root grown
for 3 days on water and then transferred to 1 µM haloxyfop methyl
(A) or water (B). Samples were taken 24 hr after treatment. v =
vacuole, arrows denote bits of cytoplasm engulfed by vacuoles, Nu =
nucleus. Bars = 2 µm. Courtesy of K. C. Vaughn.

Haloxyfop-methyl also strongly inhibits root growth of corn at concentrations (1 μm) that have no effect on soybean root growth [135]. The effect on growth precedes effects on respiration rate or ATP content.

Diclofop-methyl strongly retards adventitious root development in both wild oats and wheat when applied in nutrient solution [132]. Root elongation and reduction in mitotic index, however, are more sensitive to diclofop-methyl in wild oats than in wheat (Table 9), showing significant inhibition within 24 hr at submicromolar concentrations. Histological examination of treated roots indicated that the developing central cylinder of wild oat roots, close to the apex, was particularly sensitive to this herbicide [132].

Donald et al. [139] extended this work and found that wild oat root morphology was altered when as little as 10 to 50 nM diclofop-methyl was added to nutrient media in which the roots were bathed. They also observed a decrease in shoot growth in response to root application of herbicide. They concluded that diclofop-methyl caused a loss of apical dominance in the roots.

3. Auxin Antagonism

The first indication that the activity of the PCAs might be related to auxin action in sensitive plants came from attempts to mix PCAs and phenoxyalkanoic acid herbicides (e.g., 2,4,-D) for broad spectrum weed control. The mixture resulted in severe decrease in efficacy of the PCAs but not in 2,4-D [50,133,140,141]. For instance, in oats, 2,4-D and 2,3,6-TBA partially reversed the chlorosis and growth inhibition caused by diclofop-methyl [133,142] (Fig. 5).

Shimabukuro et al. [137], examining the effects of diclofop and diclofop-methyl on auxin-stimulated oat and wheat coleoptile segments, found that, at 10 μM, both herbicides inhibited the IAA-induced coleoptile growth. The inhibition by diclofop-methyl could be partly overcome with higher concentrations of auxin. In subsequent work it was demonstrated that diclofop-methyl could inhibit IAA-induced acidification of the medium in peeled coleoptile segments from both oat and wheat but only at a concentration of 100 μM [143]. The relative efficacy of diclofop-methyl in inhibiting a IAA-stimulated coleoptile elongation is wild oat > oat > wheat > corn [144] (Fig. 6).

The basis for the observed antagonism between 2,4-D and diclofop-methyl may be due to an altered metabolism of the diclofop-methyl. Todd and Stobbe [145] found that the de-esterification of diclofop-methyl to its parent acid was inhibited in vivo by 2,4-D in wild oats. This resulted in a decreased translocation of diclofop to meristems which allowed the plants to recover from contact injury. The auxin-like herbicide MCPA also inhibited basipetal translocation of diclofop in wild oat [144]. When assayed in vitro, however, the esterase responsible for the conversion of diclofop methyl to diclofop

TABLE 9

The Effect of Root-Applied Diclofop on the Mitotic
Index of Adventitious Root Tips of Wheat and
Wild Oat Treated for 8, 12, and 24 hr

Concentration (μM)	Mitotic index (%)[a]		
	8 hr	12 hr	24 hr
Wheat			
0.00	8.4 a	9.1 a	8.9 a
0.15	7.5 a	8.7 a	7.4 a
0.30	7.0 a	7.8 a,b	6.4 a,b
1.50	6.3 a	7.1 a,b	4.3 b,c
3.00	6.6 a	5.1 b	2.8 c
Wild oat			
0.00	7.6 a	8.5 a	8.1 a
0.15	5.4 a,b	3.2 b	5.8 b
0.30	4.0 b,c	2.6 b,c	2.3 c
1.50	3.4 b,c	1.6 b,c	1.6 c
3.00	2.7 c	1.1 c	0.5 c

[a]Values within a column for each species followed
by the same letter are not significantly different
at the 1% level according to Duncan's multiple
range test.
Source: From Ref. 132.

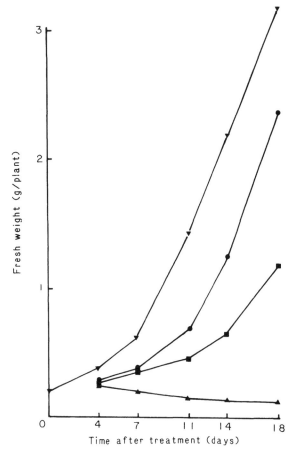

FIG. 5   Increase in fresh weight of oat seedlings after foliar treatment with diclofop-methyl at 3 kg/ha.   ▼ = no diclofop methyl, ▲ =
diclofop-methyl, ■ = pretreated with 2,4-D, ● = pretreated with
2,3,6-TBA.   (Redrawn from Ref. 133.)

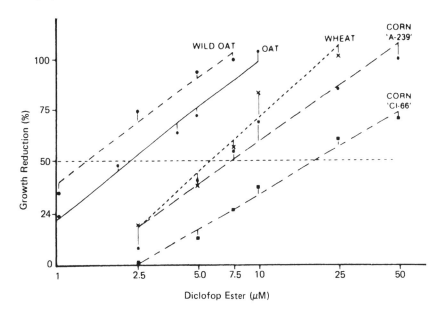

FIG. 6   Percent reduction of IAA-simulated coleoptile growth of wild oat, oat, wheat, and two maize (corn) varieties as influenced by diclofop methyl. (From Ref. 144.)

was not inhibited by 2,4-D [146]. This finding suggests that the effect of 2,4-D on the esterase may be a secondary effect.

Taylor and Loader carefully examined the antagonism between diclofop-methyl and a number of related phenoxyalkanoic acid herbicides [133]. They found the rate of re-esterification of diclofop was increased in tissue treated with 2,4-D or 2,3,6-TBA, resulting in an accumulation of nontoxic conjugates. This enhanced detoxification occurred even when diclofop-methyl was applied as a foliar spray to plants previously treated with 2,4-D through the roots.

Recent studies by Shimabukuro et al. [147] have demonstrated that the antagonistic interaction between diclofop-methyl and 2,4-D is reciprocal and dependent on the relative concentration of the two herbicides, both in corn (susceptible) and soybean (resistant). The antagonism of 2,4-D on diclofop-methyl activity was not due to reduction in diclofop-methyl absorption or an increase in detoxication. They concluded that this antagonism requires the presence of both compounds at the molecular site of action.

In structure-activity studies of antagonism of diclofop-methyl activity with compounds with auxin-like activity, the degree of antagonism and the level of growth-promoting activity of the

auxin-like compound have not positively correlated well [52,133].
It was concluded that antagonism is related to structure—not to
auxin activity.   These findings indicate that herbicides with auxin-
like activity interact with PCA herbicides at a site of action unre-
lated to the site of the auxin-like herbicides—presumably at one of
the ring binding sites.   An experiment in which the antagonisms of
PCAs of widely varying activity on the activity of a phenoxy alkanoic
acid herbicide has not been conducted.   Such an experiment would
help to determine if the PCA-auxin antagonism is related to the
mechanism of action or is simply an interesting phenomenon resulting
from partial structural similarities.   In summary, this partial structural similarity can lead to antagonism by two mechanisms: (a)
competition for the molecular site of PCA action and (b) auxin-like
compound induction of metabolic activity which also detoxifies PCAs.

4.   Inhibition of Fatty Acid Synthesis

Work from the laboratory of Hoppe implicates the inhibition of
fatty acid biosynthesis is the mechanism of action of the PCA herbicides.   In his early work, Hoppe [127] found that diclofop-methyl
greatly decreased levels of phospholipids (primarily phosphatidylethanolamine and phosphatidylcholine) in maize root tips.   Hoppe
and Zacher [148] found that diclofop-methyl inhibited the incorporation of [^{14}C]acetate into lipids in maize root tips.   Inhibition could
be observed after as little as 1 hr preincubation in the herbicide.
In maize leaves [149], Hoppe found inhibition of incorporation of
[^{14}C]acetate into lipids within 30 min after treatment with diclofop-
methyl.   Furthermore, the effect of low concentrations (0.1 μM) of
diclofop-methyl on root growth could be reversed by the addition of
exogenous fatty acids.   Ikai et al. [118] recently found lipid synthesis to be inhibited by levels of quizalofop that have no effect on
protein or RNA synthesis.

Extending this study to include a range of plants of differential
sensitivity to diclofop-methyl, Hoppe observed that fatty acid synthesis was inhibited with a I_{50} of 0.1 μM after a 0.5 to 4 hr pretreatment with the herbicide [150].   Fatty acid biosynthesis was not
inhibited in beans, sugar beets, or soybeans while wheat was as
sensitive as wild oats, maize, and barnyardgrass (Fig. 7).   Unlike
oats, maize, and barnyardgrass, however, wheat recovered from
fatty acid synthesis inhibition by diclofop-methyl.   Synthesis of
many types of lipids is severely inhibited by diclofop-methyl in maize
[148] (Fig. 8).   Sterols, however, are unaffected.

Diclofop and other PCA herbicides inhibit fatty acid synthesis
in chloroplasts isolated from maize, but not from bean [151] (Fig. 9).
In chloroplasts, the inhibition was similar to that in whole leaves
with regard to the activity of different metabolites and stereo-
specificity.   Plastidic fatty acid synthesis was 100 times more

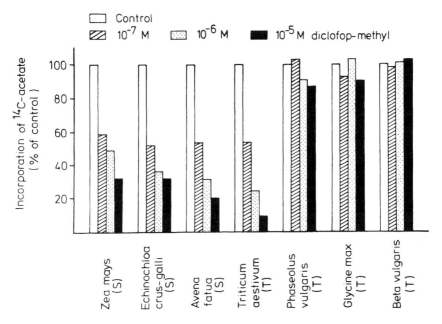

FIG. 7   Effect of diclofop-methyl on the incorporation of [^{14}C]ace-
tate into lipids of leaf pieces from sensitive (S) and tolerant (T)
plant species [From Ref. 150.]

sensitive to D-diclofop (I_{50} = 9 × 10^{-8}M) than to L-diclofop.   The
sensitivity, selectivity, and response time of lipid synthesis inhibition
in vivo and in vitro strongly implicate this process as a primary site
of action of the PCA herbicides.

The mechanism of inhibition of lipid synthesis by PCAs is not
clear, however.   Thus far, this effect has only been demonstrated
at the organelle level.   The herbicide could be inhibiting lipid synthesis by direct inhibition of an enzyme or by perturbation of a
membrane system intimately involved in lipid synthesis.   Recent work
with haloxyfop has indicated that lipid synthesis may be inhibited
at a location near where metabolic intermediates enter both the
Kreb's cycle and lipid synthesis [152].

5. *Miscellaneous Physiological Effects*

Although numerous papers on the effects of PCA herbicides on
such processes as photosynthesis exist, it is generally thought that
these effects are secondary or tertiary.   For example, flamprop-
isopropyl and benzoylprop-ethyl did not significantly reduce photosynthetic rates in wild oats until 3 and 9 days, respectively, after

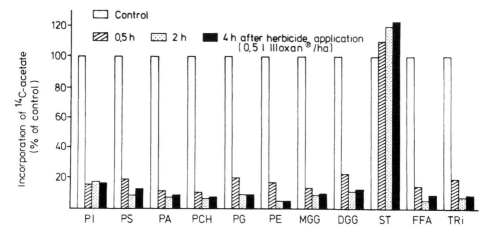

FIG. 8 Effect of diclofop-methyl on the incorporation of [14C]acetate into lipids from leaves of maize. PI = phosphatidylinositol; PS = phosphatidylserine; PA = phosphatidic acid; PCH = phosphatidylcholine; PG = phosphatidylglycerol; PE = phosphatidylethanolamine; MGG = monogalactosyldiglycerides; DGG = digalactosyldiglycerides; ST = sterols; FFA = free fatty acids; TRI = triglycerides [from Ref. 150].

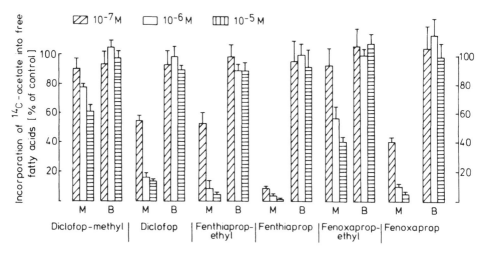

FIG. 9 Effect of phenoxy-phenoxypropionic acid derivatives and structurally related compounds on the incorporation of [14C]acetate into free fatty acids by isolated maize (M) or bean (B) chloroplasts. Isolated chloroplasts were incubated under illumination with [14C] acetate and the herbicides. Incubation was terminated after 1 hr by lipid extraction. 14C-labeled free fatty acids were isolated by TLC, scraped into scintillation vials, and measured in a liquid scintillation spectrometer (mean values ± SD of four replicates from two experiments) (from Ref. 151).

treatment [153]. However, flamprop-isopropyl reduces partitioning into starch within 1 day and benzoylprop-ethyl reduced partitioning of fixed carbon into starch within 2 hr of treatment.

Diclofop-methyl greatly inhibited Ca^{2+} uptake by maize mitochondrial through a mechanism other than interference with respiration [13]. Gauvrit [14], however, has shown that several membrane-related effects of benzoylprop-ethyl and flamprop-isopropyl are not caused by the free acids, but are due to the esters, which are presumably nonherbicidal.

There is no effect of diclofop-methyl on protein synthesis and only a very weak effect on nucleic acid synthesis at treatment levels (5 µM) that reduce lipid synthesis in maize root tips by about 70% [154]. Benzoylprop-ethyl up to 0.1 mM had no effect on DNA synthesis in wild oats in which growth was strongly affected [155]. Peregoy and Glenn [156] found fluazifop-butyl to inhibit protein synthesis at 1 and 10 µM in corn coleoptiles and soybean hypocotyls, respectively. However, the responses were attributed primarily to indirect effects on uptake of radiolabeled precursors. Similar results were obtained with nucleic acid synthesis. Thus, there are no published data indicating direct effects of PCAs on protein or nucleic acid synthesis.

6. Summary

In summary, the data suggest the following mode of action for most PCA herbicides. In postemergent application, the herbicide contacts the leaves and most of it is absorbed. In sensitive plants, a significant percentage of applied PCA is hydrolyzed into the parent acid which is phloem and xylem mobile. The acid moves toward and accumulates in meristematic regions. Disruption of cell division and cell elongation in the meristems results in stunted growth, presumably through membrane alterations and/or inhibition of fatty acid synthesis. When sufficient herbicide has accumulated in the meristematic regions, the occurrence of wholesale membrane disruption and cellular autolysis eventually results in plant death or the inability of the plant to compete with tolerant or resistant plant species.

Presently, it is not possible to assign unequivocally one specific mechanism of action to the PCAs. As with most herbicides, it may be a combination of effects which result in phytotoxicity. Undoubtedly, most of the effects of the PCA herbicides appear to be secondary and are a consequence of intracellular membrane disruption. Membrane changes at the cellular level can account for all of the observed phototoxic symptoms exhibited in membrane leakage and inhibition of growth. Although the molecular mechanism of action has not yet been firmly established, inhibition of fatty acid synthesis can also explain all of the phytotoxic effects, including most of the membrane effects. Until the effects of membrane alterations on fatty acid synthesis and vice versa are examined in detail in both sensitive and resistant plants, it will be difficult to designate unequivocally the primary sequence of events resulting from PCA activity.

C. Selectivity

One of the attractive features of the PCA herbicides is their selectivity. As mentioned earlier, they are phytotoxic to a wide range of weed species, while several important crop species are in- sensitive. The biochemical bases of these selectivities will be dis- cussed in this section.

There is no good evidence that selectivity of PCA herbicides is related to differential absorption. In many studies resistant species have taken up more PCA herbicide than susceptible species [e.g., 81,100,103,106]. Interpretation of translocation studies is difficult because in only a few studies have the researchers determined the identify of the translocated radioactive material. Generally, however, translocation is low in both sensitive and resistant species and the differences observed do not suggest differential translocation as a mechanism of tolerance or resistance. This has been the conclusion of those studying diclofop-methyl-tolerant dicots vs. susceptible monocots [98,117] and tolerant vs. susceptible monocots [100,112, 157]. Kells et al. [103] also found no difference in translocation of fluazifop-butyl in tolerant soybeans and susceptible quackgrass.

Differential metabolic activation or degradation can account for some selectivity. There is evidence that a higher level of esterase results in faster hydrolysis of benzoylprop-ethyl to phytotoxic benzoylprop in susceptible oats than in tolerant wheat [63]. As mentioned in Section II.A, the basis for tolerance of some species appears to be inactivation of the hydrolyzed PCA ester through aryl- hydroxylation and/or phenolic conjugation. However, in some cases, such as tolerant bushbeans [158], the greater tolerance compared to sensitive species could not be related to patterns of degradation.

Recent work in Hoppe's laboratory strongly supports the view that PCA selectivity is often at the site of action of lipid biosynthesis. It was found that lipid synthesis in leaves of three tolerant species (*Beta vulgaris*, soybeans, and *Phaseolus vulgaris*) was insensitive to diclofop-methyl, while lipid synthesis in maize, barnyardgrass, wild oat, and wheat were sensitive [150]. Although wheat was sensitive initially, it recovered, apparently due to detoxication of the herbi- cide. Further investigation revealed that this insensitivity of lipid synthesis in resistant species extends to the plastid level [151]. These results were virtually identical for three PCA herbicides (diclofop, fenthiaprop, and fenoxaprop) and their esters.

Thus, the basis for selectivity of PCA herbicides cannot be attributed to one particular mechanism, even with only one herbicide.

Much more research will have to be done to give an accurate picture of the relative importance of detoxication versus site of action mechanisms. At present, however, the work of Hoppe [150,151] suggests that insensitivity at the site of action is more important than detoxication.

Little has been published on the generation of PCA resistance in susceptible weed populations by sustained utilization of PCA herbicides. A diclofop-methyl-resistant biotype of Wimmera grass (*Lolium rigidim*) has occurred in Australia after only four years of diclofop-methyl use [159]. This biotype is cross-resistant to the other PCA herbicide fluazifop-butyl and to chlorsulfuron, a sulfonylurea, but not to oxyfluorfen, a *p*-nitro-substituted diphenylether [160,161]. No studies of physiological mechanisms of resistance of this biotype has been published, however, the lack of correlation between cross resistance and mechanism of action indicates that it metabolizes PCAs.

IV. CONCLUSIONS

The PCA herbicides represent a wide range of chemical structures that are reflected in the hundreds of patents on PCA herbicides applied for over the last 15 years. PCA herbicides can be generally defined as alkanoic acids (generally propanoic) with more than one ring structure attached to an asymmetric, noncarbonyl carbon. Commercialized PCA herbicides are oxyphenoxy (phenoxyphenoxy, pyridinyloxyphenoxy, quinoxalinyloxyphenoxy, and benzoxazolyloxphenoxy) and benzoyl-*N*-phenylphenoxy propanoic acid esters. Although the acid is the herbicidally active form of these herbicides, the ester is absorbed much more readily and is more "rainfast." Once inside the plant, it is rapidly hydrolyzed to the acid. This hydrolysis both bioactivates the herbicide and transforms it to a more water-soluble, translocatable form.

Although the mechanism(s) of action of the PCA herbicide is (are) not well understood, there is increasing evidence that in many species PCAs inhibit lipid synthesis. Lipid synthesis in several insensitive species is not affected by PCA herbicides. In at least some other species metabolism of the PCA may be the cause for tolerance or insensitivity to the PCAs. Auxin and auxin-like herbicides often lower the activity of PCA herbicides and in some cases, vice versa. These effects have been related to metabolism, translocation, and site of action. Thus, there is no clear picture of how PCAs work and there may be several mechanisms of action, resistance, and antagonism with auxin-like compounds.

The PCAs are generally environmentally safe, because of: (a) low volatility and mobility in soil, (b) low toxicological thresholds,

(c) low levels needed for efficacious control of weeks in field environments, and (d) reasonably short half life in soil and water. These are generalities based on a very small amount of published data. Our understanding of the behavior and fate of the PCAs in organisms, soil, and water is very limited compared to many other classes of herbicides.

If the number of recent patents are an accurate indication of the future of herbicide use, the PCAs will become an even more important herbicide class. We expect that many gaps in the understanding of the herbicide class will be closed in the next decade.

ACKNOWLEDGMENTS

We thank W. C. Koskinen for his help in arriving at a name for this herbicide class. The colleagues that provided preprints of their work and criticisms of this chapter are most appreciated by the author. We especially thank K. C. Vaughn for providing unpublished photomicrographs and M. Gohbara for his insight into the structural requisites for PCA activity. We take full responsibility for all interpretations and points of view expressed in this chapter.

REFERENCES

1. C. Fedtke, *Biochemistry and Physiology of Herbicide Action*, Springer-Verlag, Berlin, 1982.
2. S. Fujisawa, *Japan Pestic. Info.*, *39*:19 (1981).
3. K. Kobayashi and K. Ichinose, *Weed Res.* (Japan), 29, 38 (1984).
4. S. Matsunaka, in *Herbicides: Chemistry, Degradation and Mode of Action*, Vol. 2 (P. C. Kearney and D. D. Kaufman, eds.), Dekker, New York, 1975, pp. 709-739.
5. F. M. Ashton and A. S. Crafts, *Mode of Action of Herbicides*, Wiley Interscience, New York, 1981.
6. J. D. Banting, *Can. Plains Proc.*, *12*, 11 (1984).
7. *Herbicide Handbook of the Weed Science Society of America*, 5th ed., WSSA, Champaign, IL, 1983.
8. Dupont, Technical data bulletin on Assure herbicide (1985).
9. Anonymous, Application to register quizalofop as a common name of a pest control agent to American National Standard K62.1-1979 (1985).
10. Dow Chemical Co., Technical data sheet on DOWCO 453 ME, 1983.
11. ICI Americas, Inc., Technical data sheet on Fusilade 2000.
12. C. Hertel and D. Marmé, *FEBS Lett.*, *152*, 44 (1983).

13. C. Hertel and D. Marmé, *Pestic. Biochem. Physiol.*, *19*, 282 (1983).
14. C. Gauvrit, *Pestic. Biochem, Physiol.*, *21*, 377 (1984).
15. Anonymous, *Fed. Reg.*, *51*(19), 3598 (1986).
16. Anonymous, *Fed. Reg.*, *48*, 19022 (1983).
17. R. Handte, M. Koch, H. Bieringer, and G. Horlein, *Z. Naturforsch.*, *37B*, 912–922, (1982).
18. G. Sakata, K. Makino, Y. Kawamura, and T. Ikai, *Nippon Noyako Gakkaishi*, *10*, 61 (1985).
19. R. M. Scott, A. J. Sampson, and D. Jordan. *Proc. 1976 Br. Crop Protect. Conf. Weeds*, *2*, 723 (1976).
20. R. J. Jones, *Can. Plains Proc.*, *12*, 79 (1984).
21. H. J. Nestler and H. Bieringer, *Z. Naturforsch.*, *35B*, 366 (1980).
22. G. Sakata, K. Makino, K. Morimoto, T. Ikai, and S. Haseke, *Nippon Noyako Gokkaishi*, *10*, 69 (1985).
23. J. W. Dicks, J. W. Slater, and D. W. Bewick, *Proc. 1985 Br. Crop Protect. Conf. Weeds*, *7*, 271, 1985.
24. D. W. A. Barrett and P. B. Sutton, *Proc. 1985 Br. Crop Protect. Conf. Weeds*, *7*, 231, 1985.
25. G. Sakata, K. Makino, K. Kusano, J. Satow, T. Ikai, and K. Suzuki, *Nippon Noyako Gokkaishi*, *10*, 25 (1985).
26. W. A. Kleschick, Dow Chem. Co., U.S. Patent 4,532,328 (1985).
27. J. D. Gaynor and D. C. MacTavish, *Analyst*, *107*, 700 (1982).
28. P. K. Johnstone, I. R. Minchinton, and R. J. W. Truscott, *Pestic. Sci.*, *16*, 159 (1985).
29. A. E. Smith, *J. Agric. Food Chem.*, *27*, 428 (1979).
30. A. E. Smith, *J. Chromatogr.*, *129*, 309 (1978).
31. A. E. Smith and L. J. Milward, *J. Agric. Food Chem.*, *31*, 633 (1983).
32. H-B. Lee and A. S. Y. Chau, *J. Assoc. Off. Anal. Chem.*, *66*, 1322 (1983).
33. H-B. Lee and A. S. Y. Chau, *J. Assoc. Off. Anal. Chem.*, *66*, 651 (1983).
34. R. Stringham and B. R. Bennett, *J. Assoc. Off. Anal. Chem.*, *66*, 1207 (1983).
35. P. D. Bland, *J. Assoc. Off. Anal. Chem.*, *67*, 499 (1984).
36. A. E. Smith, *J. Agric. Food Chem.*, *33*, 972 (1985).
37. H. F. Taylor, M. P. C. Loader, and S. J. Norris, *Ann. Appl. Biol.*, *103*, 311 (1983).
38. A. I. Hsiao and A. E. Smith, *Weed Res.*, *23*, 231 (1983).
39. M. Schwalbe, E. Dorn, and K. Beyermann, *J. Agric. Food Chem.*, *32*, 734 (1984).
40. D. D. Buhler and O. C. Burnside, *Weed Sci.*, *32*, 574 (1984).

41. M. M. Schreiber, G. F. Warren, and P. L. Orwick, *Weed Sci.*, *27*, 679 (1979).

42. W. L. Barrentine and C. G. McWhorter, *Proc. Sou. Weed Sci. Soc.*, *39*, 533 (1986).

43. J. E. Dale, *Weed Res.*, *25*, 231 (1985).

44. P. W. Stahlman, *Weed Sci.*, *32*, 59 (1984).

45. P. A. O'Sullivan and K. J. Kirkland, *Weed Sci.*, *32*, 285 (1984).

46. F. A. Qureshi and W. H. Vanden Born, *Weed Sci.*, *27*, 202 (1979).

47. F. A. Qureshi and W. H. Vanden Born, *Can. J. Plant Sci.*, *59*, 87 (1979).

48. W. A. Olson and J. D. Nalewaja, *Weed Sci.*, *29*, 566 (1981).

49. J. R. Campbell and D. Penner, *Weed Sci.*, *30*, 458 (1982).

50. C. Hall, L. V. Edgington, and C. M. Switzer, *Weed Sci.*, *30*, 672 (1982).

51. R. A. Fletcher and D. M. Drexler, *Weed Sci.*, *28*, 363 (1980).

52. H. F. Taylor, M. P. C. Loader, and S. J. Norris, *Weed Res.*, *23*, 185 (1983).

53. W. A. Dorentzio and R. F. Norris, *Weed Sci.*, *27*, 539 (1979).

54. P. A. O'Sullivan, *Can. J. Plant Sci.*, *60*, 1255 (1980).

55. C. N. Giannopolitis and T. J. Strouthopoulos, *Weed Res.*, *19*, 213 (1979).

56. I. N. Morrison, B. D. Hill, and L. G. Dushnicky, *Weed Res.*, *19*, 385 (1979).

57. P. A. O'Sullivan and W. H. Vanden Born, *Weed Res.*, *20*, 53 (1980).

58. P. A. O'Sullivan, H. A. Freisan, and W. H. Vanden Born, *Can. J. Plant Sci.*, *57*, 117 (1977).

59. J. R. Moyer, R. D. Dryden, and P. N. P. Chow, *Can. J. Plant Sci.*, *59*, 351 (1979).

60. K. K. Hatzios, *Weed Res.*, *24*, 249 (1984).

61. R. H. Shimabukuro, W. C. Walsh, and R. A. Hoerauf, *J. Agric. Food Chem.*, *27*, 615 (1979).

62. B. D. Hill, E. H. Stobbe, and B. L. Jones, *Weed Res.*, *18*, 149 (1978).

63. B. Jeffcoat and W. N. Harries, *Pestic. Sci.*, *4*, 891 (1973).

64. B. Jeffcoat and W. N. Harries, *Pestic. Sci.*, *6*, 283 (1975).

65. C. Fedtke and R. R. Schmidt, *Weed Sci.*, *17*, 233 (1977).

66. T. R. Roberts, *Pestic. Biochem. Physiol.*, *7*, 378 (1977).

67. S. G. Gorbach, K. Kuenzler, and J. Asshauer, *J. Agric. Food Chem.*, *25*, 507 (1977).

68. K. I. Benyon, T. R. Roberts, and A. N. Wright, *Pestic. Biochem. Physiol.*, *4*, 98 (1974).

69. K. I. Benyon, T. R. Roberts, and A. N. Wright, *Pestic. Sci.*, *5*, 429 (1974).

70. B. T. Grayson and S. Stokes, *Pestic. Sci.*, *9*, 595 (1978).

71. R. H. Shimabukuro, in *Weed Physiology*, Vol. II, *Herbicide Physiology* (S. O. Duke, ed.), CRC Press, Boca Raton, FL, 1985, pp. 215–240.

72. B. D. Hill and E. H. Stobbe, *Weed Res.*, *18*, 223 (1978).

73. K. Gorecka, R. H. Shimabukuro, and W. C. Walsh, *Physiol. Plant.*, *53*, 55 (1981).

74. W. C. Akey and I. N. Morrison, *Weed Sci.*, *31*, 247 (1983).

75. B. Jeffcoat, W. N. Harries, and D. B. Thomas, *Pestic. Sci.*, *8*, 1 (1977).

76. T. R. Roberts, *Pestic. Sci.*, *8*, 463 (1977).

77. A. J. Dutton, T. R. Roberts, and A. N. Wright, *Chemosphere*, *3*, 195 (1976).

78. J. B. Pillmoor, T. R. Roberts, and J. K. Gaunt, *Pestic. Sci.*, *13*, 129 (1982).

79. P. Hendley, J. W. Dicks, T. J. Monaco, S. M. Slyfield, O. J. Tummon, and J. C. Barrett, *Weed Sci.*, *33*, 11 (1985).

80. D. Coupland, *Proc. 1985 Br. Crop Protect. Conf. Weeds*, *7*, 317 (1985).

81. D. D. Buhler, B. A. Swisher, and O. C. Burnside, *Weed Sci.*, *33*, 291 (1985).

82. R. Grover, *Can. Plains Proc.*, *12*, 119 (1984).

83. J. D. Gaynor, *Can. J. Soil Sci.*, *64*, 283 (1984).

84. A. E. Smith, *J. Agric. Food Chem.*, *27*, 1145 (1979).

85. E. J. Hitchings and T. R. Roberts, *Pestic. Sci.*, *10*, 1 (1979).

86. E. J. Hitchings and T. R. Roberts, *Pestic. Sci.*, *11*, 591 (1980).

87. K. W. Smith and G. R. Stephenson, 11th Workshop on Chemistry and Biochemistry of Herbicides *11*, 113 (1982).

88. A. E. Smith, *J. Agric. Food Chem.*, *25*, 893 (1977).

89. R. Martens, *Pestic. Sci.*, *9*, 127 (1978).

90. P. G. Bosio, K. E. Elgar, B. L. Mathews, A. P. Woodbridge, and A. N. Wright, *Pestic. Sci.*, *13*, 1 (1982).

91. P. B. Bosio, E. R. Cole, B. L. Mathews, A. P. Woodbridge, and A. N. Wright, *Pestic. Sci.*, *13*, 63 (1982).

92. J. V. Crayford, P. A. Harthoorn, and D. H. Hutson, *Pestic. Sci.*, *7*, 559 (1976).

93. C. T. Bedford, J. V. Crayford, D. H. Hutson, and D. E. Wiggins, *Xenobiotica*, *8*, 383 (1978).

94. A. E. Smith, *J. Agric. Food Chem.*, *33*, 483 (1985).

95. S. K. Harrison and L. M. Wax, *Weed Sci.*, *34*, 81 (1986).

96. C. E. G. Mulder and J. D. Nalewaja, *Weed Sci.*, *27*, 83 (1979).

97. A. J. Cessna and R. Grover, *Water Studies Inst. Symp.*, *8*, 7 (1982).

98. P. F. Boldt and A. R. Putham, *Weed Sci.*, *28*, 474 (1980).

99. G. R. Gillespie and S. D. Miller, *Weed Sci.*, *31*, 658 (1983).

100. B. G. Todd and E. H. Stobbe, *Weed Sci.*, *25*, 382 (1977).
101. G. D. Wills and C. G. McWhorter, *Aspects Appl. Biol.*, *4*, 283 (1983).
102. J. E. Carr, L. G. Davies, A. H. Cobb, and K. E. Pallett, *Ann. Appl. Biol.*, *108*, 115 (1986).
103. J. J. Kells, W. F. Meggitt, and D. Penner, *Weed Sci.*, *32*, 143 (1984).
104. S. K. Harrison and L. M. Wax, *Weed Sci.*, *34*, 185 (1986).
105. C. G. McWhorter and G. D. Wills, *Weed Sci.*, in preparation (1987).
106. C. G. McWhorter, *Weed Sci.*, *29*, 87 (1981).
107. R. J. Hamilton, A. W. McCann, P. A. Sewell, and G. T. Merrall, in *The Plant Cuticle* (D. F. Cutler, K. L. Alvin, and C. E. Price, eds.), Academic Press, New York, 1982, pp. 303–313.
108. D. Coupland, *Ann. Appl. Biol.*, *108*, 353 (1986).
109. W. A. Dortenzio and R. F. Norris, *Weed Sci.*, *28*, 534 (1980).
110. H. Walter, W. Koch, and F. Muller, *Weed Res.*, *20*, 325 (1980).
111. P. Whitehouse, P. J. Holloway, and J. C. Caseley, in *The Plant Cuticle* (D. F. Cutler, K. L. Alvin, and C. E. Price, eds.), Academic Press, New York, 1982, pp. 315–330.
112. A. Jacobson and R. H. Shimabukuro, *Physiol. Plant.*, *54*, 34 (1982).
113. J. D. Nalewaja, G. A. Skrzypczak, and G. R. Gilespie, *Weed Sci.*, *34*, 564 (1986).
114. J. D. Nalewaja and G. A. Skrzypczak, *Weed Sci.*, *34*, 572 (1986).
115. F. A. Qureshi and W. H. Vanden Born, *Can. J. Plant Sci.*, *59*, 93 (1979).
116. W. Olson and J. D. Nalewaja, *Weed Sci.*, *30*, 59 (1982).
117. C. Hall, L. V. Edgington, and C. M. Switzer, *Weed Sci.*, *30*, 676 (1982).
118. T. Ikai, K. Suzuki, K. Hattori, and H. Igarashi, *Proc. 1985 Br. Crop. Protect. Conf. Weeds*, *7*, 163 (1985).
119. F. Kimura, R. Nishiyama, K. Fujikawa, I. Yokomichi, T. Haga, and N. Sakashita, *Proc. 8th Pac. Weed Sci. Soc. Conf.*, *8*, 433 (1981).
120. J. L. Wilhm, W. F. Meggitt, and D. Penner, *Weed Sci.*, *34*, 333 (1986).
121. A. G. Brezeanu, D. G. Davis, and R. H. Shimabukuro, *Can. J. Bot.*, *54*, 2038 (1976).
122. R. A. Hoerauf and R. H. Shimabukuro, *Weed Res.*, *19*, 293 (1979).
123. D. G. Davis and A. Brezeanu, *Can. J. Bot.*, *57*, 2006 (1979).
124. K. C. Vaughn, unpublished results.

125. J. C. Crowley and G. N. Prendeville, *Can. J. Plant Sci.*, *59*, 275 (1979).
126. N. F. O'Leary, J. T. O'Donovan, and G. N. Prendeville, *Can. J. Plant Sci.*, *60*, 773 (1980).
127. H. H. Hoppe, *Z. Pflanzenphysiol.*, *100*, 415 (1980).
128. D. E. Moreland, *Annu. Rev. Plant Physiol.*, *31*, 597 (1980).
129. A. S. Cohen and I. N. Morrison, *Pestic. Biochem. Physiol.*, *16*, 110 (1981).
130. J. P. Wright and R. H. Shimabukuro, *Plant Physiol.*, *67*, S-524 (1982).
131. W. J. Lucas, C. Wilson, and J. P. Wright, *Plant Physiol.*, *74*, 61 (1984).
132. I. N. Morrison, M. G. Owino, and E. H. Stobbe, *Weed Sci.*, *29*, 426 (1981).
133. H. F. Taylor and M. P. C. Loader, *Outlook Agric.*, *13*(2), 58 (1984).
134. B. Jeffcoat and W. N. Harries, *Pestic. Sci.*, *6*, 283 (1975).
135. J. W. Gronwald, *Weed Sci.*, *34*, 196 (1986).
136. P. N. P. Chow and D. E. LaBerge, *J. Agric. Food Chem.*, *26*, 1134 (1978).
137. M. A. Shimabukuro, R. H. Shimabukuro, W. S. Nord, and R. A. Hoerauf, *Pestic. Biochem. Physiol.*, *8*, 199 (1978).
138. H. H. Hoppe, *Weed Res.*, *20*, 371 (1980).
139. W. W. Donald, R. V. Parke, and R. H. Shimabukuro, *Physiol. Plant.*, *54*, 467 (1982).
140. H. F. Taylor, M. P. C. Loader, and S. J. Norris, *Weed Res.*, *24*, 185 (1983).
141. P. A. O'Sullivan, H. A. Friesen, and W. H. Vanden Born, *Can. J. Plant Sci.*, *57*, 117 (1977).
142. R. A. Fletcher and D. M. Drexler, *Weed Sci.*, *28*, 363 (1980).
143. M. A. Shimabukuro, R. H. Shimabukuro, and W. C. Walsh, *Physiol. Plant.*, *56*, 444 (1982).
144. W. A. Olson, J. D. Nalewaja, G. L. Schroeder, and M. E. Duysen, *Weed Sci.*, *29*, 597 (1981).
145. B. G. Todd and E. H. Stobbe, *Weed Sci.*, *28*, 371 (1980).
146. B. D. Hill, B. G. Todd, and E. H. Stobbe, *Weed Sci.*, *28*, 725 (1980).
147. R. H. Shimabukuro, W. C. Walsh, and R. A. Hoerauf, *Plant Physiol.*, *80*, 612 (1986).
148. H. H. Hoppe and H. Zacher, *Z. Pflanzenphysiol.*, *106*, 287 (1982).
149. H. H. Hoppe, *Z. Pflkrankh. Pflschutz.*, *9*, 187 (1981).
150. H. H. Hoppe, *Pestic. Biochem. Physiol.*, *23*, 297 (1985).
151. H. H. Hoppe and H. Zacher, *Pestic. Biochem. Physiol.*, *24*, 298 (1985).
152. H.-Y. Cho, J. M. Widholm, and F. W. Slife, *Weed Sci.*, *34*, 496 (1986).

153. E. Bergmannova and L. Taimr, *Weed Res.*, *25*, 347 (1985).
154. H. H. Hoppe, *Z. Pflanzenphysiol.*, *102*, 189 (1981).
155. B. Kowalczyk, J. C. Caseley, and C. C. McCready, *Aspects Appl. Biol.*, *4*, 235 (1983).
156. R. S. Peregoy and S. Glenn, *Weed Sci.*, *33*, 443 (1985).
157. A. Jacobson, R. H. Shimabukuro, and C. McMichael, *Pestic. Biochem. Physiol.*, *24*, 61 (1985).
158. H. Zacher and H. H. Hoppe, *Z. Pflkrankh. Pflschultz.*, *S-H. IX*, 179 (1981).
159. J. Heap and R. Knight, *J. Austral. Inst. Agric. Sci.*, *48*, 156 (1982).
160. J. Gressel, in *Pesticide Resistence*, National Academy Press, Washington, D. C., 1986, pp. 54–73.
161. I. Heap and R. Knight, *Aust. J. Agric. Res.*, *37*, 149 (1986).

Chapter 3

SULFONYLUREAS

ELMO M. BEYER, JR., MICHAEL J. DUFFY, JAMES V. HAY, and
DAVID D. SCHLUETER

Agricultural Products Department, E. I. du Pont de Nemours &
Company, Inc., Wilmington, Delaware

I. INTRODUCTION

A. History and Development

1. History

The herbicidal properties of sulfonylureas were first reported in 1966 [1]. These early sulfonylureas were derivatives of the triazine herbicides and generally exhibited herbicidal symptoms and levels of activity comparable to the parent compounds, for instance, atrazine.

Work in the area of herbicidal sulfonylureas was dormant until the mid-1970s, when George Levitt of DuPont noted that sulfonylurea I exhibited weak plant growth regulant activity at 2.0 kg/ha [2,3]. Synthesis of compounds related to I, with various substituents on both phenyl rings failed to increase growth regulant activity. In June 1975, Levitt prepared sulfonylurea II, which contained an aminopyrimidine in place of an aniline.

(I)

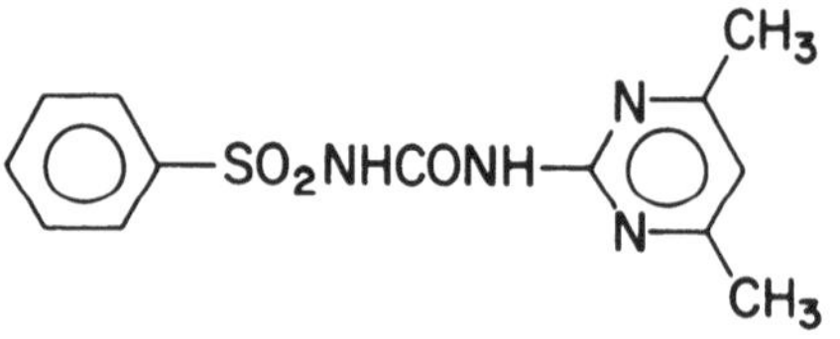

(II)

Compared to I, II displayed very high biological activity at 2.0 kg/ha, thus stimulating additional interest and activity in the area.

Through the continued efforts of Levitt, with the support and encouragement of Luckenbaugh, Farney, and Schlaf of DuPont, sulfonylureas were soon produced having up to 100 times the activity of conventional herbicides. Thus began one of the most exciting breakthroughs in the field of herbicide research in several decades. With their unprecedented activity, plus the seemingly endless number of possible structural variations, the sulfonylureas soon became one of the largest herbicide research programs in the history of the agricultural chemicals industry.

DuPont's first sulfonylurea patent application was published in October 1977 [4]. The patent literature indicates that some time elapsed before other agricultural chemical companies began to actively synthesize and test sulfonylureas. Work at Ciba-Geigy led to the first non-DuPont sulfonylurea herbicide patent application, published in January 1982 [5]. By the end of 1985, as many as 14 agricultural chemical companies had entered the field as evidenced by patent publications. By June 1987, 230 U.S. sulfonylurea herbicide patents had been issued, of which DuPont held 169 or 74%.

2. Chemistry

The sulfonylurea herbicides are represented by the general structure III. The molecule is composed of three distinct parts: an aryl group, the sulfonylurea bridge, and a nitrogen-containing heterocycle.

$$
\text{ARYL} \mid \text{BRIDGE} \mid \text{HETEROCYCLE}
$$

(III)

When the aryl portion is a phenyl group, the highest herbicidal activity occurs when this group contains a substituent *ortho* to the bridge. Although many types of electron-donating and electron-withdrawing *ortho* substituents potentiate herbicidal activity, there are some, such as *ortho* carboxyl and hydroxyl substituents, that do not. Sulfonylureas containing aryl groups other than a phenyl are also biologically active. Examples of these include a thiophene [6-8], furan [6,8], pyridine [9], or naphthalene group [6,10]. With these aryl groups, as with a phenyl group, the presence of a substituent *ortho* to the bridge enhances activity.

Sulfonylureas usually exhibit the highest herbicidal activity when the heterocyclic portion of the molecule is a *symmetrical* pyrimidine or *symmetrical* triazine containing lower alkyl or lower alkoxy substituents. Other nitrogen heterocycles, such as triazoles [11,12], *asymmetrical* triazines [6], fused-ring pyrimidines [13,14], and pyridines [15] also produce herbicidal sulfonylureas. These compounds, however, generally have lower activity.

Sulfonylureas with an unmodified bridge, as shown in (III), are usually the most active, but compounds with modified bridging groups, such as $SO_2NHC(S)NH$ [6], $OSO_2NHCONH$ [16], $SO_2NHCON(CH_3)$ [17], or $CH_2SO_2NHCONH$ [18], are active also. As expected, the level of activity of each modified bridge sulfonylurea is also dependent on the type of aryl and heterocycle in the molecule.

3. Biology

Sulfonylureas are potent inhibitors of plant growth. While seed germination is not usually affected, subsequent root and shoot growth are severly inhibited in sensitive seedlings. Growth inhibition is very rapid, with visual symptoms usually apparent within 1 to 2 days in rapidly growing plants.

Depending on the plant species, dose, and environmental conditions, a variety of secondary plant responses often develop. These include enhanced anthocyanin formation, loss of leaf nyctinasty, abscission, vein discoloration, terminal bud death, chlorosis, and necrosis. These secondary effects are often slow to develop, with plant death sometimes not occurring until a week or more following treatment. Sulfonylureas are taken up readily by both the roots and foliage, and once inside the plant they are translocated via the xylem and phloem.

The potency of the sulfonylurea herbicides is unprecedented. Conventional herbicides, such as alachlor and atrazine, have use rates of 0.5–2.0 kg/ha. Sulfonylureas, on the other hand, have use rates as low as 0.002 kg/ha.

4. Commercial and Developmental Compounds

Tables 1 and 2 list the commercial and developmental sulfonylurea herbicides, the crops on which they are used, and recommended use rates.

First commercialized in 1982, Glean Herbicide (chlorsulfuron) is effective on a broad spectrum of weeds and is safe to use on such cereal grains as wheat, barley, oats, rye, and triticale [19,20]. Use rates as low as 4 g a.i./ha control many important broadleaf weeds and some grasses in these crops. Telar Herbicide was commercialized in 1984 and is chlorsulfuron for industrial, or noncrop, uses.

Oust Herbicide (sulfometuron methyl), first sold commercially in 1982, is an industrial, or noncrop, herbicide [21,22]. It controls both broadleaves and grasses on railroad and highway rights-of-way. Sulfometuron methyl is particularly effective on both seedling and rhizome johnsongrass, giving season-long control. In areas where bermudagrass and johnsongrass coexist, sulfometuron methyl can be used to selectively remove the johnsongrass, thus "releasing" the desirable bermudagrass.

Ally Herbicide (metsulfuron methyl) effectively controls a wide spectrum of broadleaf weeds in wheat and barley. Metsulfuron methyl is also sold as a selective herbicide for cereals under the trade names Allie (France) and Gropper (West Germany). It was introduced commercially in the United Kingdom in the spring of 1984, in France in 1985 [23,24], and in West Germany in 1986. Metsulfuron methyl complements chlorsulfuron, and a mixture of the two was introduced in the United Kingdom in the fall of 1984, under the trade name Finesse Herbicide. A similar mixture was introduced in the United States in 1986. Escort Herbicide was commercialized in 1986 in the United States and is metsulfuron methyl for industrial, or noncrop, uses.

Classic Herbicide (chlorimuron ethyl) was introduced under EUP in 1984 for postemergence use in soybeans [22,25] and commercialized in 1986 in the United States. Chlorimuron ethyl effectively controls many of the problem broadleaf weeds in soybeans, including cocklebur, morning glory, and sicklepod at postemergence rates of 8 to 13 g a.i./ ha. Combinations with metribuzin (Canopy Herbicide) and linuron (Gemini Herbicide) were commercialized in 1986 for preemergence applications.

Londax Herbicide (bensulfuron methyl) was introduced in 1984 for use in direct-seeded and transplanted rice and was first sold commercially in Thailand in 1985 [18,26]. Bensulfuron methyl controls many problem broadleaf weeds and sedges in rice. In combination with thiocarbamate grass herbicides, bensulfuron methyl provides total season-long weed control with good crop safety.

Harmony Herbicide (formerly DPX-M6316) was introduced in 1986 for postemergence use in cereals [8,27]. Harmony controls a wide range of broadleaf weeds and is especially effective in controlling wild garlic. Due to its rapid degradation in soil, Harmony provides maximum rotational crop flexibility. Sensitive rotational crops can be planted in Harmony-treated soil within 30 days; whereas, plantings in Glean-treated soil may be restricted for a year or longer depending on the soil, climate, and use rate.

In 1985, Ciba—Geigy announced the development of Logran Herbicide (triasulfuron) for cereal crops, and Nissan is reportedly developing a sulfonylurea herbicide (NC-311) for rice. Patent publications indicate that other companies are also actively involved in developing new sulfonylurea herbicides.

TABLE 1

Sulfonylurea Herbicides

Structural formula	Chemical name	Trade name(s) and common name
	2-chloro-N-[4-methoxy-6-methyl-1,3,5-triazin-2-yl)-aminocarbonyl]benzenesulfonamide	Glean/Telar (chlorsulfuron)
	methyl 2-[[(4,6-dimethyl-pyrimidin-2-yl)aminocarbonyl]aminosulfonyl]benzoate	Oust (sulfometuron methyl)
	methyl 2-[[[(4-methoxy-6-methyl-1,3,5-triazin-2-yl)aminocarbonyl]aminosulfonyl]benzoate	Ally/Allie/Gropper/Escort (metsulfuron methyl)

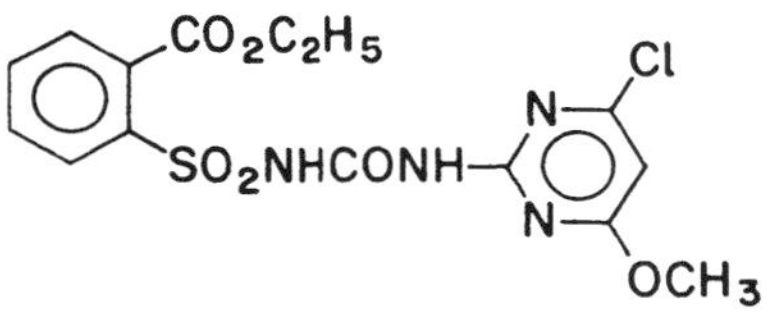

Classic
(chlorimuron ethyl)

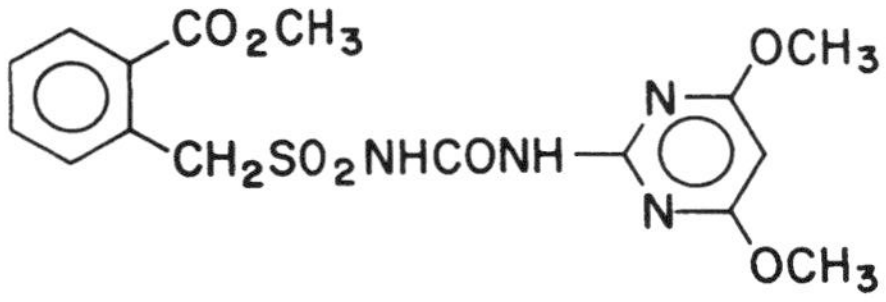

ethyl 2-[[(4-chloro-6-methoxy-
pyrimidin-2-yl)aminocarbobnyl]-
aminosulfonyl]benzoate

Londax
(bensulfuron methyl)

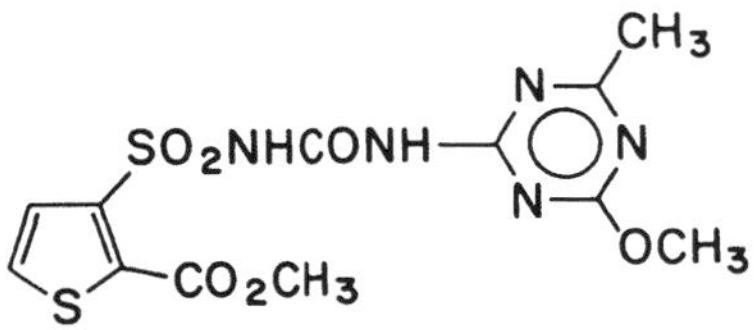

methyl 2-[[[(4,6-dimethoxy-
pyrimidin-2-yl)aminocarbonyl]-
aminosulfonyl]methyl]benzoate

Harmony
(DPX-M6316)

methyl 3-[[[(4-methoxy-6-methyl-
1,3,5-triazin-2-yl)aminocarbonyl]-
aminosulfonyl]-2-
thiophenecarboxylate

TABLE 2

Use and Application Rate of
Sulfonylurea Herbicides

Sulfonylurea	Use	Rate (g ai/ha)
Chlorsulfuron	Cereals	4–26
Chlorsulfuron	Noncrop	13–158
Sulfometuron methyl	Noncrop	70–840
Metsulfuron methyl	Cereals	1.8–8
Metsulfuron methyl	Noncrop	2.6–126
Chlorimuron ethyl	Soybeans	8–13
Bensulfuron methyl	Rice	20–75
DPX-M6316	Cereals	9–52.5

B. Physical and Chemical Properties

Selected physical and chemical properties of commercial and developmental sulfonylurea herbicides are listed in Table 3 [28–33]. Typically, sulfonylureas exhibit moderate melting points, low octanol-water partition coefficients at neutral pH, very low vapor pressures, and low to moderate water solubilities at neutral pH.

Because sulfonylureas are weak acids (see page 127), pH greatly affects their water solubility and partition coefficient. For example, the water solubility of chlorsulfuron is 60 ppm at pH 5, while at pH 7 it is 7000 ppm. This is because as the pH increases, the proportion of the water-soluble anionic form increases. Generally, sulfonylureas containing a triazine heterocycle (e.g., chlorsulfuron) are more water soluble than those containing a pyrimidine heterocycle (e.g., sulfometuron methyl).

The effect of pH on the octanol-water partition coefficient (K_{OW}) is opposite to its effect on water solubility. As pH increases, more of the sulfonylurea exists in its water-soluble, anionic form and, accordingly, partitioning into octanol decreases. Chlorsulfuron, for example, has a K_{OW} of 5.5 at pH 5, but only 0.046 at pH 7. As a rule, the K_{OW} of pyrimidinyl sulfonylureas are higher than triazinyl sulfonylureas.

The dissociation constants for sulfonylureas are covered in Sec. I.E.

TABLE 3

Physical and Chemical Properties of Sulfonylurea Herbicides

	Melting point	Molecular weight	Dissociation constant (pKa)	Partition coefficient (octanol-water, at 25°C)		Vapor pressure (mmHg at 25°C)	Water solubility (ppm at 25°C)	
				pH 5	pH 7		pH 5	pH 7
Chlorsulfuron	174–178°C	357.78	3.6	5.5	0.046	2.3×10^{-11}	60	7000
Sulfometuron methyl	203–205°C	364.39	5.2	15	0.31	5.5×10^{-16}	8	70
Metsulfuron methyl	158°C	381.37	3.3	1.0	0.014	2.5×10^{-12}	1100	9500
Chlorimuron ethyl	186°C	414.83	4.2	320	2.3	3.7×10^{-12}	11	1200
Bensulfuron methyl	185–188°C	410.40	5.2	155	4.1	2.1×10^{-14}	2.9	120
DPX-M6316	186°C	387.40	4.0	3.3	0.027	1.3×10^{-10}	260	2400[a]

[a] At pH 6.

C. Toxicological Properties

All six of the sulfonylurea herbicides discussed in this review have low acute oral, dermal, and inhalation toxicities in rats and/or rabbits. The acute oral LD_{50} value of these compounds in rats was >4100 mg/kg. By comparison, the acute oral LD_{50} of table salt in rats is 3000 mg/kg [34]. Eye and skin exposure on rabbits and guinea pigs generally caused only mild to moderate irritation and no sensitization. The subject sulfonylureas are not mutagenic or teratogenic, and they exhibit low toxicity to fish, wildlife, and honey bees. Subchronic feeding and reproduction studies, as well as chronic feeding studies in rats, mice, and dogs are also favorable. This low toxicity, combined with the very low application rates of the sulfonylureas, makes them especially attractive from an environmental and human health standpoint.

A summary of key findings from many of the completed toxicity studies is presented in Table 4. Additional toxicological results on the active ingredient and formulations can be found in the Technical Bulletin for each product [29–33].

Nontarget microorganisms involved in nitrogen fixation were found to be unaffected by chlorsulfuron [35] as was the growth of three strains of *Rhizobium* and *Azotobacter chroococcum*. The ability of *Rhizobium* strains to infect their respective host plants and the ability of *Azotobacter* to reduce acetylene were also unaltered.

D. Synthesis

Several methods exist for the synthesis of herbicidal sulfonylureas (IV). This section summarizes the synthetic routes most often followed.

$$ArSO_2NHCONHHet \qquad\qquad (\underline{IV})$$

1. Reaction of Sulfonylisocyanates with Amines

The preferred route for commercial manufacture of sulfonylureas is the coupling of an aryl sulfonylisocyanate with a heterocyclic amine.

$$ArSO_2NCO + H_2NHet \longrightarrow ArSO_2NHCONHHet \qquad (1)$$

By combining a solution of the aryl sulfonylisocyanate with a suspension of the heterocyclic amine in an inert organic solvent, such as xylene, the sulfonylurea precipitates as a fine crystalline solid [2,3]. The sulfonylisocyanates are easily prepared by the reaction of phosgene with a sulfonamide, in the presence of an alkyl isocyanate, in an inert organic solvent at 120–140°C [36].

2. Reaction of Carbamic Acid Phenylesters with Sulfonamides

Meyer and Föry [37] demonstrated that heterocyclic carbamic acid phenylesters react with sulfonamides in the presence of base at room temperature to give the corresponding sulfonylurea and phenol.

$$ArSO_2NH_2 + C_6H_5OCONHHet \xrightarrow{\text{base}}$$

$$ArSO_2NHCONHHet + C_6H_5OH \qquad (2)$$

3. Reaction of Sulfonylcarbamic Acid Phenylesters with Amines

In a modification of the above method, Meyer and Föry [37] showed that arylsulfonyl carbamic acid phenylesters react with heterocyclic amines at elevated temperatures (e.g., 100°C) to give the corresponding sulfonylurea and phenol.

$$ArSO_2NHCOOC_6H_5 + H_2NHet \xrightarrow{\Delta}$$

$$ArSO_2NHCONHHet + C_6H_5OH \qquad (3)$$

E. Chemical Reactions

Sulfonylureas undergo several chemical reactions but only those reactions that are relevant to their normal use as herbicides (e.g., chemical stability, uptake, soil residual properties) are discussed.

1. Salt Formation

Sulfonylureas are weak acids, and in aqueous media they exist as an equilibrium mixture of the neutral undissociated form (V) and

TABLE 4

Toxicology Properties of Sulfonylurea Herbicides

Acute toxicity studies	Chlorsulfuron	Sulfometuron methyl
Oral LD$_{50}$ in rat	5546 mg/kg (male) 6293 mg/kg (female)	>5000 mg/kg
Dermal LD$_{50}$ in rabbit	>3400 mg/kg	>2000 mg/kg
Skin irritation in guinea pig	Negative	Mild to none
Skin sensitization in guinea pig	Negative	Negative
Eye irritation in rabbit	Mild; reversible	Mild to minimal; reversible
Inhalation 4-hr LC$_{50}$ in rat	>5.9 mg/L	>5 mg/L

Chronic Toxicity Studies

Oral feeding-

NOEL in rat (2-yr)	100 ppm	50 ppm
NOEL in mouse	500 ppm (2-yr)	(18-mo in prog.)
NOEL in dog		200 ppm

Reproduction, teratogenicity, mutagenicity

Reproduction

NOEL in rat	500 ppm (3-gen)	500 ppm (2-gen)

Metsulfuron methyl	DPX-M6316	Bensulfuron methyl	Chlormuron ethyl
>5000 mg/kg	>5000 mg/kg	>5000 mg/kg	4102 mg/kg (male) 4236 mg/kg (female)
>2000 mg/kg	>2000 mg/kg	>2000 mg/kg	>2000 mg/kg
Mild	Negative	Negative	Mild
Negative	Negative	Negative	Negative
Moderate to severe; reversible	Mild; reversible	Negative	Negative
>5.3 mg/L	>7.9 mg/L	>5.0 mg/L	>5.0 mg/L
500 ppm	25 ppm	750 ppm	250 ppm
5000 ppm (18-mo)	7500 ppm (18-mo)	2500 ppm (18-mo)	125 ppm (18-mo)
500 ppm (male) 5000 ppm (female)	750 ppm	750 ppm	250 ppm
500 ppm (2-gen)	2500 ppm (2-gen)	7500 ppm (2-gen)	250 ppm (2-gen)

TABLE 4

(continued)

Acute toxicity studies	Chlorsulfuron	Sulfometuron methyl
Teratogenicity		
NOEL in rat	2500 ppm	1000 ppm
NOEL in rabbit	25 mg/kg	300 mg/kg
Mutagenicity	Negative in 5 assays	Negative in 4 assays
Aquatic & Wildlife Toxicity		
Rainbow trout 96-hr LC_{50}	>250 ppm	>12.5 ppm
Bluegill sunfish 96-hr LC_{50}	>300 ppm	>12.5 ppm
Daphnia 48-hr LC_{50}	370 ppm	>12.5 ppm
Mallard duck Oral LD	>5000 mg/kg	>5000 mg/kg
8-day dietary LC_{50}	>5000 ppm	>5000 ppm
Bobwhite quail 8-day dietary LC_{50}	>5000 ppm	>5620 ppm
Honey Bee LD_{50}	>25 µg/bee	>12.5 µg/bee

the negatively charged dissociated form (VI) [Eq. (1)]. The dissociation constants (pKas) for the sulfonylurea herbicides were determined from their solubility in water as a function of pH [38] and found to range from 3.3 to 5.2 (Table 3). Thus, they are

Metsulfuron methyl	DPX-M6316	Bensulfuron methyl	Chlorimuron ethyl
40 mg/kg	200 mg/kg	500 mg/kg	30 mg/kg
25 mg/kg	200 mg/kg	300 mg/kg	15 mg/kg
Negative in 5 assays; Equivocal in 1	Negative in 5 assays	Negative in 6 assays	Negative in 4 assays
>150 mg/L	>100 mg/L	>150 mg/L	>1000 mg/L
150 ppm	>100 mg/L	>150 mg/L	>100 mg/L
>150 ppm	1000 mg/L	>100 mg/L	1000 mg/L
>2510 mg/kg	>2510 mg/kg	>2510 mg/kg	>2510 mg/kg
>5620 ppm	>5620 ppm	>5620 ppm	>5620 ppm
>5620 ppm	>5620 ppm	>5620 ppm	>5620 ppm
>25 µg/bee	>12.5 µg/bee	>12.5 µg/bee	>12.5 µg/bee

about as acidic as acetic acid (pKa = 4.75). Generally, sulfonylureas containing a triazine heterocycle, such as chlorsulfuron, metsulfuron methyl, and DPX-M6316, are stronger acids (lower pKas) than sulfonylureas containing a pyrimidine heterocycle, such as sulfometuron methyl, chlorimuron ethyl, and bensulfuron methyl.

$$ArSO_2NHCONHHet \rightleftharpoons [ArSO_2NCONHHet]^{\ominus} + H^{\oplus} \qquad (1)$$

$$(\underline{V}) \qquad\qquad\qquad (\underline{VI})$$

As weak acids, sulfonylureas form stable metal salts, (VII). This occurs when the sulfonylurea is treated with an alkali or alkaline earth hydroxide or carbonate [Eq. (2)]. Similarly, stable ammonium and substituted ammonium salts are formed with ammonia and aliphatic amines [6].

$$ArSO_2NHCONHHet \xrightarrow{MOH} [ArSO_2NCONHHet]^{\ominus} \bullet M^{\oplus} + H_2O$$

$$(\underline{VII}) \qquad\qquad\qquad\qquad (2)$$

The nitrogen-containing heterocycle present in the sulfonylurea molecule results in the formation of acid-addition salts with suitable acids, such as p-toluenesulfonic acid or trichloroacetic acid [Eq. (3)] [6].

$$ArSO_2NHCONHHet \xrightarrow{Cl_3CCO_2H} ArSO_2NHCONHHet \cdot Cl_3CCO_2H$$

$$(3)$$

2. Chemical Hydrolysis

Sulfonylurea herbicides undergo hydrolysis in aqueous media at a rate which is a function of pH and temperature. Comparative hydrolysis half-lives for chlorsulfuron, chlorimuron ethyl, metsulfuron methyl, bensulfuron methyl, and sulfometuron methyl in buffered, sterilized aqueous solutions are listed in Table 5 [28–31,39]. These data clearly show that hydrolysis is much more rapid under acidic conditions.

The predominant hydrolysis reaction of sulfonylureas under mildly acidic conditions (e.g., pH 5), is cleavage of the sulfonylurea bridge. The reaction products are the sulfonamide, the heterocyclic amine, and carbon dioxide [Eq. (4)]. Depending on their structures, the sulfonamide and amine may undergo further hydrolytic degradation [40,41].

TABLE 5

Hydrolysis Half-Lives of Selected Sulfonylurea Herbicides as a
Function of pH and Temperature

Temperature (°C)	Compound	Half-life in days			
		pH 5	pH 6	pH 7	pH 8
25	Metsulfuron methyl	33	—	—	—
	Sulfometuron methyl	18	—	—	—
	Bensulfuron methyl	11	—	143	—
35	Chlorsulfuron	6.0	53	256	208
	Chlorimuron ethyl	2.4	17	64	89
	Sulfometuron methyl	1.8	6	15	20
45	Chlorsulfuron	1.7	14	51	58
	Chlorimuron ethyl	0.6	4	14	18
	Sulfometuron methyl	0.4	1	6	7
	Metsulfuron methyl	2.1	—	33	—
55	Chlorsulfuron	0.5	4	10	12
	Chlorimuron ethyl	0.2	1	3	4
	Sulfometuron methyl	0.1	0.3	2	2

$$ArSO_2NHCONHHet \xrightarrow{\text{H}_2\text{O}} ArSO_2NH_2 + H_2NHet + CO_2 \qquad (4)$$

Since sulfonylureas are weak acids, they exist primarily in the anionic form [structure VI in Eq. (1)] in aqueous solutions at neutral or basic pH's. In this form, the sulfonylureas are much less subject to hydrolysis. For example, the hydrolytic half-life at 35°C for chlorsulfuron at pH 8 is 208 days, whereas it is only 6.0 days at pH 5.

Under very alkaline conditions (pH >10), the rate of hydrolysis of sulfonylureas typically increases because of a base-catalyzed reaction mechanism [39,42].

A mathematical model has been developed to calculate the hydrolysis rate constant at any pH and temperature [39,43]. The model assumes that, over the pH range of 3 to 11, the observed hydrolysis rate is the sum of three simultaneous reactions: (1) hydrolysis of the neutral sulfonylurea, (2) hydrolysis of the sulfonylurea anion, and (3) hydrolysis of the sulfonylurea anion by hydroxide ion.

F. Formulations

Sulfonylureas are generally formulated as wettable powders or water-dispersible granules. Granules are often preferred for their ease of handling and measuring. The low solubilities of sulfonylureas in tap water and many organic solvents, precludes such common formulations as emulsifiable concentrates.

Wettable powders are prepared by mixing and grinding the active ingredient with dispersing agents, wetting agents, and diluents. Granules are prepared by granulation of premixes formulated similarly to wettable powders.

Both the wettable powder and granule formulations of sulfonylureas disperse easily in spray tanks and can easily be applied in normal spray volumes with conventional equipment. In addition, the sulfonylureas are particularly compatible with low and ultra-low spray volumes. These sulfonylurea formulations can be easily tank mixed with a variety of commercial herbicides. Addition of a nonionic surfactant to the sulfonylurea spray solution often enhances herbicidal activity (see page 181).

G. Methods of Analysis

1. Mass Spectrometry

Mass spectrometry is used extensively for structural confirmation and identification of sulfonylureas. By using electron impact ionization, many lower mass fragment ions such as the heterocyclic amine and the isocyanate of the sulfonamide are detected. However, the spectra contain no molecular ions. "Softer" ionization techniques, such as thermospray or chemical ionization, produce protonated molecular ions with much less fragmentation. With the use of these "softer" techniques, fragment ions will typically include protonated and adduct ions corresponding to the sulfonamide, the heterocyclic amine, and the heterocyclic urea of the sulfonylurea.

Shalaby recently described a method for connecting a liquid chromatograph and mass spectrometer, via thermospray and direct liquid interfaces, for the analysis of chlorsulfuron and sulfometuron methyl [44–46]. With the direct liquid interface, one can control the extent of fragmentation and relative abundance of the protonated molecular ion by adjusting either the source pressure or the source

temperature. This approach has been especially useful for the identification of sulfonylurea metabolites from plants, soils, and animal
tissues.

2. Chromatography

Technical grade sulfonylureas and their formulations are most
conveniently assayed by high performance liquid chromatography
(HPLC). Separations in the reverse-phase mode are typically performed on a Zorbax-ODS column with a mobile phase of acetonitrile
(25–45% by volume) and water with a pH adjusted between 2.2–3.0
with phosphoric acid. Acidification of the mobile phase ensures that
the sulfonylurea will be in the neutral, undissociated form during
the analysis. The samples are dissolved in a mixture of acetonitrile
and water, containing approximately 0.01–0.025 N ammonium hydroxide. The base is needed to achieve the desired solubility and to
stabilize the sulfonylurea prior to analysis.

Normal-phase HPLC assay methods have been developed as well.
These typically use Zorbax-SIL columns and a mobile phase containing several percent acetic acid and a few tenths percent water in
methylene chloride. Samples for analysis are dissolved in methylene
chloride, and the eluting compounds are detected with a fixed- or
variable-wavelength ultraviolet detector. Absorptivity values of
each compound at their wavelength maximum and at 254 nm are listed
in Table 6.

Gas chromatography, a powerful analytical tool for many agricultural chemicals, is not particularly useful for the analysis of sulfonylurea herbicides because of their low volatility and thermal instability.

3. Residue Analysis

The very low use rates and relatively rapid dissipation characteristics of the sulfonylurea herbicides complicate residue and
metabolite analyses by necessitating extremely sensitive and selective
detection methods. Methods have been reported for the determination of residues of chlorsulfuron in soil [47,48] and cereals [49,50],
for metsulfuron methyl in soil and crops [51], and for sulfometuron
methyl in soil and water [52]. The final measurement in all these
procedures employs normal-phase liquid chromatography with photo-
conductivity detection to achieve the necessary sensitivity and selectivity. Detection limits of 0.01–0.05 parts per million in plant and
animal tissues, and 0.2 parts-per-billion in soil have been demonstrated, but conducting routine analyses at these levels is very
difficult.

Kelly et al. [53] reported on the determination of chlorsulfuron
in crude soil extracts using an enzyme-linked immunosorbent assay.
This technique is rapid, inexpensive, relatively specific and sensitive; detection limits were estimated to be about 0.4–1.2 parts-per-
billion.

TABLE 6

Ultraviolet Spectral Data of Sulfonylurea Herbicides

Compound	Wavelength maximum (nm)	Absorptivity at wavelength maximum	Absorptivity at 254 nm
Chlorsulfuron	222	63	28
Sulfometuron methyl	228	68	12
Metsulfuron methyl	220	68	8
DPX-M6316	220	50	28
Bensulfuron methyl	230	46	27
Chlorimuron ethyl	232	54	19

4. Bioassay

Bioassay methods have been developed to determine the residue level of many herbicides in soil and water. Santelman [54] has compiled a list of plant species that have been used as bioassay indicators for various herbicides. Variables controlling the response of bioassay indicators have been reviewed by Appleby [55].

Advantages of using a bioassay for determining sulfonylurea herbicide residues include:

1. The extreme sensitivity of certain plants to specific sulfonylureas resulting in a highly sensitive determination.
2. Expensive instrumentation is not required.
3. Only the biologically active material is measured.
4. Pre-extraction for soil and water analysis generally is not necessary.

Disadvantages of the method include:

1. Assay results are usually only semiquantitative.
2. Bioassay indicators normally have a very limited response range, making it necessary to use several different indicator species to analyze a broad concentration range.

TABLE 7

Relative Sensitivity of Common Rotational Crops to
Chlorsulfuron Soil Residues

Tolerant	Moderately sensitive	Sensitive	Very sensitive
Wheat (winter/spring)	Sorghum	Flax	Lentils
Durum	Soybeans	Corn	Sugar beets
Rye	Peas	Sunflower	Onions
Oats	Safflower	Mustard	
Triticale	Millet (proso/pearl)	Rape	
Barley (winter/spring)	Dry beans	Potatoes	
.	Ryegrass	Alfalfa	
	Bluegrass	Millet (setaria)	
	Guar		
	Mungbeans		
	Cotton		

3. Separate calibration curves must be prepared for each sulfonylurea and plant species [56].
4. Bioassay procedures are relatively slow, requiring from several
   days [57] to several weeks [58,59].

Field bioassays are sometimes conducted to determine whether
a particular rotational crop can be grown on fields previously treated
with sulfonylureas in locations where soil and climatic conditions restrict breakdown, for example, low rainfall, high soil pH, low temperature. In the United States, the current Glean herbicide label
requires that a field bioassay be conducted before Glean-treated
fields are planted to sensitive rotational crops. The relative

sensitivity of common rotational crops to chlorsulfuron residues is
given in Table 7 [60].

Laboratory and greenhouse plant bioassays have been used to
study chlorsulfuron's mobility [61,62], efficacy [58,63], and the influence of soil and environmental factors on its dissipation and movement [61,62,64–66]. The corn root bioassay, initially developed at
DuPont [67], as well as other plant bioassays using white mustard
(*Sinapis alba* L.) [61,62,66], alfalfa [68], lettuce [64], pea [65],
rapeseed [69], carrot [69], corn [63], sugar beet [69], and sorghum
[58] have proven invaluable in sulfonylurea research.

II. DEGRADATION PATHWAYS

A. Degradation in Plants

Plant metabolism studies have been conducted with radiolabeled
sulfonylureas to study the metabolism of the parent compound and
to characterize the residue in all plant parts used for food or feed.
When possible, short-term studies are conducted at higher than
recommended use rates to facilitate the identification of initial metabolites while long-term greenhouse or field studies are conducted at the
maximum recommended use rate. The extended studies focus on the
fate of the parent compound and early metabolites from the time of
normal application to harvest.

1. Metabolism of Chlorsulfuron

The metabolism of the cereal herbicide chlorsulfuron has been
studied in wheat [70] and in tolerant broadleaves [71]. The metabolic pathways in these two plant groups are very different.
Metabolism in tolerant broadleaves, such as flax and black nightshade, occurs via hydroxylation on the methyl group of the heterocycle and conjugation with a sugar. In wheat, barley, and other
tolerant grasses, chlorsulfuron first undergoes hydroxylation on the
phenyl ring followed by conjugation with glucose (Fig. 1). The
metabolites formed are nontoxic and herbicidally inactive. Rapid inactivation or detoxification of sulfonylureas by crops has been found
to be the basis of crop tolerance (see page 169).

2. Metabolism of Metsulfuron Methyl

Metabolism of the cereal herbicide metsulfuron methyl proceeds
via the same basic mechanism in both field-grown wheat and barley,
namely, hydroxylation of the phenyl ring and conjugation with
glucose. Other significant metabolites identified in these studies
include the hydroxymethyl analog of metsulfuron methyl, methyl 2-
(aminosulfonyl)benzoate, 2-(aminosulfonyl)benzoic acid, saccharin,

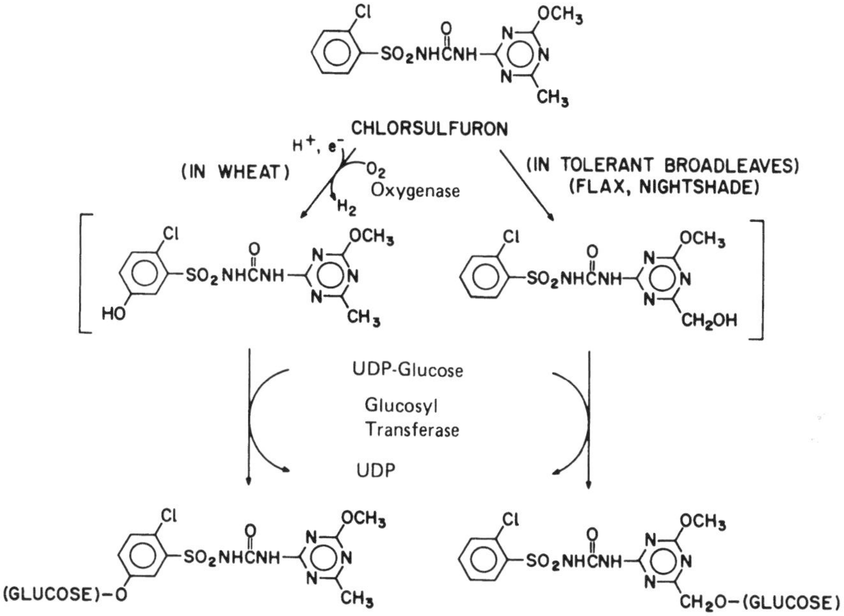

FIG. 1   Proposed metabolic pathway of chlorsulfuron in wheat and
tolerant broadleaves.

4-methoxy-6-methyl-1,3,5-triazin-2-amine, and 6-hydroxymethyl-4-
methoxy-1,3,5-triazin-2-amine.   Methyl 2-(aminosulfonyl)-4-hydroxy-
benzoate was detected in foilage samples from barley.   The proposed
metabolic pathway for metsulfuron methyl in wheat and barley is
summarized in Figure 2.

3.   Metabolism of DPX-M6316

    Metabolism of DPX-M6316 occurs rapidly in wheat and soybeans
with half-lives ($t_{1/2}$) of 3 to 4 and 5 to 6 hours, respectively.   Data
from laboratory and greenhouse studies using excised wheat leaves
suggest that at least three different metabolic pathways may be
operative.   A proposed scheme, based on preliminary studies with
[thiophene-2-14C] DPX-M6316, is shown in Figure 3.

4.   Metabolism of Chlorimuron Ethyl

    In soybeans, conjugation with homoglutathione is a major
metabolic pathway of chlorimuron ethyl (Fig. 4).   Homoglutathione
is the naturally occurring analogue of glutathione in soybeans.

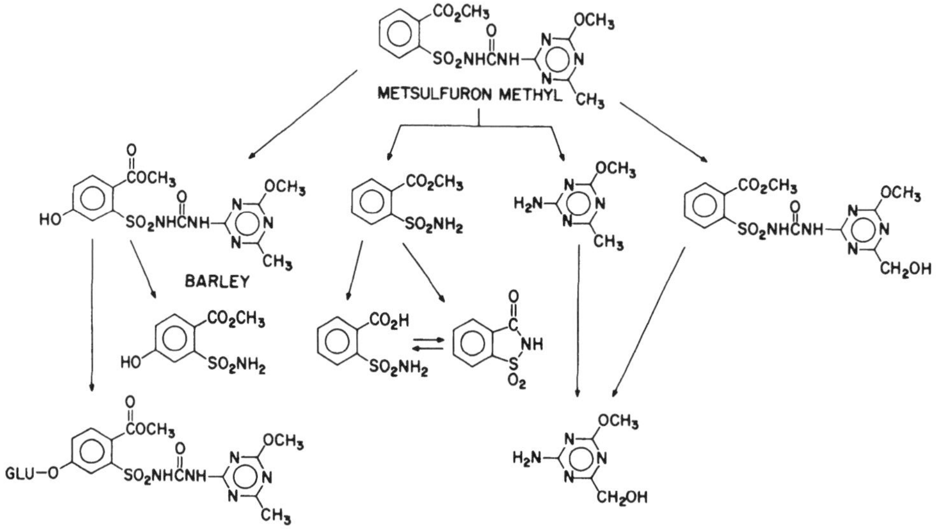

FIG. 2   Proposed metabolic pathway of metsulfuron methyl in wheat and barley.

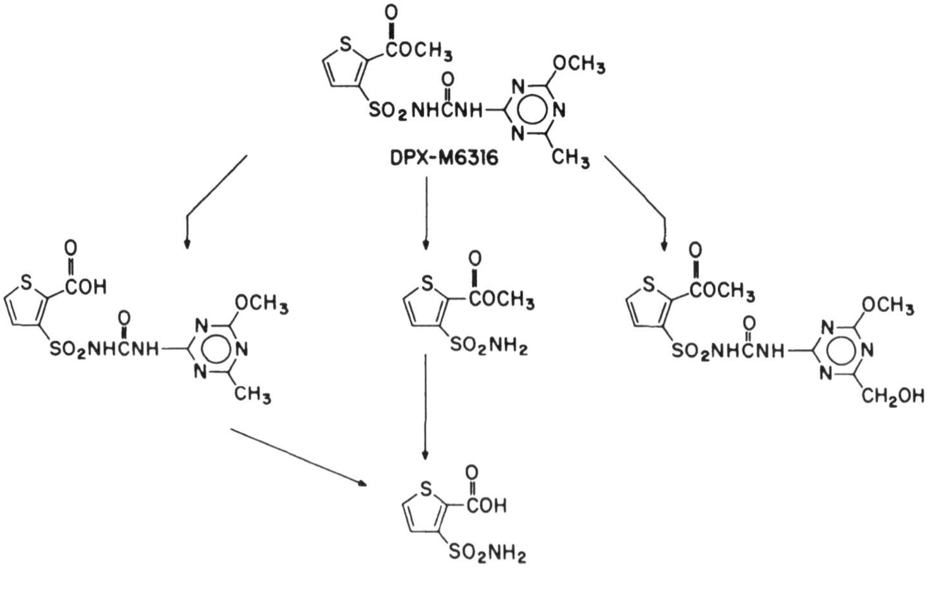

FIG. 3   Proposed metabolic pathway of DPX-M6316 in wheat.

Acifluorfen [72], the active ingredient in Blazer herbicide, and metribuzin [73], the active ingredient in Lexone and Sencor Herbicides, are metabolized in soybeans via this same mechanism. Recent data indicate that the homoglutathione conjugate undergoes further degradative metabolism to a cysteinyl conjugate. Chlorimuron ethyl also undergoes deesterification, followed by hydrolysis of the carboxylic acid analog to the sulfonamide and aminopyrimidine. These routes of metabolism are summarized in Figure 4.

5. Metabolism of Bensulfuron Methyl

In rice, bensulfuron methyl rapidly undergoes hydroxylation of one of the methoxy groups on the pyrimidine ring. This reaction, which probably involves a mixed function oxidase, presumably proceeds via an unstable $-O-CH_2-OH$ intermediate. A similar O-demethylation reaction has been reported for the metabolism of chlomethoxynil in rice [74]. Several other metabolites have been isolated and identified in rice harvested from a simulated paddy treated with a granular

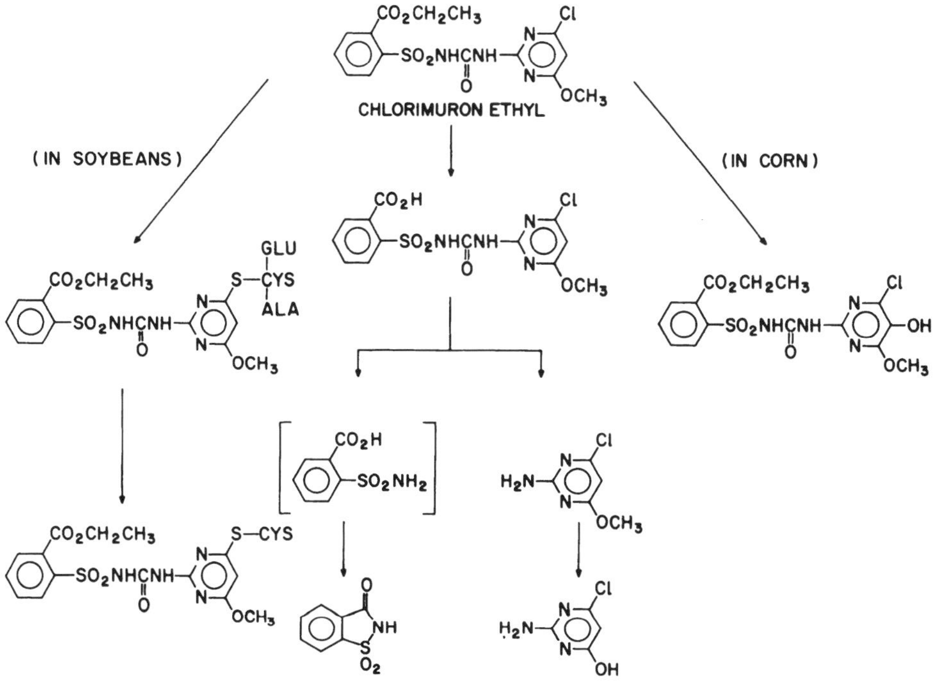

FIG. 4   Proposed metabolic pathway of chlorimuron ethyl in soybeans and corn.

formulation of either [phenyl-^{14}C(U)]bensulfuron methyl or [pyrimidine-2^{14}C]bensulfuron methyl. In this long-term study, the rice was continuously exposed, to parent compound and other radiolabeled compounds (i.e., soil metabolites, hydrolysis products) formed in the sediment and water during the course of the experiment. Radioactivity extracted from the foilage and grain (other than bensulfuron methyl and its O-demethylated analogue) was characterized as:

   Normal hydrolysis products, methyl 2-(aminosulfonyl-methyl) benzoate and 2-amino-4,6-dimethoxypyrimidine
   1H-2,3-benzothiazin-4-(3H)-one,2,2 dioxide
   4,6-dimethoxy-2-ureido-pyrimidine
   The carboxylic acid analog of bensulfuron methyl
   A very small percentage of the 4-hydroxy analogue of bensulfuron methyl

The proposed metabolic pathways of bensulfuron methyl in rice are summarized in Figure 5.

B. Degradation in Animals

   Metabolism studies in a mammalian species (normally the rat) are conducted to produce data on the absorption, distribution, excretion, and metabolism of the test compound. These tests are intended to fortify the understanding of the safety of the chemical in consideration of its intended uses and anticipated human exposure [75]. Male

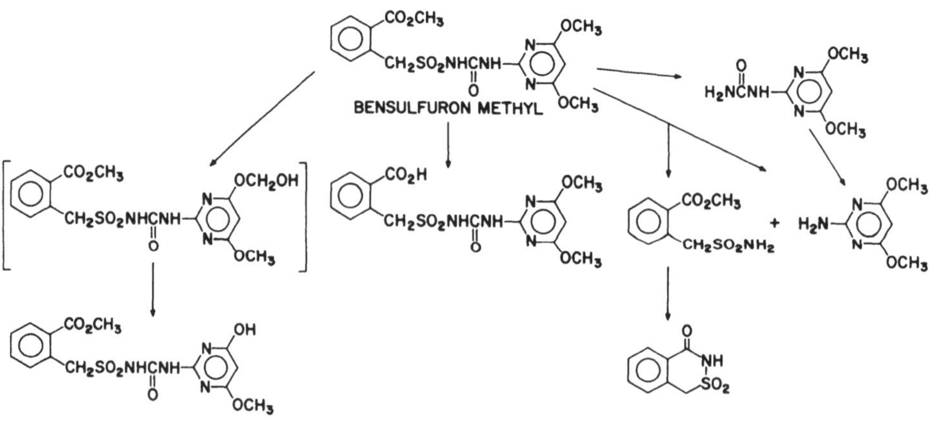

FIG. 5   Proposed metabolic pathway of bensulfuron methyl in rice.

and female rats receive a single oral dose of the radiolabeled test compound by gastric intubation at a low (no-effect) level, both with and without preconditioning, and at a high level, preferably one producing a toxic or pharmacological effect. The animals are sacrificed after more than 90% of the dosed radioactivity is excreted.

Additional studies in ruminants (usually goats) are conducted whenever residues are present in plants used for food or feed. In lactating goat studies, the animal is typically dosed twice daily for 5 to 7 consecutive days with the radiolabeled test compound and sacrificed approximately 24 hours after the final dose.

1. Metabolism of Chlorsulfuron

In the rat metabolism study with [phenyl-^{14}C(U)]chlorsulfuron, the excreted radioactivity was predominantly intact chlorsulfuron (approximately 85%) with smaller amounts of 2-chlorobenzenesulfonamide. Low levels of at least two unidentified polar metabolites were also detected. By contrast, radioactivity excreted by the goat was mostly the N-glucuronic acid conjugate of unaltered chlorsulfuron; the remainder was characterized as intact chlorsulfuron and 2-chlorobenzenesulfonamide.

2. Metabolism of Metsulfuron Methyl

In similar animal studies with metsulfuron methyl, most of the excreted radioactivity, recovered from both the rat and goat, was intact metsulfuron methyl. The composition of the small amount of remaining radioactivity suggests that hydrolysis of the *ortho*-carbomethoxy group and sulfonylurea bridge occurred as well as O-demethylation and hydroxylation of the 5-methyl triazine group. A proposed metabolic pathway for this sulfonylurea herbicide, based on the metabolites identified in the urine and feces, is presented in Figure 6.

3. Metabolism of DPX-M6316

The metabolism of DPX-M6316 has been studied in the rat. As in the rat studies with chlorsulfuron and metsulfuron methyl, most of the dosed ratioactivity was excreted in the urine and feces and at least 70 to 75% of this amount was recovered as intact DPX-M6316. Hydrolysis of the o-carbomethoxy group, O-demethylation on the heterocyclic ring, and hydrolysis of the sulfonylurea bridge appear to be the primary degradative mechanisms. A proposed metabolic pathway for DPX-M6316 in rats is presented in Figure 7.

4. Metabolism of Bensulfuron Methyl

Unlike chlorsulfuron, metsulfuron methyl, or DPX-M6316, bensulfuron methyl is metabolized extensively by the rat and goat. In

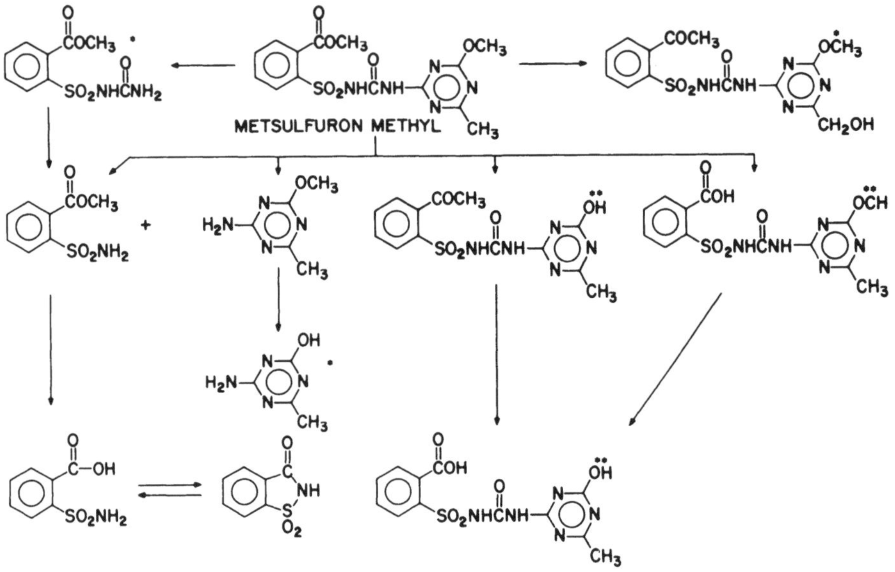

*Compounds identified only in extracts from the rat metabolism studies.
**Compounds identified only in extracts from the goat metabolism studies.

FIG. 6   Proposed metabolic pathway of metsulfuron methyl in the rat and goat.

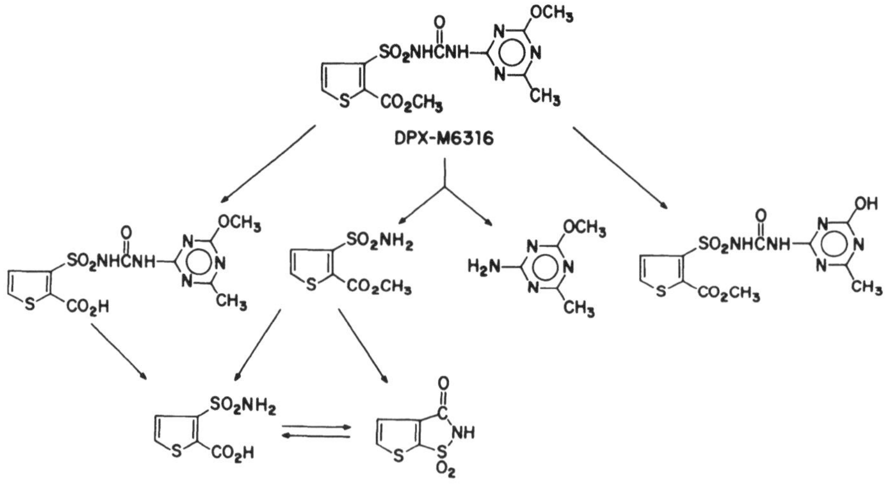

FIG. 7   Proposed metabolic pathway of DPX-M6316 in the rat.

the rat studies, parent bensulfuron methyl comprised less than 2% of
the total radioactivity in excreta from the low-dose-level animals and
typically only 4 to 11% in the high-dose-level animals. The major
radiolabeled metabolites, extracted from the excreta and identified by
mass spectrometry, include the mono O-demethylated analogue of
bensulfuron methyl, the 5-hydroxy analog of bensulfuron methyl,
and the 5-hydroxy analog of the pyrimidine amine. A glucuronide
or sulfate ester conjugate of the O-demethylated compound was identified in the rat urine. Minor radiolabeled metabolites were identified
as the carboxylic acid analog of bensulfuron methyl, the pyrimidine
urea, the sulfonamide hydrolysis product, and its corresponding
homosaccharin analog. Proposed metabolic pathways, based on these
identifications, are summarized in Figure 8.

5. Metabolism of Chlorimuron Ethyl

Metabolism of [pheny-^{14}C(U)]chlorimuron ethyl and [pyrimidine-
2-^{14}C]chlorimuron ethyl was studied in the rat. Chlorimuron ethyl,
like bensulfuron methyl, is extensively metabolized. The predominant
metabolic pathways involve hydroxylation reactions on the pyrimidine

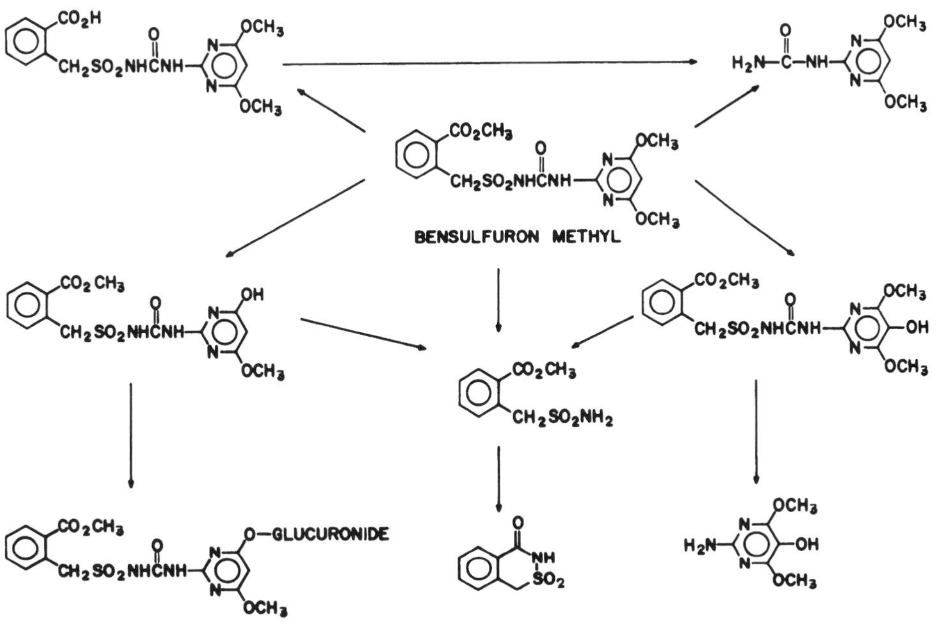

FIG. 8   Proposed metabolic pathway of bensulfuron methyl in the
rat and goat.

ring, the phenyl ring, and the o-carboethoxy group. Some hydrolysis of the sulfonylurea bridge occurs as well. These reactions are summarized in Figure 9. Unlike the other sulfonylurea herbicide studies, the dosed radioactivity was excreted somewhat more slowly and concentrations in the organs and tissues at sacrifice, although still very low, were somewhat higher than those found in the studies with other sulfonylurea herbicides.

6. Summary

The following general conclusions can be drawn from these studies:

1. The extent of metabolism is dependent on both the compound and the species. For example, chlorsulfuron is eliminated from the rat primarily as the intact sulfonylurea, whereas in the goat it is metabolized and excreted as a conjugate.
2. The dosed radioactivity is excreted very rapidly in the rat.
3. When little metabolism occurs, most of the parent compound and metabolites (85-87%) appear in the urine.

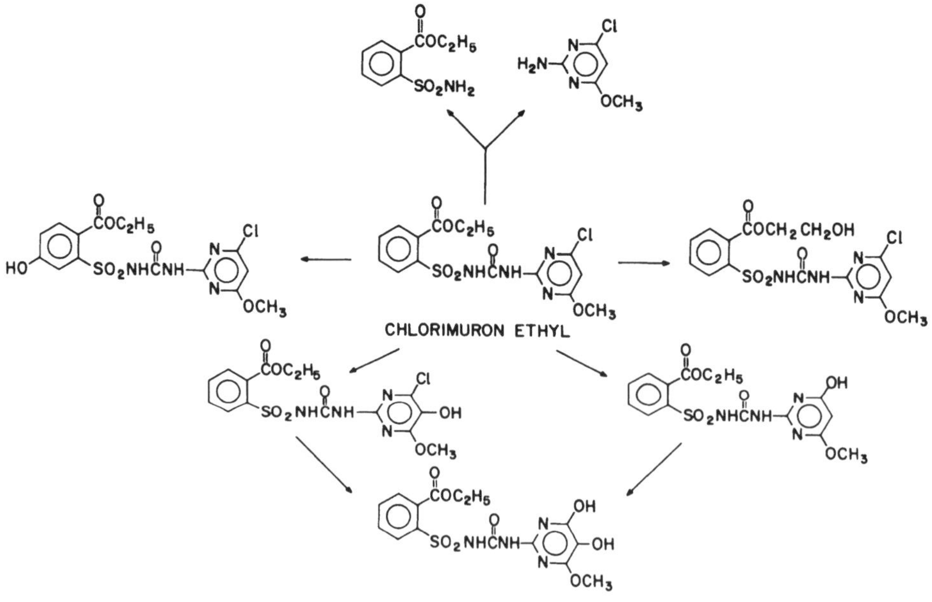

FIG. 9   Proposed metabolic pathway of chlorimuron ethyl in the rat.

4. No significant accumulation of the sulfonylurea herbicide or its
    metabolites occurs in the goat's milk or in any of the rat or
    goat tissues.

C. Degradation in Soil

The most important pathways of sulfonylurea degradation/dissi-
pation in soil are chemical hydrolysis and microbial breakdown.
Photolysis and volatilization are relatively minor processes. To
assess the relative importance of the two major processes, degradation in sterilized (no microbial activity) and nonsterilized soil is
compared. Because only chemical hydrolysis occurs in sterile soil,
the difference in the rate of breakdown under sterile and nonsterile
conditions can be used to estimate the relative contributions of chemical hydrolysis and microbial breakdown. The factors having the
greatest influence on chemical hydrolysis and microbial breakdown include temperature, pH, soil moisture, and soil organic matter.

1. Chemical Hydrolysis

The major hydrolysis pathway in soil is via cleavage of the
sulfonylurea bridge to give the corresponding sulfonamide and heterocyclic amine. This pH-sensitive cleavage parallels that observed for
hydrolysis in aqueous solution [cf Sect. I.E.2], that is, with increasing pH the rate of chemical hydrolysis in soil decreases. As
illustrated in Figure 10, Joshi et al. [76] observed that the half-life
of chlorsulfuron in sterilized soil at 30°C increases from 3 to approximately 45 weeks or 15-fold as the pH of the soil increases from 5.9
to 8.0.

The same trend of slower chemical hydrolysis with increasing
pH was observed by Fredrickson and Shea [58,77]. In a nonsterile,
acidic soil (Sharpsburg silty clay loam, pH 5.6, 2.4% organic matter)
the degradation half-life of chlorsulfuron at 25°C increased from 1.5
weeks to more than 9 weeks after it was amended with calcium carbonate to increase the pH to 7.5.

The expected effect of soil pH on degradation rate was also
observed by M. M. Joshi, H. M. Brown, and A. T. Van (DuPont,
unpublished data). On average, the rate of hydrolysis of chlorsulfuron was more than six times faster at a soil pH of 5.7 than at
one of 7.5 (Table 8). Using four air-dried soils from Colorado,
Zimdahl et al. [78] similarly observed that as the soil pH increased
the rate of chlorsulfuron degradation decreased. Anderson and
Barrett [79] likewise demonstrated that chlorsulfuron degradation
slowed as the pH of a sandy loam soil was increased from 6.1 to 7.2
by amendment with wheat residue.

Not surprisingly, both chemical hydrolysis and microbial breakdown are enhanced by warmer temperatures. For example,

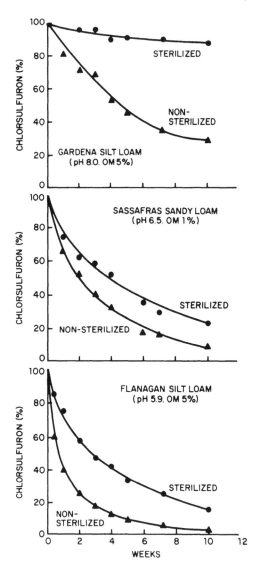

FIG. 10   Chlorsulfuron degradation in soil at 30°C [from Ref. 76].

TABLE 8

Half-Life (Weeks) of Chlorsulfuron Degradation in Soil at Several Temperatures Under Laboratory Conditions

Temperature (°C)	Rochelle, IL silt loam (pH = 5.7; 4.9% OM)		Fargo, ND silt loam (pH = 7.5; 5.7% OM)	
	Sterile	Nonsterile	Sterile	Nonsterile
40	1.0	0.40	8.0	1.1
35	2.4	0.65	14.0	2.0
30	4.3	1.2	33.0	7.8
20	14.0	4.0	69.0	32.0

M. M. Joshi, H. M. Brown, and A. T. Van (DuPont, unpublished data) found that by increasing the temperature of two silt loam soils from 20 to 40°C they could significantly increase the rate of degradation of chlorsulfuron under both sterile and nonsterile conditions (Table 8). The magnitude of this increase ranged from 9- to 30-fold.

Walker and Brown [64] obtained similar results in a nonsterile sandy loam soil (pH 6.5, 1.2% organic matter). The half-life of chlorsulfuron increased 6.5-fold (from 1.3 weeks to 8.5 weeks) as the temperature was lowered from 30 to 10°C while the soil moisture was maintained at 75% of holding capacity. When the soil moisture was reduced to 37% of holding capacity, the same temperature dependence was observed, but in the drier soil the rate of degradation was reduced about 50%.

Similar effects of moisture and temperature were observed by others. Anderson and Barrett [79] found that by increasing the temperature from 20 to 40°C chlorsulfuron persistence decreased. Increasing soil water content increased degradation of chlorsulfuron in a sandy loam soil but had no effect in a loam soil. After adding chlorsulfuron to an air-dried Rago silty clay loam soil (pH 7.7, 1.3% organic matter), Zimdahl et al. [78] adjusted the moisture from 25% to 75% of holding capacity and then incubated the soils at temperatures ranging from 10 to 40°C. Decreasing the soil moisture at a constant temperature resulted in a slower rate of chlorsulfuron degradation. When the soil moisture was held constant, decreasing the temperature had a similar effect. For example, the chlorsulfuron half-life increased 3.7-fold (from 8.9 to 32.7 weeks) as the temperature was dropped from 40 to 10°C. The fact that the soil was air-dried prior to its use may explain the unusually long half-lives found for chlorsulfuron in these experiments. Soil microorganisms were probably adversely affected, resulting in less microbial breakdown than might otherwise have occurred [58,65,76].

2. Microbial Degradation

Joshi, Brown, and Romesser [65,76,80] were the first to demonstrate the very important role that microorganisms play in the degradation of sulfonylureas. They showed that chlorsulfuron degraded significantly faster in nonsterile, microbially active soil than in soil that had been sterilized by exposure to ethylene oxide. Chlorsulfuron was thoroughly mixed with sterile and nonsterile Gardena silt loam soil (pH 8.0, 5.0% organic matter), Sassafras sandy loam soil (pH 6.5, 1.0% organic matter) and Flanagan silt loam soil (pH 5.9, 5.0% organic matter) to give a final concentration of 100 ppb. Treated soil was incubated at 30°C for 10 weeks. During this period, aliquots were removed and analyzed (Fig. 10). Loss of herbicide was followed by means of a pea bioassay (T. T. Obrigawitch and C. J. Peter, DuPont, unpublished data) or high performance liquid chromatography (HPLC) analysis using radiochemical detection. Microbial breakdown accounted for 79% of the initial degradation in the nonsterile Flanagan soil, 50% in Sassafras soil, and 91% in Gardena soil. Similarly, Fredrickson [58] observed microbial breakdown of chlorsulfuron in a Sharpsburg silty clay loam soil (pH 5.6, 2.4% organic matter) with 15% moisture, but he found that soil microbes accounted for only 25% of the initial degradation of chlorsulfuron in this soil. Reinoculation of sterilized soil with a microbial suspension from nonsterilized soil restored most of its ability to degrade chlorsulfuron [76].

Several groups of microorganisms, including actinomycetes, fungi, and bacteria, have been isolated from soil and shown to actively metabolize chlorsulfuron, metsulfuron methyl, and chlorimuron ethyl [65,76,80]. These microorganisms have also been demonstrated to metabolize sulfometuron methyl, bensulfuron methyl, and DPX-M6316 (M. M. Joshi and A. T. Van, DuPont, unpublished data).

The metabolic pathways of three of these microorganisms have been reported [65,80]. *Streptomyces griseolus*, a soil actinomycete, rapidly metabolizes chlorsulfuron via conversion of the methoxy group to a hydroxyl group and hydroxylation of the methyl group (Fig. 11). Metsulfuron methyl contains the same heterocycle as chlorsulfuron and undergoes the same metabolic reactions. In addition, metsulfuron methyl undergoes deesterification of the *ortho*-carbomethoxy group on the aryl portion, resulting in the formation of the free acid (Fig. 11). *Aspergillus niger* and *Penicillium* sp., two soil fungi, appear to catalyze hydrolysis of the sulfonylurea bridge, yielding

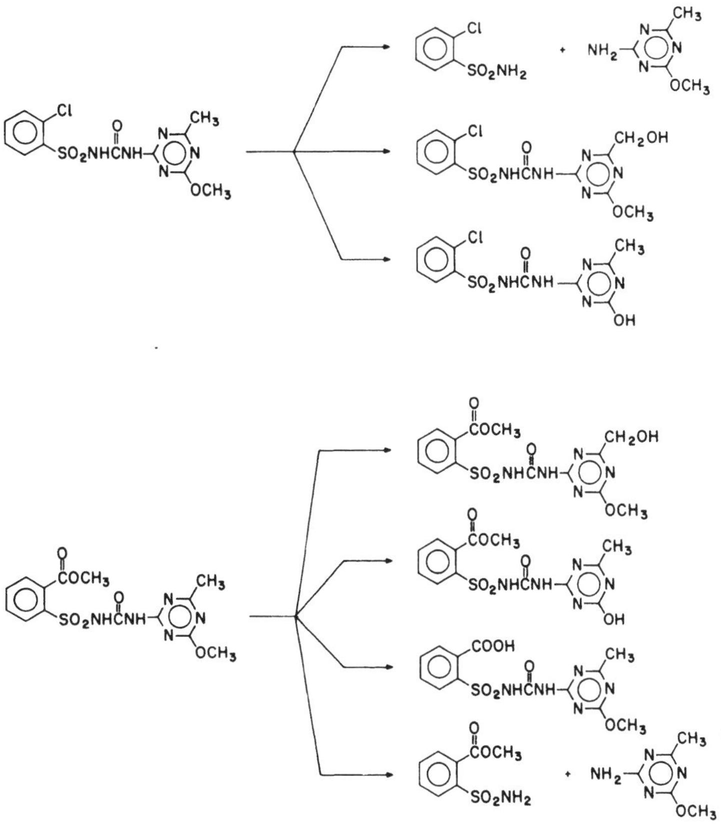

FIG. 11   Metabolic pathways of chlorsulfuron and metsulfuron methyl by soil microorganisms.

the corresponding sulfonamide and heterocycle [80]. These fungi are also able to convert the methoxy group to a hydroxyl group on the heterocycle, but this is a relatively minor pathway (Fig. 11).

3.   Photolysis

Photodegradation studies are conducted to assess the importance of photolysis as a degradation mechanism. Natural sunlight or artificial light sources, which simulate sunlight, are used as irradiation sources. Control samples, maintained in darkness at constant temperature, are similarly analyzed to determine the significance of photolysis relative to hydrolysis and/or microbial degradation.

In the aqueous photolysis studies, the sulfonylurea, dissolved in sterilized aqueous solutions buffered at the most hydrolytically stable pH, is irradiated for up to 30 days under natural sunlight or 15 days continuous exposure under the artificial lights. In general, the importance of photodegradation depends on the extent of overlap between the emission spectrum of the light source and the ultraviolet-visible absorption spectrum of the sulfonylurea. For example, metsulfuron methyl, which does not absorb ultraviolet radiation at wavelengths in the solar emission spectrum, degraded at essentially the same rate in the irradiated and dark control studies. Chlorimuron ethyl had a first half-life at pH 9 (the most hydrolytically stable pH) of 31 to 43 days under the simulated sunlight source. DPX-M6316, on the other hand, which absorbs radiation in the lower wavelength region of natural sunlight, degraded with a first half-life (corrected for hydrolysis) of approximately 5 days at a pH of either 5, 7, or 9.

In the photodegradation studies with soil, a thin layer of soil (typically 0.4–1.0 mm) is spread on glass plates and the test compound is uniformly applied to the soil surface prior to irradiation. Metsulfuron methyl, when exposed to natural sunlight, degraded at essentially the same rate as it did in the dark ($t_{1/2}$ = 6 days). Degradation of DPX-M6316 on soil was only slightly faster in the irradiated samples ($t_{1/2}$ = 12 days) vs. the dark controls ($t_{1/2}$ = 14 days). Chlorimuron ethyl, on the other hand, degraded more than twice as fast when the soil was exposed to natural sunlight ($t_{1/2}$ = 20 days vs. 43 to 46 days for the dark controls).

4. Volatilization

Sulfonylureas are very nonvolatile with vapor pressures at 25°C ranging from 1.3×10^{-10} to 5.5×10^{-16} mmHg at 25°C (Table 3). In laboratory studies where these radiolabeled sulfonylureas were applied to soil and aged for up to one year in a closed biometer flask, there was no evidence of the formation of volatile radiolabeled compounds other than labeled carbon dioxide formed via microbial breakdown. Volatilization will thus be insignificant in the dissipation of sulfonylureas following application.

D. Dissipation Under Field Conditions

1. Chlorsulfuron

Sulfonylurea degradation under field conditions occurs at rates that are similar to, and often faster than, conventional soil-active herbicides. Palm et al. [81], when summarizing worldwide field experience, reported that the half-life of chlorsulfuron in soil is usually between one and two months. Similarly, Campion [82] found the half-life of chlorsulfuron in Australia generally to be between one and one-and-a-half months. Under typical growing conditions in Newark, Delaware, on a Keyport silt loam soil (pH 6.0, 1.4% organic matter), chlorsulfuron exhibits a half-life of about one month (Table 9). This fairly rapid rate of dissipation is comparable to that of metribuzin, somewhat shorter than that of linuron, and about one-half that of bromacil or diuron at the same location.

The disappearance of chlorsulfuron under field conditions in the spring was studied at fourteen locations by J.C-Y. Han and C. Rapisarda (DuPont, unpublished data). Several stainless steel cylinders, 10 cm in diameter, were driven into the ground at each site. Radiolabeled chlorsulfuron was then applied to the bare soil surface at the relatively high rate of 100 g/ha. The cylinders were removed from the field at various times and the amount of intact chlorsulfuron determined by thin layer chromatography of soil extracts, followed by liquid scintillation counting. Half-lives ranged from 2 to 13 weeks (Table 9), with the slowest rates of degradation occurring in the highest pH soils.

Walker and Brown [64] found a chlorsulfuron half-life of 5.7 weeks following a late April application at a rate of 30 g/ha to a sandy loam soil (pH 6.5, 1.2% organic matter) in Wellesbourne, United Kingdom. Predictions of chlorsulfuron dissipation were made by using a computer program [83] and actual weather data for the site. Reasonable agreement was obtained with the measured values at each time interval and predictions of chlorsulfuron dissipation were extended to several worldwide locations using this computer model.

Chlorsulfuron half-lives ranging from 2.6 (Mississippi, 1979) to 7.6 weeks (Holland, 1980) were projected. The estimated 6.4 week half-life in Saskatchewan using 1980 weather data is consistent with the measured three-site average half-life of 4.3 weeks (Table 9) determined by J. C-Y. Han and C. Rapisarda (DuPont, unpublished data). Walker and Brown [64] concluded from their studies that during the summer growing season the most rapid disappearance of chlorsulfuron would be expected in the warmer, wetter regions of North America. Intermediate rates of loss would occur in the warm but drier climates of Colorado, Alberta, and Saskatchewan, and the slowest degradation would occur in the cooler climates of northern Europe. Furthermore, they indicated that during the winter, some degradation would be expected at the relatively milder European locations, whereas little to no breakdown would be anticipated in the frozen soil of Canada.

Smith and Hsiao [84] determined the amount of radiolabeled chlorsulfuron remaining 45 and 95 weeks after treatment of 660 g/ha to a sandy loam soil at White City, Saskatchewan. They found 5–8% intact chlorsulfuron at 45 weeks and 2% at 95 weeks. This is equivalent to half-lives of 10 to 17 weeks. Since the temperature during the growing season is much warmer than either the 45- or 95-week

TABLE 9

Half-Life of Chlorsulfuron Under Field Conditions

Location	Soil type	pH	Organic matter (%)	Half-life (weeks)
Clay Center, NE	Silt loam	5.6	3.8	4
Newark, DE	Silt loam	6.0	1.4	4
Swift Current, Saskatchewan	Silt loam	6.1	1.0	2
Saskatoon, Saskatchewan	Silty clay loam	6.2	3.4	4
Versailles, France	Loam	6.4	1.9	3
Carrington, ND	Loam	6.6	4.4	3
Rochelle, IL	Silt loam	6.7	7.4	3
Akron, CO (Site 1)	Silt loam	6.9	1.0	5
Stettler, Alberta	Sandy loam	6.9	4.4	12
Rosetown, Saskatchewan	Clay loam	7.3	2.3	7
Fargo, ND	Silty clay loam	7.6	5.3	8
Fisher Branch, Manitoba	Silty clay loam	7.8	6.4	9
Kimberly, ID	Silt loam	8.0	1.3	13
Akron, CO (Site 2)	Silt loam	8.2	1.2	11

average, the half-life during the summer months would be expected to be much shorter than these values.

In related work, Smith and Hsiao [84] applied chlorsulfuron at 40 g/ha to a clay soil at Regina, Saskatchewan, and to the sandy loam soil at White City in May 1981, 1982, and 1983. Prior to the 1982 application, 15-16% of the chlorsulfuron applied the previous year was found in the top 10 cm of the soil at each location. One year after the 1982 and 1983 greatments, only 4 to 6% of the chlorsulfuron remained. These results suggest that at these sites there is little potential for herbicide accumulation resulting from repeat applications.

Despite the relatively rapid dissipation of chlorsulfuron from the soil, there have been reports of continued weed control [85] and crop injury [61,62,68,69,86-88] after the first full growing season following treatments of up to 120 g/ha. This continuation of residual herbicidal activity is caused by the unprecedented difference in sensitivity to chlorsulfuron among different crop and weed species (Fig. 12, Table 7). For example, Sweetser et al. [70] have found wheat to be greater than 1000 times more tolerant to chlorsulfuron than the extremely sensitive sugar beet. Thus, despite breakdown

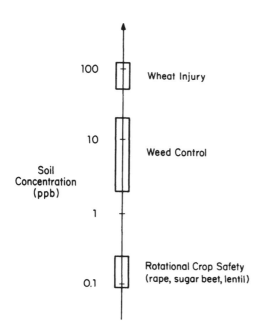

FIG. 12   Glean plant sensitivity.

of 99% of applied chlorsulfuron (6.3 half-lives), the extremely low soil residue that remains can cause injury to a highly sensitive rotational crop like sugar beet. Sugar beet root growth can be affected by concentrations of chlorsulfuron in the soil as low as 0.1 to 1 ppb.

Observations from field rotational crop studies (Table 10) confirm this unprecedented difference in sensitivity. Although Hageman [86] found no injury to sugar beet two years after a spring application of chlorsulfuron at a rate of 125 g/ha in Rosemount, Minnesota, the same treatment caused 84% injury to sugar beet at Crookstown, Minnesota. Most probably, the much faster degradation of chlorsulfuron in the wetter, acidic soil at Rosemount resulted in dissipation to a nonphytotoxic residual concentration. Brewster and Appleby [87] observed no sugar beet stand reduction 26 months after application of 35 g/ha chlorsulfuron to an acidic silt loam soil at Corvalis, Oregon, but they did see a 37% reduction in foilage fresh weight 6 weeks after planting. In a sandy clay soil at Turku, Finland, Junnila [69] observed no injury to beets 24 months after a 20 g/ha application of chlorsulfuron, but 12 months after the second of two annual applications of 10 g/ha he found 82% injury when planting into soil from the top 5 cm of the profile. In contrast, only 15% injury was observed when using soil from the 5 to 10 or 10 to 20 cm layers. It is important to recognize that this soil residual activity, which may continue for more than two years following some applications, is due to the very high sensitivity of sugar beets, and several other rotational crops such as oilseed rape to chlorsulfuron and not to a failure of the herbicide to degrade appreciably in the soil.

Shorter rotational crop intervals are possible when more tolerant crops are planted after application of sulfonylurea herbicides. Hageman [86] observed no injury to sunflower two years after a 250 g/ha treatment of chlorsulfuron at Rosemount. Foy and Mersie [68] saw no injury to corn and soybean 10 and 11 months after application of up to 120 g/ha chlorsulfuron to an acidic silt loam soil in Virginia. Peterson and Arnold [88] found no injury to corn, sorghum, soybean, sunflower, or flax 12 months after treatment with up to 68 g/ha chlorsulfuron at Watertown, South Dakota. At Redfield, South Dakota, using the same species and experimental design, all crops were injured significantly 12 months after application of the lowest rate (17 g/ha). After 24 months, crop injury ranged from <10% at 17 g/ha to 20-50% at 68 g/ha. Corn and sorghum were most susceptible while flax showed the least susceptibility. Differences in carryover at the two locations were attributed to the lower soil pH at Watertown.

The timing of application also affects the duration of biological activity in the soil. Nilsson [61,62] studied chlorsulfuron dissipation

at five locations in Sweden at rates from 5 to 40 g/ha. He found
that injury to white mustard persisted for 3 to 4 months following
spring applications and for 9 to 11 months after autumn treatments
at the same rates. Since little, if any, chlorsulfuron degradation
occurs during the cold winter months, the duration of herbicidal
activity is somewhat longer following autumn treatments.

2. Metsulfuron Methyl

Under field conditions in Europe, Canada, and the United
States, Doig et al. [23] found that the half-life for metsulfuron
methyl ranged from one week to one month. Royrvik [66] studied
the disappearance of metsulfuron methyl in the summer growing
season in Denmark and found the average half-life to be one month.
This was comparable to that observed for chlorsulfuron at these
locations.

These field soil dissipation observations are in good agreement
with laboratory results. Anderson [89] has shown that metsulfuron
methyl degrades in the soil to herbicidally inactive products at rates
slightly faster than chlorsulfuron under identical conditions.

The rate of degradation of sulfonylurea herbicides in soil is
fastest in warm, moist, light-textured, low pH soils, and slowest in
cold, dry, heavy, high pH soils. The sometimes long residual
activity that has been indicated by replanting highly sensitive
species is caused primarily by the very high susceptibility of the
rotational crops to the herbicide and not to a slow (compared with
conventional herbicides) rate of dissipation.

E. Mobility in Soil

The mobility of sulfonylurea herbicides (predominantly chlorsulfuron) has been the topic of numerous laboratory [90-92], and
field [61,62,69] studies. These laboratory studies typically involve
either measuring the elution of the test compound through a uniformly packed soil column as water is continuously added (i.e., soil
leaching), or measuring its migration along a uniform thin layer of
soil coated on a glass plate after development with water (i.e., soil
thin-layer chromatography, TLC).

Soil thin-layer chromatography (TLC) R_f values on four soils
(Table 11 and 12) show that the mobility of a particular sulfonylurea
herbicide generally increases with increasing soil pH and decreasing
soil organic matter [T. M. Priester, DuPont, unpublished data]. A
positive correlation of soil pH with the TLC mobility of chlorsulfuron
was also observed by Nicholls and Evans [90] and Mersie and Foy
[91]. The latter authors also reported mobility was negatively
correlated with soil organic matter.

TABLE 10

Rotational Crop Studies with Chlorsulfuron

Ref.	Location	Soil type	Soil pH	Organic matter (%)	Rate [g ai/ha]	Time after application (months)	Indicator species	Injury
[88]	Watertown, SD	Brookings silty clay	5.3	3.5	17	12	Corn, flax, soybean sorghum & sunflower	None
						24		None
					34	12		None
						24		None
					68	12		None
						24		None
[85]	Alberta, Canada	Sandy loam	5.4	7.4	40	12	Lamb's-quarters, Tartary buckwheat White cockle, rough cinquefoil, white clover, dandelion, stink weed	>80% for all

[87]	Corvalis, OR	Woodburn silt loam	5.8	2.8	35	14	Sugar beet	No stand reduction, 54% reduction fresh wt.
						18	Sugar beet	No stand reduction, 8% reduction fresh wt.
						26	Sugar beet	No stand reduction, 37% reduction fresh wt.
[86]	Rosemount, MN	Waukegan silt loam	6.1	5.0	60	24	Sugar beet	None
					125	12	Sugar beet	70%
						24	Sugar beet	None
					125	12	Sunflower	46%
						24	Sunflower	None
					250	12	Sugar beet	86%
						24	Sugar beet	28%
					250	12	Sunflower	60%
						24	Sunflower	None
[68]	VA	Duffield-Ernest silt loam	6.5	2.5	<120	10	Corn	None
						11	Soybean	None

TABLE 10

(continued)

Ref.	Location	Soil type	Soil pH	Organic matter (%)	Rate [g ai/ha]	Time after application (months)	Indicator species	Injury
[88]	Redfield, SD	Great Bend silty clay loam	6.5	2.5	17	12 24	Corn, flax, soybean, sorghum & sunflower	40–60% <10%
					34	12 24		55–80% 10–30%
					68	12 24		80–90% 20–50%
[85]	Alberta, Canada	Silt loam	7.1	13.7	50	12	Canada thistle	76% reduction # shoots
					100	12	Canada thistle	94% reduction # shoots
					150	12	Canada	94% reduction # shoots
[86]	Crookston, MN	Bearden silt loam	8.3	2.8	125	12 24	Sugar beet Sugar beet	94% 84%
					250	12 24	Sugar beet Sugar beet	100% 99%

Ref	Location	Soil				Crop	Result
[69]	Turku, Finland	Sandy clay	—	—	20	24	Beet	None
					10 + 10	12	Beet	82%, 0– 5 cm 15%, 5– 10 cm 15%, 10– 20 cm
					20 + 20	12	Beet	92%, 0– 5 cm 55%, 5– 10 cm 42%, 10– 20 cm
[61]	Ugerup and Lamma, Sweden	Sand, clay	—	—	40	4– 11	White mustard	Observed
	Ultuna, Robacks-dalen and Vojakkala, Sweden	Clay, sand-silt, fine sand	—	—	40	>11	White mustard	Observed
[62]	Ultuna, Sweden	Clay	—	—	5	3	White mustard	Observed
						9	White mustard	Observed
					10	4	White mustard	Observed
						11	White mustard	Observed
					20	4	White mustard	Observed
						11	White mustard	Observed

TABLE 11

Mobility of Sulfonylurea Herbicides

Compound	Soil thin-layer chromatography R_f value[a]			
	Woodstown sandy loam[b]	Cecil sandy loam[b]	Flanagan silt loam[b]	Keyport silt loam[b]
Chlorsulfuron	0.90 (5)[c]	0.65 (4)	0.59 (3)	0.52 (3)
Metsulfuron methyl	0.88 (4)	0.74 (4)	0.70 (4)	0.58 (3)
Sulfometuron methyl	0.84 (4)	0.59 (3)	0.26 (2)	0.21 (2)
Chlorimuron ethyl	0.71 (4)	0.59 (3)	0.41 (3)	0.18 (2)
Bensulfuron methyl	0.46 (3)	0.30 (2)	0.06 (1)	0.05 (1)
DPX-M6316	0.92 (5)	0.73 (4)	0.44 (3)	0.49 (3)

[a] The R_f value is the ratio of the migration of the compound relative to the water solvent front.
[b] Selected soil characteristics are listed in Table 12.
[c] U.S. EPA classification

U.S. EPA classification	Range of R_f values	Mobility descriptor
1	0.0-0.09	Immobile
2	0.10-0.34	Low mobility
3	0.35-0.64	Intermediate mobility
4	0.65-0.89	Mobile
5	0.90-1.0	Very mobile

TABLE 12

Soil Characterization[a]

Parameter	Woodstown sandy loam	Cecil sandy loam	Flanagan silt loam	Keyport silt loam
% Sand (0.05–2.0 mm)[b]	60	61	2	12
% Silt (0.002–0.05 mm)[b]	33	21	81	83
% Clay (<0.002 mm)[b]	7	18	17	5
% Organic matter	1.1	2.1	4.3	7.5
pH	6.6	6.5	5.4	5.2
Cation exchange Capacity, meg/100 g	5.3	6.6	21.1	15.5
Origin of soil	Dover, DE	Raleigh, NC	Rochelle, IL	Newark, DE

[a]Soil analyses were performed at the Soil Testing Laboratory, College of Agricultural Sciences, University of Delaware, Newark, DE.

[b]Mechanical analysis, to determine the sand, silt, and clay content, was conducted after removal of the organic matter by wet oxidation.

In soil column leaching studies, Mersie and Foy [91] showed
that the depth to which chlorsulfuron penetrated the soil was in the
same order as the soil TLC R_f values. Using a column packed with
a silt loam soil, Nilsson [92] found that chlorsulfuron moved vertical-
ly with rising capillary water. Therefore, during periods of net upward flow of soil water, chlorsulfuron might reenter the root zone
from deeper in the soil profile where it had penetrated during earlier
periods of net downward water flow.

Mobility under field conditions is similarly a function of soil
characteristics (e.g., percent organic matter, pH, soil type, porosity,
etc.). Environmental factors such as rate of compound application,
time of the year, rainfall and soil temperature also influence mobility.
Chlorsulfuron has been characterized as a relatively mobile compound
by field studies conducted in Sweden [61,62], Finland [69], and in
Canada and the United States. The reappearance of chlorsulfuron
residues in upper layers of soil, previously containing nondetectable
levels, has been attributed to its upward mobility by capillary action
[61,62,69,92]. The appearance of chlorsulfuron in control plots adjacent to chlorsulfuron-treated plots, has been cited as evidence for
the lateral movement of chlorsulfuron [61]. In spite of their relatively high mobilities, chlorsulfuron and the other sulfonylurea herbicides,
are not expected to pose groundwater contamination problems because
of their exceptionally low use rates, low toxicities, and their relatively rapid soil degradation/dissipation characteristics.

III. MODE OF ACTION

A. Initial Observations

1. Growth Inhibition Studies

Studies of the mode of action of the sulfonylurea herbicides
were first undertaken at DuPont in 1978 by T. B. Ray. Recogniz-
ing their potent growth inhibition properties, Ray [93,94] examined
the dose dependence and time course of this response in sensitive
corn seedlings. Concentrations of chlorsulfuron as low as 2.8 nM
(1 ppb) significantly inhibited root growth, and higher concentrations effectively reduced shoot growth within 2 to 4 hours of treatment. These results indicated a rapid effect of chlorsulfuron on
either cell division or cell enlargement.

Studies using a variety of plant hormone bioassays ruled out
cell enlargement as a sensitive process, since auxin-, cytokinin-, or
gibberellin-induced cell enlargement was unaffected by chlorsulfuron
concentrations that completely stopped growth (28 μm, 10 ppm). In
marked contrast, cell division was highly sensitive. For example,
in *Vicia faba* L. root tips, the average number of cell division or
mitotic figures per 100 cells was reduced 86% following a 24-hour

treatment at 2.8 μM chlorsulfuron. Subsequent studies used [3H] thymidine incorporation into DNA as a highly sensitive measurement of cell division. With this technique, the threshold concentration for chlorsulfuron inhibition of cell division in corn, pea, and Jerusalem artichoke (*Helianthus tuberosus* L.) tissues ranged from 2.8 to 28 nM and inhibition was readily apparent in 2 to 4 hours.

A more detailed analysis by Rost of dividing pea root cells [95] revealed that chlorsulfuron blocked their progression from G_2 to M (mitosis) in the cell division cycle and secondarily reduced movement from G_1 to S (DNA synthesis). No aberrant mitotic figures were observed and, in agreement with the results of Ray [93,94], there was no change in the distribution pattern of the mitotic stages (prophase through telophase). These early results suggested that chlorsulfuron was a potent and rapid inhibitor of plant cell division but did not directly interefere with the mitotic apparatus.

Several pieces of evidence [96] suggest that the inhibitory effect of chlorsulfuron on cell division is not due to a direct effect on DNA synthesis or the synthesis of nucleosides. These include:

1. Chlorsulfuron had no effect on DNA synthesis when added directly to isolated plant nuclei even though it was a powerful inhibitor of [3H]thymidine incorporation when added to intact cells.
2. Chlorsulfuron had no effect on the in vitro activity of DNA polymerase obtained from corn roots or *Micrococcus leteus* or on the in vitro activity of thymidine kinase extracted from the roots of etiolated barley or corn seedlings.
3. Nucleoside precursors of DNA, when added exogenously to corn roots, did not overcome the chlorsulfuron inhibition of [3H]thymidine incorporation.

2. Metabolism Studies

Ray [94] found protein synthesis ([14C]leucine incorporation) unaffected, and RNA synthesis ([3H]uridine incorporation) inhibited by only 28% under conditions (2.8 μM chlorsulfuron, 6 hr) where cell division in corn roots was reduced 80–90%. Using pea roots, isolated leaf cells of soybean, or leaf cells of *Phaseolus vulgaris* L., other workers [95,97,98] also found marginal to no effects on protein and RNA synthesis during the first few hours following chlorsulfuron treatment (2.8 μM).

Because of the urea moiety and the triazine ring in the structure of chlorsulfuron, some workers initially thought sulfonylureas might act by blocking photosynthesis [98]. A variety of methods, including oxygen evolution from isolated chloroplasts [93,94], $^{14}CO_2$ fixation, and analysis of leaf fluorescence in whole leaves [93,94,97], have ruled this out as a primary effect of the sulfonylureas. Likewise, respiration has been eliminated as an initial target [93,94]. One report [97] suggests that chlorsulfuron inhibits lipid synthesis in isolated soybean leaf cells at concentrations as low as 0.1 μM. The significance of this observation is unknown since this effect was not observed in navy bean leaf cells [98].

3. Plant Hormone Studies

    In addition to the cell enlargement studies cited above, which clearly demonstrate that chlorsulfuron does not directly block the growth-promoting action of auxins, cytokinins, and gibberellins, the possible intervention of ethylene in chlorsulfuron action has been evaluated by several workers. Suttle and Schreiner [99] found that treatment of soybean seedlings with 1 μg of chlorsulfuron per plant inhibited growth and greatly stimulated anthocyanin formation and phenylalanine ammonia lyase activity. Because ethylene production was also greatly stimulated, and because the gas is known to cause plant responses similar to those caused by the herbicide, experiments were conducted to determine whether ethylene might be the intervening causative agent. The presence, however, of a 3- to 4-day lag period between chlorsulfuron application and an increase in ethylene production led the authors to rule out ethylene as an important primary factor. A similar conclusion was reached about chlorsulfuron-enhanced phenolic content in sunflower seedlings [100].

    These and other studies [Beyer, DuPont, unpublished data] have demonstrated that ethylene is not involved in the primary action of the sulfonylurea herbicides; however, "stress-ethylene," formed in some plant species as a result of the phytotoxic action of the sulfonylureas, can contribute to the secondary symptoms that develop following herbicide treatment. For example, Hageman and Behrens [101] found that the defoliation of velvetleaf caused by chlorsulfuron treatment was due to enhanced ethylene production.

B. Discovery of the Site of Action

1. Studies with Bacteria

    Although the initial studies of Ray [93,94] and Rost [95] established that sulfonylureas such as chlorsulfuron are rapid and potent inhibitors of cell division, these and other studies [96–98] identified no sensitive biochemical process that could account for this inhibition. A major advance in identifying the target site of action of the sulfonylurea herbicides came from studies involving bacteria. Since bacteria and plants share many common biochemical pathways, LaRossa and Schloss [102] reasoned that bacteria might provide an expedient means of localizing the site of sulfonylurea action. Accordingly, several bacterial species were selected for their sensitivity to sulfometuron methyl. In the presence of valine, sulfometuron methyl inhibited the growth of wild-type *Salmonella typhimurium*, and this inhibition was reversed by isoleucine, but not by 18 other common amino acids. These and other reversal experiments suggested that sulfonylureas inhibit some step in the biosynthesis of branched-chain amino acids. In a series of elegant experiments using sulfometuron methyl-resistant mutants, the site of action of sulfometuron methyl was identified as the enzyme, acetolactate synthase.

Acetolactate synthase (ALS; EC 4.13.18), also known as aceto-hydroxyacid synthase (AHAS), is a key enzyme in the branched-chain amino acid biosynthetic pathway of bacteria, fungi, and higher plants. The enzyme requires thiamine pyrophosphase and Mg^{2+}, as well as FAD, even though the reactions catalyzed by this enzyme involve no net oxidation or reduction. ALS catalyzes: (a) the condensation of two molecules of pyruvate to form CO_2 and α-aceto-lactate, which leads to valine and leucine synthesis, and (b) the condensation of one molecule of pyruvate with α-ketobutyrate to form CO_2 and α-aceto-α-hydroxybutyrate, which leads to isoleucine formation.

Since the initial work of LaRossa and Schloss [102], a series of papers by these and other DuPont scientists [103–105] has characterized the molecular and genetic details of ALS inhibition in bacteria. In *S. typhimurium* and *Escherichia coli*, there are several isozymes of acetolactate synthase, each encoded by a separate gene. In contrast, only a single gene encoding for ALS has been detected in yeast [S. C. Falco, DuPont, unpublished data]. Isozyme II from *S. typhimurium* [102] and isozyme III from *E. coli* [103] have been shown to be sensitive to sulfometuron methyl; isozyme I is insensitive [103]. Valine is a feedback inhibitor of isozyme I, and only when the insensitive isozyme I is inhibited by valine, does sulfometuron methyl inhibit the growth of *S. typhimurium* [102].

Sulfometuron methyl inhibition of acetolactate synthase isozyme II from *S. typhimurium* exhibits slow, tight-binding kinetics with an initial K_i of 660 ± 60 nM and a final steady-state K_i of 65 ± 25 nM [102]. Current evidence suggests that sulfometuron methyl binds tightly but reversibly to the ALS-FAD-TPP-Mg^{2+}-decarboxylated pyruvate complex and competes for the second pyruvate binding site [104]. Recombinantly-produced *S. typhimurium* acetolactate synthase isozyme II has been purified and partially characterized [105]. While normally an extremely labile enzyme, recent work has demonstrated that in the absence of thiamine pyrophosphate and metal, the enzyme-FAD complex is quite stable. Mutants of *S. typhimurium* have been obtained that are insensitive to sulfometuron methyl because of a modified ALS isozyme II [102].

2. Studies with Plants

Based on the initial work of LaRossa and Schloss [102], Ray [106] determined that a similar site of action exists in plants. Crude preparations of the ALS enzyme from pea shoots revealed that it was strongly inhibited by chlorsulfuron; in fact, the plant enzyme was significantly more sensitive than the bacterial and yeast ALS enzymes. Inhibition was detectable at concentrations as low as 2.8 nM (1 ppb) paralleling the threshold established for plant cell division inhibition. The I_{50} value, or the concentration of chlorsulfuron inhibiting the pea ALS by 50% in a 30-minute assay, was 21 nM, while that for sulfometuron methyl was 15 nM (Table 13). This is in contrast to values of 65 and 120 nM for sulfometuron methyl inhibition of $S.$ $typhimurium$ [102] and yeast [113,114] ALS, respectively. These results are consistent with the observation that plants are generally much more sensitive to sulfonylurea herbicides than are bacteria and yeast. The kinetics of inhibition of pea ALS by sulfonylureas, as with ALS isozyme II from $S.$ $typhimurium$, were biphasic with inhibition increasing with time.

The ALS from two broadleaf plants and three grasses were all highly sensitive to chlorsulfuron with I_{50} values ranging between 7 and 28 nM (Table 13) [106,107,110]. Although wheat is highly tolerant to chlorsulfuron, the herbicide is a very potent inhibitor of the ALS from this crop with an I_{50} value of 19 to 22 nM. Similarly, none of the other selective sulfonylurea herbicides (e.g., metsulfuron methyl, DPX-M6316, bensulfuron methyl, chlorimuron ethyl) owe their crop tolerance to differential ALS sensitivity. As discussed below, tolerance is due to the rapid inactivation or detoxification of the sulfonylurea by the crop.

Chaleff and Ray [108,109], using cell culture selection techniques, regenerated tobacco mutant plants containing a single, semidominant nuclear gene mutation that conferred greater than a 100-fold level of resistance to chlorsulfuron. These mutant plants also possessed an ALS enzyme that was much less sensitive to inhibition by this herbicide (I_{50} >8000, Table 13). Through a series of genetic crosses, Chaleff and Mauvais [110] demonstrated that all normal, sensitive segregants possessed a chlorsulfuron-sensitive ALS, whereas all homozygous mutants contained a highly resistant form of the enzyme. Heterozygotes contained an ALS with an intermediate degree of resistance to chlorsulfuron. Thus, resistance at the whole plant level paralleled resistance at the ALS enzyme level.

Collectively, the biochemical evidence of Ray [106,107], together with the genetic evidence of Chaleff and Mauvais [110], provides unequivocal proof that ALS is the primary target site of action of the sulfonylureas. Interestingly, American Cyanamid's new imidazolinone herbicides are also plant ALS inhibitors, although they are less potent inhibitors than the sulfonylureas (Table 13) [111,112].

Homogeneous preparations of higher plant ALS have not been reported. Until such material is available, information on the molecular
mechanism of sulfonylurea action in plants will have to be extra-
polated from data obtained with the bacteria and yeast ALS.

3. Studies with Yeast

Falco and co-workers [113,114] used yeast (*Saccharomyces
cerevisiae*) as an eukaryotic model to gain additional insight into the
action of the sulfonylureas at the cellular and molecular level. Sulfometuron methyl at 3 μg/ml inhibited the growth of yeast on minimal
media, but as with bacteria and higher plants, the addition of
branched chain amino acids reversed this inhibition. The isolation
and characterization of several dozen mutants resistant to sulfometuron methyl revealed that most mutations conferring high levels
of resistance were in the *ILV*2 gene, which encodes for ALS. The
ALS from highly resistant mutants was not inhibited by >30 μg/ml
of sulfometuron methyl, while the wild-type ALS was very sensitive
having an I_{50} of 45 ng/ml (0.12 μM). Genetic and biochemical studies of sulfonylurea resistant ALS mutants showed that their resistance
is caused by a single dominant nuclear gene mutation and that the
resistant ALS enzyme activity cosegregates with the resistant pheno-
type. ALS activity increased 4- to 6-fold upon the introduction into
sensitive yeast of a multicopy plasmid containing the wild-type *ILV*2
gene, and resulted in a similar increase in sulfometuron methyl
tolerance. These data provide strong biochemical and genetic evidence
that ALS is also the target enzyme of sulfometuron methyl in yeast.
Yeasts, like bacteria, are generally less sensitive than plants to the
sulfonylurea herbicides.

The DNA sequence of the yeast *ILV*2 gene, which codes for
yeast ALS, has been recently reported for both sulfonylurea-resistant and sensitive yeast [115]. The ALS gene from the resistant
yeast contains a single DNA base change, a G:C to A:T, which results in the substitution of a serine for a proline in the amino acid
sequence of yeast ALS [116]. This single amino acid substitution in
an enzyme containing 687 amino acid units is sufficient to confer
high levels of resistance to sulfometuron methyl. The availability of
such resistant structural genes is the first step toward genetically
engineering sulfonylurea-resistant crops and such advances now
appear imminent.

C. Basis of Selectivity

1. Studies with Chlorsulfuron

Studies to elucidate the biochemical basis of crop tolerance to
chlorsulfuron were initiated at DuPont by P. B. Sweetser in 1979.

TABLE 13

Inhibition of Acetolactate Synthase (ALS) by Sulfonylurea (SU) and Imidazolinone (IM) Herbicides

Herbicide	Type	Active ingredient	Source	I_{50} (nM)[a]	Reference
Oust	SU	Sulfometuron methyl	Bacteria	65	102
			Yeast	120	114, 115
			Pea	15	108
			Wheat	13	108
			Wild oats	7	108
			Wild mustard	9	108
Glean	SU	Chlorsulfuron	Pea	21	107
			Wheat	19, 22	107, 108
			Wild oats	16	108
			Wild mustard	11	108
			Tobacco (sensitive)	28, 7, 14	107, 108, 111
			Tobacco (resistant)	>8,000	111
Ally	SU	Metsulfuron methyl	Pea	14	108
			Wheat	30	108

Harmony	SU	DPX-M6316	Wild oats	15	108
			Wild mustard	13	108
			Pea	75	108
			Wheat	74	108
			Wild oats	46	108
			Wild mustard	32	108
Londax	SU	Bensulfuron methyl	Pea	64	108
			Rice	16	108
			Barnyardgrass	15	108
Classic	SU	Chlorimuron ethyl	Pea	6	108
			Soybean	8	108
			Morningglory	7	108
Arsenal	IM	Imazapyr	Pea	9,000	Unpubl., Ray
			Corn	12,000 3,000	112, 113
			Bacteria (S. typhimurium)	>100,000	Unpubl., Schloss
Scepter	IM	Imazaquin	Pea	3,000	Unpubl., Ray
			Corn	3,400	112

TABLE 13

(continued)

Herbicide	Type	Active ingredient	Source	I_{50} (nM) [a]	Reference
Scepter (continued)			Bacteria (*S. typhimurium* ALS II)	20,000	Unpubl., Schloss

[a]Concentration of the herbicide required to inhibit ALS by 50%. See references for assay procedures.

Differences in sensitivity to chlorsulfuron of up to 4000-fold were observed between highly tolerant plants, such as wheat, barley, and wild oats, and highly sensitive broadleaf plants, such as mustard, sugar beet, soybean, and cotton [70]. These large differences in sensitivity could not be explained in terms of differences in penetration or translocation; nor could they be explained by differences in the sensitivities of the ALS enzymes from these plants to chlorsulfuron [106,107].

Metabolism studies, however, indicated a highly positive correlation between tolerance and the rate of [14C]chlorsulfuron metabolism. In highly sensitive sugar beets, only 3% of the [14C]chlorsulfuron was metabolized in the leaf after 24 hours, whereas in tolerant wheat leaves, 95% was metabolized during the same period. Similarly, dramatic differences were observed between the rate of [14C]chlorsulfuron metabolism in other tolerant grasses (barley, wild oats, annual bluegrass) and sensitive broadleaves (cotton, soybean, mustard). This led Sweetser et al. [70] to conclude that metabolism might account for differences in tolerance if the products of metabolism were herbicidally inactive.

This idea was tested by combining high performance liquid chromatography (HPLC) with enzymatic hydrolysis and mass spectrometric (MS) analysis to identify chlorsulfuron metabolites [70]. A major 14C metabolite was detected in [14C]chlorsulfuron-treated wheat leaves, and treatment of the isolated metabolite with β-glucosidase yielded a less polar product. This product was subsequently identified as the 5-hydroxyphenyl analogue of chlorsulfuron and confirmed by comparing HPLC retention times and spectral results with a synthetic reference standard. These and other results established that chlorsulfuron is first metabolized in wheat and other tolerant grasses to the 5-hydroxyphenyl intermediate and then rapidly conjugated with glucose (Fig. 1). No 5-hydroxychlorsulfuron was detected in wheat extracts, indicating that the final conjugation step is very rapid. Synthesis and whole-plant testing of the conjugate demonstrated that it is herbicidally inactive. Thus, it was concluded that rapid metabolism is the basis of chlorsulfuron selectivity.

Erbes [117] showed that an oxygenase and glucosyltransferase are involved in this two-step detoxification process in wheat. Using $^{18}O_2$, he found that the oxidation of chlorsulfuron involves the incorporation of an atom of oxygen from molecular oxygen and that the subsequent conjugation step involves the incorporation of [14C]glucose from uridine-5'-diphospho-[14C]glucose. Although attempts to isolate a cell-free system with oxygenase activity were unsuccessful, the presence of an active glucosyltransferase for conjugation was readily demonstrated in extracts from wheat shoots. Consistent with the weak whole-plant herbicidal activity of the glucose conjugate, chlorsulfuron was over 200 times more potent an inhibitor of ALS than the conjugate [117].

Broadleaf plants, such as flax and black nightshade, show considerable tolerance to chlorsulfuron compared to the highly sensitive velvetleaf. Independently, Hageman and Behrens [118] and Hutchison et al. [71] studied the basis of this tolerance. Hageman and Behrens found that sensitive velvetleaf was over 20,000 times more susceptible to chlorsulfuron than tolerant eastern black nightshade. Differences in spray retention, absorption, and translocation were inadequate to explain such vast differences. As with tolerant grasses [70], however, eastern black nightshade metabolized chlorsulfuron much more rapidly than velvetleaf. After 72 hours, 81% of the [^{14}C]chlorsulfuron was metabolized in nightshade, compared to only 7% metabolism in sensitive velvetleaf. Two metabolites were detected by TLC, but no attempt was made to identify them.

Similarly, Hutchison et al. [71], working with another species of black nightshade, and with flax, found that chlorsulfuron is metabolized to two metabolites. Unlike chlorsulfuron metabolism in wheat, the site of attack in these two tolerant broadleaf species was on the triazine rather than the phenyl ring. Mass spectral analysis and synthesis provided structural proof that one metabolite was the 4-hydroxy-methyltriazine analogue of chlorsulfuron and that the second, more polar metabolite was the sugar conjugate where the sugar is attached via the 4-hydroxymethyl group on the triazine ring (Fig. 1). In contrast to wheat, which only accumulates the glucose conjugate, in flax and nightshade the 4-hydroxymethyl-triazine accumulates to an even greater extent than its conjugate. This intermediate is considerably less herbicidal, however, than chlorsulfuron at the whole-plant level and is a 4-fold weaker inhibitor of acetolactate synthase [117]. Conjugation of the metabolite results in an additional ninefold decrease in ALS inhibitory activity. Thus, rapid metabolism with a concomitant loss of herbicidal activity supports the view that metabolic detoxification is the basis for flax and nightshade tolerance.

2. Studies with Other Sulfonylureas

With the exception of the resistant tobacco mutants (see Section III.E.2), rapid metabolism is the principal factor responsible for the differential responsiveness of plants to the sulfonylureas. The pathway or route of sulfonylurea detoxification can vary greatly depending on the herbicide and crop involved. Such differences are summarized in Figure 13.

Metsulfuron methyl, like chlorsulfuron, is detoxified in wheat by hydroxylation at the 5-position on the phenyl ring, followed by rapid conjugation with glucose [Sweetser (DuPont), unpublished data]. DPX-M6316, on the other hand, is detoxified in wheat by amidase attack on the $-C(=O)-N-$ bond of the bridge which is the same mechanism barley uses to inactivate chlorsulfuron and

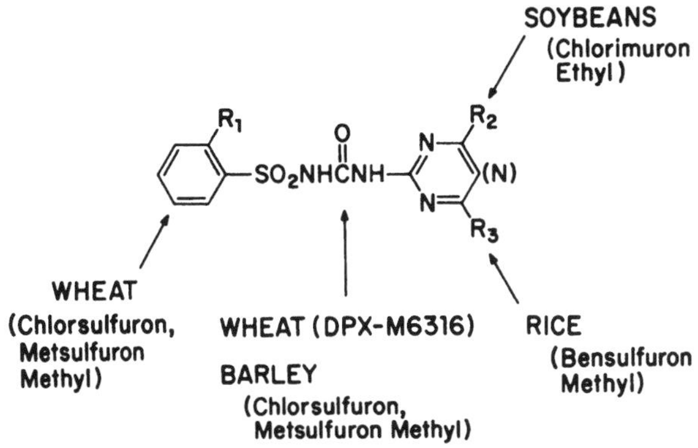

FIG. 13   Sites of attack by various crops leading to sulfonylurea inactivation or detoxification.

metsulfuron methyl [Sweetser (DuPont), unpublished data].   Brown [119] has shown that soybeans render chlorimuron ethyl inactive by displacing the chlorine at the R_2 position with homoglutathione.   The half-life of chlorimuron ethyl in excised soybean leaves is only 2 to 4 hours compared with greater than 30 hours in sensitive morningglory and cocklebur.   Takeda et al. [120] have recently shown that rice modifies the pyrimidine portion of bensulfuron methyl, converting the 6-methoxy substituent at the R_3 position to a hydroxyl group.   This metabolite is approximately 3000 times less active as an ALS inhibitor in rice than the parent herbicide.   Interestingly, *Indica* rice metabolizes the herbicide more rapidly than *Japonica* rice, and this difference appears to be responsible for the greater tolerance of the *Indica* varieties.

Based on current evidence, the chemical and biological rules for sulfonylurea tolerance include (a) the presence of a metabolizable site in the molecule, (b) rapid metabolism with a half-life of only a few hours, and (c) the formation of products with greatly reduced herbicidal potency.

D.   Uptake and Translocation

1.   Theory of Phloem Mobility

The sulfonylureas are weak acids with acid dissociation constants (pKas) ranging from 3.3 to 5.2 (Table 3) [38].   As with 2,4-D and other ionizable molecules, the neutral or uncharged form of the molecule is much more lipophilic than the ionized, anionic form.   The octanol-water partition coefficients of the sulfonylureas are about

50 to 100 times higher at pH 5.0 than at pH 7.0 (Table 3). Accordingly, at the lower pH they are much more permeable to cell membranes because a greater proportion of the molecules are in the undissociated or protonated state.

Based on these physiochemical properties, the sulfonylureas should be phloem mobile since they fit the classical "weak acid" theory of phloem mobility, whereby acidic molecules are partitioned into the phloem as a result of an acid-trapping mechanism. Operationally, acidic molecules are "trapped" by the marked changes in permeability that occur as the molecule crosses the physiological pH gradient between the apoplastic cell wall area (pH 5.5–6.0) and the alkaline phloem sap (pH 8.0). That is, as sulfonylurea molecules reach the acidic cell wall environment either via foliar or root uptake, some of the molecules are in the neutral, highly permeable form, and are able to cross the phloem tube/companion cell plasma membrane and enter the phloem. Upon entering the alkaline phloem environment, the molecule dissociates and becomes "trapped" as the relatively impermeant anionic form. Once "trapped" the sulfonylurea will move systemically by "mass flow" action with sucrose and other phloem solutes.

2. Leaf Explant Studies

Lichtner [121] has demonstrated phloem mobility of the sulfonylureas using an excised soybean leaf. [14C]-labeled sulfonylureas, together with [3H]sucrose, were applied as droplets to soybean leaves, then the phloem exudate was collected from the cut end of the petiole which had been placed in 20 mM ethylene diamine tetraacetic acid (EDTA) to enhance exudation. To facilitate uptake, the surface of the leaf was abraded with carborundum. The amount of [14C]sulfonylurea transported over a 6-hour period, compared to [3H]sucrose, [14C]2,4-D, and [14C]monuron, is shown in Table 14. Thirty percent of the applied sucrose was translocated via the phloem versus 4–10% of the sulfonylureas, 2% of the 2,4-D, and none of the monuron. These results clearly indicate the relatively high degree of phloem mobility of the sulfonylureas.

3. Whole-Plant Studies

Sweetser et al. [70] evaluated the uptake and translocation of [14C]chlorsulfuron in four sensitive broadleaf plants and three tolerant grasses. When [14C]chlorsulfuron was applied for 24 hours as droplets (10–20% acetone plus 0.2% Tween-20) to the upper leaf surface of 1- to 3-week-old seedlings, as much as 56–98% of the [14C] chlorsulfuron penetrated the leaf. Depending on the plant species, 1.1–17.6% of the total [14C] applied moved out of the treated region and was recovered in the leaves and stems, while only 0.1–4.0% was recovered in the roots.

TABLE 14

Phloem Transport of Sulfonylurea Herbicides

Compound	Log P_{ow}	pKa	Amount translocated	
			Pmole/6 hr	% of Total applied[a]
Metsulfuron methyl	1.7	3.3	55 ± 9	4
Chlorsulfuron	2.2	3.6	48 ± 19	4
Sulfometuron methyl	1.4	5.2	170 ± 20	10
Sucrose	—	—	42,000 ± 9,000	30
2,4-D	2.8	3.0	28 ± 10	2
Monuron	2.0	—	0	0

[a]All compounds applied at a concentration of 135 μM, except Monuron (200 μM) and Sucrose (25 mM).
Source: From Ref. 121.

Petersen and Swisher [122] have analyzed the uptake and translocation properties of foliarly applied and root-applied [14C]chlorsulfuron in Canada thistle. Of the 14C applied to the leaf, 39% was absorbed after 48 hours, and of this amount, 10% was translocated out of the treated leaf while 29% remained in the leaf. Only 1% of the translocated 14C was recovered in the roots. When applied to the nutrient solution, 16% was absorbed by the roots, and of this amount, 6% remained in the roots while 2% moved to the root buds and 8% went to the shoots.

Similar distribution patterns have been reported by others. Bestman and Vanden Born [123] found that with tartary buckwheat, 40% of foliarly applied [14C]chlorsulfuron was absorbed, with 37% recovered in the treated leaf and only 3% translocated from the leaf after 120 hours. In Canada thistle, Devine and Vanden Born [124] recovered 5 and 11% of the applied radioactivity in the roots and upper shoots, respectively, after 144 hours, when three-fourths of the applied 14C had been absorbed by the leaf. Smaller amounts were absorbed and translocated in perennial sow thistle.

While the sulfonylureas are clearly phloem mobile, the data currently available provide only a cursory view of the uptake and translocation properties of the sulfonylureas. Uptake seems to vary considerably depending on the method of application, the formulation, and the plant species evaluated. Generally, foliar uptake is between 40 and 80% of that applied. Between 1 and 20% is translocated to shoots and only 0.1–5% to roots. Obviously, the extent of metabolism could greatly affect the observed ^{14}C distribution patterns, especially if the physiochemical properties of the metabolites differ greatly from those of the parent sulfonylurea.

E.   Factors Affecting Performance and Crop Tolerance

1.   Safeners

Parker et al. [125] first reported the use of safeners to extend the selectivity of the sulfonylureas. A four- to eight-fold increase in the tolerance of corn was achieved in greenhouse pot experiments by treating the seed with 1,8-naphthalic anhydride (NA) at 0.5% (w/w) before applying preemergence chlorsulfuron treatments ranging from 5 to 40 g/ha. At 5 and 10 g/ha chlorsulfuron, shoot dry weight was inhibited approximately 60% and 75%, respectively, in corn that had not been pretreated with NA, whereas the corresponding values with NA pretreatment were 10 and 15%. Similarly, NA protected sensitive sorghum and rice, but the degree of safening at 5 and 10 g/ha was not as complete as with corn. NA further protected normally tolerant wheat and barley from excessively high rates of chlorsulfuron (100–400 g/ha). Varietal differences in the degree of protection provided by NA, and the less effective safener, N,N-diallyl-2,2-dichloroacetamide (DDCA), were observed with chlorsulfuron and no protection was seen in oil seed rape, sugar beet, perennial ryegrass, dwarf bean, or onion. These results have since been confirmed and extended to metsulfuron methyl [126] and several other safeners [126–128].

While these safeners protect crops such as corn and sorghum from low rates of certain sulfonylureas, the margin of safety does not appear sufficient to provide practical weed control and no such combinations have yet been commercialized. Yuyama et al. [129] have found commercially acceptable levels of safening, however, with bensulfuron methyl and certain thiocarbamate rice herbicides, such as thiobencarb, MY-93, or S-(1-methyl-1-phenyl)piperdine-1-carbothioate, CH-83 or S-isopropyl-tetrahydro-1H-azepine-1-carbothioate, and SC-2957 (experimental herbicide, Stauffer Chemical Co.). At the relatively high rate of 100 g/ha, bensulfuron methyl reduced the dry weight of rice shoots by 23%. Combination with any of the above thiocarbamate herbicides essentially eliminated this injury in transplanted *Japonica* rice. Direct seeded *Japonica* rice was also safened by MY-93. Fortuitously, these compounds not only safen bensulfuron methyl, but they also control barnyardgrass, a weed not well controlled by bensulfuron methyl.

The mechanism of sulfonylurea safening has been studied by Sweetser and co-workers [129,130]. The mode of action is an enhancement or "turning-on" of the rate of sulfonylurea detoxification. For example, treatment of rice plants with thiobencarb or MY-93 increases the rate of bensulfuron methyl detoxification 3- to 4-fold [129]. Apparently, antidotes are not particularly effective in safening broadleaf plants because of their inability to induce or "turn-on" sulfonylurea detoxification.

2. Herbicide-Resistant Plants

Traditionally, herbicide chemistry has been tailored to kill weeds without harming the crop. When true physiological tolerance is inadequate or not present, an antidote or safener can enhance the margin of safety. Today, another complementary method is available for achieving crop safety. This method involves genetics and the techniques embodied in biotechnology such as plant cell culture and gene transfer. Chaleff and Ray [108,109] have successfully used cell selection to obtain plants with greatly increased levels of resistance to the sulfonylurea herbicides. By culturing tobacco cells in the presence of chlorsulfuron and sulfometuron methyl, they selected several herbicide-tolerant mutant cell lines that were regenerated into similarly resistant mutant plants. Genetic analysis [109] of the regenerated plants has demonstrated that resistance is inherited as a single dominant or semidominant mutation. Moreover, resistance at the biochemical level is due to an altered, less sensitive form of the sulfonylurea target enzyme, acetolactate synthase (ALS). Homozygous mutant plants that are over 100 times more resistant to chlorsulfuron than normal plants have been characterized. More recently [131] double mutants exhibiting more than 1000 times the resistance of normal cultivars have been obtained. These technological advances now make possible the tailoring of genetics to fit existing herbicide chemistry. The recent isolation of the ALS gene encoding for sulfonylurea resistance in higher plants [132] is a first step in achieving this goal.

3. Sulfonylurea Herbicide Antagonism

Chlorsulfuron, like 2,4-D and several other broadleaf herbicides, can antagonize the activity of wild oat herbicides, such as diclofop, difenzoquat, and flamprop. In field tests, O'Sullivan and Kirkland [133] found that spray tank mixtures of chlorsulfuron (20 to 40 g/ha) with diclofop resulted in a 4 to 35% reduction in wild oat control. The reduction was 0 to 19% with difenzoquat and 2 to 30% with flamprop. As also observed by these workers, this antagonism could

largely be overcome by increasing the rate of the wild oat herbicide
in the mixture. For example, increasing the rate of flamprop from
0.42 to 0.56 kg/ha essentially eliminated the antagonism caused by
20 g/ha of chlorsulfuron. This antagonism can also be overcome by
applying the wild oat herbicide and chlorsulfuron several days apart.

The degree of antagonism appears to be highly variable. Several workers [134,135], using similar or even higher rates of chlorsulfuron, have concluded that in contrast to 2,4-D, chlorsulfuron does
not typically antagonize diclofop-methyl, difenzoquat, or triallate.
These results suggest that the ratio of chlorsulfuron to wild oat
herbicide is crucial to whether antagonism is observed. Apparently
the more sensitive or susceptible the wild oat plant is to the wild
oat herbicide at the time of treatment, the less likely the chance for
antagonism. The basis of this antagonism is presently unknown.

4. Environmental Factors

As with other herbicides, many factors, such as soil type,
organic matter, moisture and temperature, can significantly influence
the performance of the sulfonylurea herbicides. Healthy, actively
growing weeds are generally more sensitive than stressed weeds.
Optimum temperatures favor not only plant growth and herbicide
sensitivity but also crop tolerance. Sweetser (DuPont, unpublished
data) found that for every 10°C drop in temperature the rate of
sulfonylurea detoxification drops by a factor of 2 to 5. Thus, ab-
normally cool temperatures can potentiate crop injury by slowing the
rate of sulfonylurea detoxification.

5. Weed Competition

Sweetser and Ackerson [136] have assessed the competitive
status of weeds treated with chlorsulfuron. Within 1 to 2 days following herbicide treatments, transpiration and water uptake were
markedly inhibited in sensitive green foxtail, speedwell, bedstraw,
morningglory, and annual rye grass, but not in tolerant wheat.
After 1 day, an 8 g/ha chlorsulfuron treatment inhibited transpiration in both morningglory and annual rye grass by 30–40%, and
after 1 week this inhibition had increased to 80%. In bedstraw,
water loss for control plants was 0.5 g/hr, but it was reduced to
only 0.06 g/hr in plants treated for 6 days with 3 g/ha chlorsulfuron.
Reductions in water loss were highly correlated with reductions in
plant growth. These studies led the authors to conclude that when
growth is arrested by chlorsulfuron, there is a rapid, concomitant
reduction in water and nutrient uptake. Competition for light is

not a significant factor, since sensitive weeds are not actively producing new foliage.

6. Surfactants

In many applications, the addition of a nonionic surfactant im-
proves the postemergence performance of the sulfonylurea herbicides.
Jensen [137] found "Citowet" (nonionic surfactant, trademark of
BASF) to be especially beneficial for improving the activity of chlorsulfuron. Similarly, Hunter [138] was able to increase the activity
of chlorsulfuron on cow cockle 4- to 8-fold by adding 0.1% of this
surfactant to the spray solution. Surfactant WK (0.5%) significantly
boosted the control of wild garlic by chlorsulfuron [139], and provided similar improvement in the activity of chlorsulfuron on kochia
and green foxtail [140]. Other workers have achieved similar success by tank mixing nonionic surfactants with chlorsulfuron [141–
143], sulfometuron methyl [144], chlorimuron ethyl [145], and metsulfuron methyl [146]. Oil adjuvants, such as crop oil concentrates,
have also been investigated, especially with chlorimuron ethyl [147].
In general, improvements in weed control have been obtained with
this soybean herbicide, but because of excessive crop injury the use
of oils is not currently recommended [31].

F. Growth Stimulation

As with many other herbicides [148] chlorsulfuron at sublethal
doses can directly promote the vegetative growth of young seedlings.
Generally, this effect occurs over a fairly narrow range of rates and
the magnitude is quite variable. This variation is apparently due to
differences in plant sensitivity at the time of treatment and to variation in treatment conditions. Drexler [135] reported that chlorsulfuron
applied at 15 g/ha increased the fresh weight of wheat 33% and that
of wild oat 89%. Parker et al. [125], in their safener studies, presented preemergence data indicating that chlorsulfuron applied at
100 g/ha stimulated the growth of barley by 20%. The data of Hall
et al. [134] also suggest a similar effect on wild oats when applied
as a postemergence treatment at 20 g/ha. Studies at DuPont (unpublished data) have confirmed the growth-stimulating effects of
chlorsulfuron on the growth of young vegetative plants. Typically,
the stimulation is small, highly variable, and biphasic in nature,
ranging from 10 to 30%. The response is generally transitory and
even under well-controlled environmental conditions, reproducibility
is often difficult to achieve. Nevertheless, when exactly the right
rate is applied, which depends on the species and sensitivity of the
tissue at the time of application, small but significant stimulation of
growth can be observed.

IV. CONCLUSIONS

Within a relatively short time the sulfonylurea herbicides have emerged as a major advance in chemical weed control technology. With their unprecedented herbicidal activity, use rates have plummeted resulting in application rates of grams rather than kilograms per hectare. The need for such low-dosage compounds with greater selectivity, environmental compatibility, and groundwater safety are important factors contributing to the rapid success of these new materials.

The high versatility of sulfonylurea chemistry has the potential for solving many of the weed control problems prevalent in the major agronomic crops, and the six sulfonylureas featured in this chapter promise to be the first of a long list of new generation products based on this chemistry. In cereals alone, there is already available a complementary group of sulfonylurea herbicides with unique weed-controlling properties and vastly differing soil residual characteristics. In addition to Glean, Ally, and Harmony, a mixture of Glean and Ally called Finesse has recently been commercialized in the United States and the United Kingdom, and another short residual material for cereals called Express is under development [149]. Beyond cereals, there is Classic for soybeans, and Londax for rice as well as a number of development candidates and mixtures currently undergoing field testing for many other major crops.

Perhaps at no other time in the history of herbicide research has so much been learned in such a short time about a new family of herbicides. Only a few years after their discovery, the site of action of the sulfonylureas was pinpointed as the enzyme, acetolactate synthase. Inhibition of this enzyme, which is needed for the production of the essential amino acid building blocks valine and isoleucine, results in rapid cessation of growth and eventual plant death. The absence of this enzyme in man and other animals helps to explain the low toxicity of the sulfonylureas.

All plants contain this target enzyme and therefore it has been of considerable interest to learn how certain crop plants escape herbicidal injury following postemergence treatment. Extensive investigations have revealed that crop tolerance is due to the ability of the crop plant to rapidly convert the herbicide to inactive products. This inactivation occurs so rapidly that the active molecule never reaches the enzyme in sufficient quantities to effectively inhibit it. A wide range of different metabolic inactivation reactions has been identified. These involve oxidative and conjugative systems that quickly render the molecule herbicidally inactive. Metabolic attack can occur on the aryl, bridge, or heterocyclic portion of the molecule. In addition to plants, microbes also modify and degrade sulfonylureas to a variety of products including the corresponding sulfonamide and the aminotriazine or aminopyrimidine. These in turn are further degraded to even simpler compounds.

Sulfonylureas degrade under field conditions at rates similar to, and often faster than, conventional herbicides. Chemical hydrolysis and microbial breakdown are the main modes of dissipation. The sulfonylureas are weak acids and under acidic soil conditions often undergo rapid dissipation by chemical hydrolysis. Under alkaline soil conditions, where rates of chemical hydrolysis are minimal, microbial breakdown is the predominant dissipation mechanism. Breakdown is generally the fastest in warm, moist, light-textured, low pH soils and slowest in cold, dry, heavy, high pH soils. The differential sensitivity of plants to the sulfonylurea herbicides can be over 1000-fold. The sometimes long residual activity that has been observed by replanting highly sensitive crops into sulfonylurea-treated soil is caused primarily by the very high susceptibility of the rotational crop and not to an inherently slow rate of dissipation.

The tailoring of sulfonylurea chemistry to control weeds without injuring the crop has given way to an alternative strategy of tailoring the crop to fit the chemistry. By obtaining mutant forms of the acetolactate synthase gene that codes for insensitive forms of the enzyme, it has been possible to embark on genetic engineering of crops with high levels of sulfonylurea resistance. Rapid advances in this area will undoubtedly continue and surely impact future weed control practices.

REFERENCES

1. H. J. Koog, Jr. (to Deutsche Gold and Silver–Scheidean–stalt Vormals) Netherlands Pat. 121,788 (1966).
2. G. Levitt, in *Pesticide Chemistry: Human Welfare and the Environment* (J. Miyamato and P. C. Kearney, eds.), Vol. 1, Pergamon Press, New York, 1983, p. 243.
3. R. F. Sauers and G. Levitt, in *Pesticide Syntheses Through Rational Approaches* (P. S. Magee, G. K. Kohn, and J. J. Mean, eds.), American Chemical Society, Washington, D.C., 1984, p. 21.
4. G. Levitt (to DuPont), Belgian Pat. 853,374 (1977).
5. W. Meyer and W. Föry (to Ciba–Geigy), South African Pat. Appl. 81/4874 (1982).
6. G. Levitt (to DuPont), U.S. Pat. 4,169,719 (1979).
7. G. Levitt (to DuPont), U.S. Pat. 4,398,939 (1983).
8. G. Levitt (to DuPont), U.S. Pat, 4,481,029 (1984).
9. G. Levitt (to DuPont), U.S. Pat. 4,435,206 (1984).
10. G. Levitt (to DuPont), U.S. Pat, 4,370,479 (1983).
11. T. P. Selby and A. D. Wolf (to DuPont), U.S. Pat. 4,421,550 (1983).

12. W. Topfl, H. Kristinsson, and W. Meyer (to Ciga–Geigy), Australian Pat. Appl. 16890/83 (1983).
13. G. Levitt (to DuPont), U.S. Pat. 4,339,267 (1982).
14. W. T. Zimmerman (to DuPont), U.S. Pat. 4,487,626 (1984).
15. G. Levitt (to DuPont), U.S. Pat. 4,293,330 (1981).
16. J. J. Reap (to DuPont), U.S. Pat. 4,191,553 (1980).
17. G. Levitt (to DuPont), U.S. Pat. 4,257,802 (1981).
18. R. F. Sauers (to DuPont), U.S. Pat. 4,420,325 (1983).
19. D. W. Finnerty, D. J. Fitzgerald, H. L. Ploeg, S. E. Schehl, R. C. Weigel, and G. Levitt, *Proc. North Cent. Weed Control Conf.*, *34* (1979).
20. G. Levitt, H. L. Ploeg, R. C. Weigel, and D. J. Fitzgerald, *J. Agric. Food Chem.*, *29*, 416 (1981).
21. J. M. Green, J. E. Harrod, J. D. Long, G. Levitt, and D. J. Fitzgerald, *Proc. South. Weed Sci. Soc. Am.*, *34*, 214 (1981).
22. G. Levitt (to DuPont), U.S. Pat. 4,394,506 (1983).
23. R. I. Doig, G. A. Carraro, and N. D. McKinley, *Proc. Br. Crop Prot. Conf.*, *3*, 20 (1983).
24. G. Levitt (to DuPont), U.S. Pat. 4,383,113 (1983).
25. H. L. Ploeg, A. D. Wolf, and J. R. C. Leavitt, *Weed Sci. Soc. Am. Abstr.*, Abstr. 48 (1984).
26. T. Yuyama, F. Takeda, H. Watanabe, T. Asami, S. Peudpaichit, J. L. Malassa, and P. Heiss, *Proc. 7th Indonesian Weed Sci. Soc. Conf.*, *1*, 89 (1984).
27. J. M. Hutchison, L. H. Hageman, and S. E. Schehl, *Weed Sci. Soc. Amer. Abstr.*, Abstr. 26 (1985).
28. Chlorsulfuron Technical Data Sheet, E. I. du Pont de Nemours & Co., Bulletin No. B-22713, August 1982.
29. Sulfometuron Methyl Technical Data Sheet, E. I. du Pont de Nemours & Co., Bulletin No. AG-361, November 1983.
30. Metsulfuron Methyl Technical Data Sheet, E. I. du Pont de Nemours & Co., Bulletin No. AG-404, April 1984.
31. Classic® Herbicide Technical Bulletin, E. I. du Pont de Nemours & Co., Bulletin No. E-74208, March 1985 and E-62928, January 1984.
32. Londax® Herbicide Technical Bulletin, E. I. du Pont de Nemours & Co., Bulletin No. E-62943, July 1985.
33. Harmony® Herbicide Technical Bulletin, E. I. du Pont de Nemours & Co., Bulletin No. E-62932, February 1985.
34. *Dangerous Properties of Industrial Materials*, 5th Edition (N. Irving Sax, ed.), Van Nostrand Reinhold Company, New York, 1979, p. 978.
35. T. Larsen, State Plant Protection Institute, Flakkenbjerg, Denmark, personal communication (1983).
36. H. Ulrich and A. A. R. Sayrgh, *Angew. Chem. (Int. Ed.)*, *78*, 761 (1966).

37. W. Meyer and W. Föry (to Ciba-Geigy), U.S. Pat. 4,419,121 (1983).

38. J. J. Dulka and M. J. Duffy, 5th Annual Mtg. of the Society of Environmental Toxicology and Chemistry, Arlington, VA (1984).

39. J. J. Dulka, A. C. Barefoot, and M. J. Duffy, 6th Annual Mtg. of the Society of Environmental Toxicology and Chemistry, St. Louis, MO (1985).

40. J. Harvey, Jr., J. J. Dulka, and J. J. Anderson, *J. Agric. Food Chem.*, *33*, 590 (1985).

41. P. L. Friedman, J. J. Anderson, and J. Harvey, Jr., *J. Agric. Food Chem.*, submitted for publication (1987).

42. F. Kurzer in *Organic Sulfur Compounds* (N. Kharasch, ed.), Pergamon Press, New York, 1961.

43. A. C. Barefoot, J. J. Dulka, and M. J. Duffy, *J. Agric. Food Chem.*, submitted for publication (1987).

44. L. M. Shalaby, *Biomed. Mass Spectrom.*, *12*, 261 (1985).

45. L. M. Shalaby, 33rd Annual Conference on Mass Spectrometry and Allied Topics, San Diego, CA, May 29-31, 1985, Paper No. WPE 5.

46. L. M. Shalaby, Chapter 12, Applications of New Mass Spectrometry Techniques in Pesticide Chemistry (Ed. J. B. Rosen), *91* in the Chemical Analysis Series, Wiley Interscience, p. 161 (1987).

47. E. W. Zahnow, *J. Agric. Food Chem.*, *30*, 854 (1982).

48. E. W. Zahnow, *J. Agric. Food Chem.*, *32*, 953 (1984).

49. R. V. Slates, *J. Agric. Food Chem.*, *31*, 113 (1983).

50. E. W. Zahnow, *L. C. Magazine*, submitted for publication (1986).

51. L. W. Hershberger, *J. Agric. Food Chem.*, submitted for publication (1986).

52. E. W. Zahnow, *J. Agric. Food Chem.*, *33*, 479 (1985).

53. M. M. Kelley, E. W. Zahnow, W. C. Petersen, and S. T. Toy, *J. Agric. Food Chem.*, *33*, 962 (1985).

54. P. W. Santelman, *Research Methods in Weed Science*, 2nd Ed. (B. Truelove, ed.), Southern Weed Science Society, 1977, p. 79-87.

55. A. P. Appleby, *Weed Sci.*, *33* (Suppl. 2), 2 (1985).

56. K. E. M. Groves and R. K. Foster, *Weed Sci.*, *33*, 825 (1985).

57. A. I. Hsiao and A. E. Smith, *Weed Res.*, *23*, 231 (1983).

58. D. R. Fredrickson, "Effects of Soil pH on the Degradation and Availability of Chlorsulfuron in Soil," University of Nebraska, Lincoln, Nebraska (1984).

59. D. W. Morishita, D. C. Thill, D. G. Flom, T. C. Campbell, and G. L. Lee, *Weed Sci.*, *33*, 420 (1985).

60. Relative Sensitivity of Common Rotational Crops to Glean® Herbicide Soil Residues, E. I. du Pont de Nemours & Co., Bulletin E-71699, March 1985.

61. H. Nilsson, *Weeds Weed Control*, 24th (1), 302 (1983).
62. H. Nilsson, *Weeds Weed Control*, 25th (1), 76 (1984).
63. W. Mersie and C. L. Foy, *Weed Sci.*, *33*, 564 (1985).
64. A. Walker and P. A. Brown, *Bull. Environ. Contam. Toxicol.*, *30*, 365 (1983).
65. M. M. Joshi, H. M. Brown, and J. A. Romesser, *Proc. 1984 WSWS Meeting*, Spokane, WA (1984).
66. J. Royrvik, Nordic Plant Protection Conference, Uppsala, Sweden (1981).
67. DPX-4189 Corn Root Bioassay, E. I. du Pont de Nemours & Co. (1980).
68. C. L. Foy and W. Mersie, *Proc. South. Weed Sci. Soc.*, *37*, 108 (1984).
69. S. Junnila, *Weeds Weed Control*, 24th (1), 296 (1983).
70. P. B. Sweetser, G. S. Schow, and J. M. Hutchison, *Pestic. Biochem. Physiol.*, *17*, 18 (1982).
71. J. M. Hutchison, R. Shapiro, and P. B. Sweetser, *Pestic. Biochem. Physiol.*, *22*, 243 (1984).
72. D. S. Frear, H. R. Swanson, and E. R. Mansager, *Pestic. Biochem. Physiol.*, *20*, 299 (1983).
73. D. S. Frear, H. R. Swanson, and E. R. Mansager, *Pestic. Biochem. Physiol.*, *23*, 56 (1985).
74. Y. Niki, S. Kuwatsuka and I. Yokomichi, *Agric. Biol. Chem.*, *40*, 683 (1976).
75. United States Environmental Protection Agency, Pesticide Assessment Guidelines, Subdivision F - Hazard Evaluation: Human and Domestic Animals, Document No. PB83-153916, October 1982.
76. M. M. Joshi, H. M. Brown, and J. A. Romesser, *Weed Sci.*, *33*, 888 (1985).
77. D. R. Fredrickson and P. J. Shea, *Weed Sci.*, *34*, 328 (1986).
78. R. L. Zimdahl, K. Thirunarayanan, and D. E. Smika, *Weed Sci.*, *33*, 558 (1985).
79. R. L. Anderson and M. R. Barrett, *J. Environ. Qual.*, *14*, 111 (1985).
80. M. M. Joshi and H. M. Brown, *Proc. WSSA Meeting*, Seattle, WA (1985).
81. H. L. Palm, J. D. Riggleman, and D. A. Allison, *Proc. Br. Crop Prot. Conf. Weeds*, *1*, 1 (1980).
82. J. G. Campion, *Aust. Weeds*, *1*, 31 (1982).
83. A. Walker and A. Barnes, *Pestic. Sci.*, *12*, 123 (1981).
84. A. E. Smith and A. I. Hsiao, *Weed Sci.*, *33*, 555 (1985).
85. P. A. O'Sullivan, *Can. J. Plant Sci.*, *62*, 715 (1982).
86. L. H. Hageman, "Investigations of Chlorsulfuron Mode of Action and Soil Persistence," University of Minnesota, Minneapolis, MN (1982).
87. B. D. Brewster and A. P. Appleby, *Weed Sci.*, *31*, 861 (1983).

88. M. A. Peterson and W. E. Arnold, *Weed Sci.*, *34*, 131 (1985).

89. R. L. Anderson, *J. Environ. Qual.*, *14*, 517 (1985).

90. P. Nicholls and A. A. Evans, *Proc. Br. Crop Prot. Conf. Weeds*, *1*, 333 (1985).

91. W. Mersie and C. L. Foy, *J. Agric. Food Chem.*, *34*, 89 (1986).

92. H. Nilsson, *Weeds Weed Control*, 26th (1) (1985).

93. T. B. Ray, *Proc. Br. Crop Prot. Conf. Weeds*, *15*, 7 (1980).

94. T. B. Ray, *Pestic. Biochem. Physiol.*, *17*, 10 (1982).

95. T. L. Rost, *J. Plant Growth Regul.*, *3*, 51 (1984).

96. T. B. Ray, *Pestic. Biochem. Physiol.*, *18*, 262 (1982).

97. K. K. Hatzios and C. M. Howe, *Pestic. Biochem. Physiol.*, *17*, 207 (1982).

98. O. T. de Villiers, M. L. Vandenplas, and H. M. Koch, *Proc. Br. Crop Prot. Conf. Weeds*, 237 (1980).

99. J. C. Suttle and D. R. Schreiner, *Can. J. Bot.*, *60*, 741 (1982).

100. J. C. Suttle, H. R. Swanson, and D. R. Schreiner, *J. Plant Growth Regul.*, *2*, 137 (1983).

101. L. H. Hageman and R. Behrens, *Weed Sci.*, *32*, 132 (1984).

102. R. A. LaRossa and J. V. Schloss, *J. Biol. Chem.*, *259*, 8753 (1984).

103. R. A. LaRossa and D. R. Smulski, *J. Bacteriol.*, *160*, 391 (1984).

104. J. V. Schloss, in *Flavins and Flavoproteins* (R. C. Bray, P. C. Engel, and S. G. Mayhew, eds.), Walter de Gruyter & Co., New York, 1984, p. 737.

105. J. V. Schloss, D. E. Van Dyk, J. F. Vasta, and R. M. Kutny, *Biochemistry*, *24*, 4952 (1985).

106. T. B. Ray, *Plant Physiol.*, *75*, 827 (1984).

107. T. B. Ray, *Proc. Br. Crop Prot. Conf. Weeds*, *1*, 131 (1985).

108. R. S. Chaleff and T. B. Ray, *Science*, *223*, 1148 (1984).

109. T. B. Ray and R. S. Chaleff, *Weed Sci.*, submitted for publication (1986).

110. R. S. Chaleff and C. J. Mauvais, *Science*, *224*, 1443 (1984).

111. D. L. Shaner, P. C. Anderson, and M. A. Stidham, *Plant Physiol.*, *76*, 545 (1984).

112. P. C. Anderson and K. A. Hibberd, *Weed Sci.*, *33*, 479 (1985).

113. S. C. Falco and K. S. Dumas, *Genetics*, *109*, 21 (1985).

114. S. C. Falco, K. S. Dumas, and R. E. McDevitt, *Molecular Form and Function of the Plant Genome*, Plenum Press, New York, 1985.

115. S. C. Falco, K. S. Dumas, and K. J. Livak, *Nucleic Acids Res.*, *13*, 4011 (1985).

116. R. W. F. Hardy and R. T. Giaquinta, *Bioassays*, *1*, 152 (1984).

117. D. L. Erbes, *Pestic. Biochem. and Physiol.*, in press
      (1987).
118. L. H. Hageman and R. Behrens, *Weed Sci.*, *32*, 162 (1984).
119. H. M. Brown, *Pestic. Biochem. Physiol.*, in press
      (1987).
120. S. Takeda, D. L. Erbes, P. B. Sweetser, J. V. Hay, and T.
      Yuyama, *Weed Research* (Japan), *31*, 157 (1986).
121. F. T. Lichtner, *Proc. Internat. Phloem Conf.* (Asilomar, CA).
      (W. J. Lucas, J. Cronshaw, eds.), Alan R. Liss Inc., New
      York, 601 (1986).
122. P. J. Petersen and B. A. Swisher, *Weed Sci.*, *33*, 7 (1985).
123. H. D. Bestman and W. H. Vanden Born, *Abstr. Weed Sci.
      Soc. Am.*, No. 203 (1983).
124. M. D. Devine and W. H. Vanden Born, *Weed Sci.*, *33*, 524
      (1985).
125. C. Parker, W. G. Richardson, and T. M. West, *Proc. Br.
      Crop Prot. Conf. Weeds*, *15*, 15 (1980).
126. W. Mersie and C. L. Foy, *Proc. South. Weed Sci. Soc.*, *37*,
      328 (1984).
127. K. K. Hatzios, *Weed Res.*, *24*, 249 (1984).
128. K. K. Hatzios, *Weed Sci.*, *32*, 51 (1984).
129. T. Yuyama, P. B. Sweetser, R. C. Ackerson, and S. Takeda,
      Weed Res. (Japan), *31*, 164 (1986).
130. P. B. Sweetser, *Proc. Br. Crop Prot. Conf. Weeds*, *1*, 1147
      (1985).
131. R. S. Chaleff, S. A. Sebastian, T. B. Ray, C. J. Mauvais,
      and B. J. Mazur, *J. Cell. Biochem.* (Suppl.), *10C*, 10 (1986).
132. B. J. Mazur, C-F. Chui, S. C. Falco, R. S. Chaleff, and C.
      J. Mauvais, *Word Biotech Rep. (USA)* 2, 97 (1985).
133. P. A. O'Sullivan and K. J. Kirkland, *Weed Sci.*, *32*, 285
      (1984).
134. C. Hall, L. V. Edgington, C. M. Switzer, *Weed Sci.*, *30*, 672
      (1982).
135. D. M. Drexler, "The Bioactivity of Chlorsulfuron," University
      of Guelph, Canada (1982).
136. P. B. Sweetser and R. C. Ackerson, *Weed Sci.*, submitted
      for publication (1986).
137. P. G. Jensen, *Weeds Weed Control*, 21st (1), 24 (1980).
138. J. H. Hunter, *Proc. North Cent. Weed Control Conf.*, *36*, 52
      (1981).
139. E. J. Retzinger, Jr. and P. A. Richard, *Louisiana Agric.*, *27*,
      14 (1983).
140. J. D. Nalewaja and Z. Woznia, *Weed Sci.*, *33*, 395 (1985).
141. P. N. P. Chow and H. F. Taylor, *Proc. Br. Crop Prot.
      Conf. Weeds*, *15*, 23 (1980).

142. P. N. P. Chow, *Proc. 16th Annual Workshop Pesticide Residual Anal. (West Canada)*, p. 177 (1981).
143. W. W. Donald, *Weed Sci.*, *32*, 42 (1984).
144. A. V. Glaser, R. H. Koester, J. E. Primus, and P. Sarin, *Proc. North Cent. Weed Control Conf.*, *36*, 81 (1981).
145. S. H. Crowder, M. T. Edwards, and L. B. Gillham, *Proc. South. Weed Sci. Soc. Conf.*, *38*, 78 (1985).
146. J. M. Balneaves and B. J. Cosslett, *Proc. 36th New Zealand Weed and Pest Control Conf.*, p. 41 (1983).
147. B. E. Norris and R. H. Walker, *Proc. South. Weed Sci. Soc. Conf.*, *37*, 70 (1984).
148. S. J. Wiedman and A. P. Appleby, *Weed Res.*, *12*, 65 (1972).
149. D. T. Ferguson, S. E. Schehl, L. H. Hageman, and G. E. Lepone, *Proc. Br. Crop Prot. Conf. Weeds*, *1*, 43 (1985).

Chapter 4

METRIBUZIN

KRITON K. HATZIOS

Department of Plant Pathology, Physiology, and Weed Science
Virginia Polytechnic Institute and State University
Blacksburg, Virginia

DONALD PENNER

Department of Crop and Soil Sciences
Michigan State University
East Lansing, Michigan

I.   INTRODUCTION: HISTORY AND DEVELOPMENT

A.   History and Use

Two decades ago Dornow et al. [1] were the first to report on
the synthesis and chemistry of a new class of heterocyclic compounds,
the 4-amino-1,2,4-triazin-5-ones.   Shortly after their report, selected
aminotriazinone derivatives were synthesized and screened as herbicides by researchers of the agrochemical division of the Farbenfab-
riken Bayer GmbH in Leverkusen, West Germany.   In 1966, the first
patent applications were made, covering several 4-amino-1,2-4-triazin-
5-ones, their influence on the growth of undesired plants, and their
selectivity on such crops as cotton, wheat, oats, and beans [2].
Following these discoveries, numerous studies on the structure–
activity relationships of substituted aminotriazinones have demonstrated the excellent herbicidal activity and selectivity of these
compounds [3–9], which are also known as asymmetrical (as) triazines.

At present, three compounds of the triazinone group have
gained attention as commercial herbicides.   Of these, metribuzin is
the most active and most widely used in the United States and in
other parts of the world [10,11].   Applied either preemergence or
early postemergence, metribuzin is very effective against annual
grasses and numerous broadleaf weeds, including some hard-to-con-
trol weeds, such as cocklebur, velvetleaf, jimsonweed, coffeeweed,
teaweed, and sicklepod [11].   Crops tolerant to metribuzin include
direct-seeded or transplanted tomato [12,13], potato [14,15], soybeans [16,17], sugarcane [18], established alfalfa [11], asparagus
[11], and carrots [11].   Metribuzin is also registered for special
use in weed management programs of the following crops: barley
(spring and winter), dryland winter wheat, established bermudagrass
turf, lentils, and dry field beans [19].   Although metribuzin was
discovered by Bayer AG in West Germany, field testing and commercial development of this herbicide in the United States and Canada
have been conducted by Mobay Chemical Corporation and DuPont.

Metamitron is less potent as a herbicide than metribuzin but
exhibits excellent selectivity in sugar beets and is widely used in
Europe [20,21].   Isomethiozin is an inactive derivative of metribuzin

that is metabolized to the phytotoxic metribuzin by susceptible plants
[22]. It has a good degree of selectivity, which makes it useful for
the control of broadleaf and grass weeds in winter or spring barley
and in spring wheat in Europe [23]. Metamitron and isomethiozin
were synthesized and developed as herbicides in Europe by Bayer
AG. The chemical structures and names of the three commercially
developed triazinone herbicides are illustrated in (1). The rest of
this chapter will deal with the chemistry, degradation, and mode of
action of metribuzin. Further information on the herbicidal properties of metamitron and isomethiozin can be found in the literature
[20–23].

METRIBUZIN   METAMITRON   ISOMETHIOZIN

[1]

B. Physical and Chemical Properties

Metribuzin is a heterocyclic, basic organic molecule. The chemical structure of this herbicide is shown in (1). To facilitate the
recognition and comparison of the parent compound and metabolites
resulting from chemical or metabolic reactions the nomenclature illustrated in (1) will be used throughout this chapter.

In acidic aqueous solutions metribuzin is protonated and ionizes,
forming cations and molecular species depending on the pH of the
solution and the pK_a of the herbicide. The pK_a of metribuzin has
been determined to be about 1.0 when measured by spectrophotometric titration [24,25]. When measured by potentiometric titrimetry
the pK_a of this herbicide was determined to be 7.1 [26]. This
discrepancy in the pK_a values of metribuzin has been attributed to
an acid-catalyzed decomposition of the molecule that occurs during
protonation when the spectrophotometric titration is used [26]. The
potentiometrically determined pK_a of metribuzin did not correspond
to protonation of the amino group [26]. The pK_a values of three
of the known metribuzin metabolites determined by spectrophotometric
titration were in good agreement with those obtained by potentiometric titration. These values were 7.3 for the deaminated (DA)

metribuzin, 10.0 for the diketo (DK) metribuzin, and 8.3 for the
deaminated diketo (DADK) metribuzin [26].

Comparisons of the spectrophotometrically determined pK_a values
of metribuzin (pK_a = 1.0) and 3-amino-as-triazine (pK_a = 3.2) indicated that metribuzin is a weaker base, probably due to the presence
of the carbonyl group at the 5-position of the heterocyclic ring [25].
Metribuzin is also a weaker base than the methylthio- or methoxy-s-
triazines but comparable to the chloro-s-triazines [25]. Additional
physical and chemical properties of the active ingredient of the herbicide metribuzin are summarized in Table 1.

C. Toxicological Properties

Extensive investigations and detailed reports on the toxicological
properties of metribuzin have been published elsewhere [11,28,29]. In
general, evidence accumulated so far indicates that metribuzin has a low
order to toxicity to fish and avian species. The 96-hr LC_{50} values for
rainbow trout and bluegill are greater than 100 ppm [28]. Oral LD_{50}
values ranging from 500 to 1000 mg/kg have been reported for several
bird species, such as bobwhite quail, mallard ducks, canaries, red-
winged blackbirds, brownhead cowbirds, and house sparrows [11,28].

The acute oral LD_{50} for technical metribuzin is 2200 mg/kg for
male rats [11]. Dermal and inhalation toxicity of metribuzin are also
low. Technical-grade or formulated (50 WP) metribuzin taped to the
abraded skin of rats for a period of 24 hr at a dosage of 20 g/kg
did not result in any toxic symptoms or death of the treated animals
[28]. Formulated metribuzin did not cause eye or skin irritation and
did not sensitize the skin of treated rats. In an inhalation study,
male or female rats survived dust treatments with an amount of
formulated metribuzin (50 WP) equivalent to 20,000 µg/liter of space
[28].

In chronic toxicity studies, rats and dogs fed dietary concentrations of metribuzin up to 100 ppm for 2 years survived without
any symptoms of adverse effects caused by the herbicide [28].

Formulated metribuzin also exhibited a low order of dermal or
inhalation toxicity in toxicological investigations with honeybees.
The acute oral LD_{50} is approximately 35 µg of formulated metribuzin
per honeybee [11,29].

Recently, Heimbach [30] reported that metribuzin had no adverse effects on the earthworm fauna of three different soils that
had been treated annually with 0.5 to 1.5 kg/ha of the herbicide
over an 11-year period.

D. Synthesis and Analytical Methods

Aminotriazinone herbicides are prepared as products of the
reaction of an appropriate carbohydrazide with an α-keto carboxylic

TABLE 1

Selected Physical and Chemical Properties of Metribuzin

Property	Data or comments
Chemical name	4-amino-6-tert-butyl-3-methylthio-as-triazin-5(4H)-one
Common name	Metribuzin
Code name	BAY 94337 (Bayer AG); DPX 2504 (DuPont)
Empirical formula	$C_8H_{14}N_4O\ S$
Molecular weight	214.3
Appearance	White, crystalline solid
Melting point	125.5–126.5°C
Density	$D_4^{20} = 1.28\ g/cm^3$
Vapor pressure	$<10^{-5}$ Torr (mmHg) at 20°C
Solubility	in H_2O: 1200 ppm at 20°C Good solvents: acetone, methanol, low solubility in paraffinic hydrocarbon solvents
Saturation vapor concentration[a]	1.3×10^{-8}
Water/air partition coefficient[b]	9.2×10^{-7}
pK_a	1.0
Stability	Resistant to diluted acids and alkali (up to pH 12.5) at 20°C

[a]Calculated from the molecular weight and the vapor pressure data [27].
[b]Calculated as a ratio of water solubility and saturation vapor concentration [27].
Source: From Refs. 11, 27, 29.

acid [1,2,31]. Thus, synthesis of metribuzin starts from thiocarbo-
hydrazide according to (2). The obtained product can be purified
by recrystallization from a 4:1 mixture of hexane and chloroform [28].

[2]

Several chromatographic methods have been used successfully
for the detection and quantitation of residues of metribuzin or its
metabolites in plant tissues, soil samples, or water [32–40]. The
lower limit of residue detection of metribuzin or its metabolites in
recovery experiments using gas chromatography procedures with a
nitrogen-specific detector (N-FID) was at 0.05 ppm for plant or soil
samples and 0.01 ppm for water samples [35,36]. High sensitivity
for the detection of metribuzin and its metabolites (0.01 ppm) was
also reported in gas chromatography studies using an ^{3}H electron
capture detector (ECD), but the selectivity of this method was limited
because many soil coextractants gave interfering peaks [33]. Use
of a flame photometric detector was also successful in detecting and
quantitating metribuzin and its metabolites [33]. An improved gas
chromatographic procedure for metribuzin and its metabolites using
an[63]Ni ECD detector has been developed by Thornton and Stanley
[34]. The sensitivity limit of this method was 0.01 ppm for metribuzin
and its metabolites.

Recently, a high performance liquid chromatography (HPLC)
procedure for the simultaneous extraction and detection of metribuzin
and its deaminated diketo (DADK) metabolite in runoff water was developed by Brown et al. [37]. Residues of metribuzin and DADK as
low as 20 ppb were detected with this method in recovery experiments, demonstrating its excellent sensitivity and resolution.

Storage of metribuzin-treated soil samples at cold temperatures (down to -37°C) for extensive periods of time prior to analysis may significantly affect residue analysis because metribuzin was shown to be degraded in such samples even at these low temperatures [40]. Webster and Reimer [40] suggested that immediate analysis following collection of samples is highly desirable. Otherwise, the results of residue analysis conducted after long periods of cold storage of soil samples should be adjusted to compensate for residue losses in such samples.

Thin layer chromatography (TLC), nuclear magnetic resonance (NMR) spectroscopy, and mass spectrometry (MS) procedures for the separation and characterization of metribuzin and its metabolites in plant tissues have been described in detail by Frear and his colleagues [38,39].

E.   Chemical Reactions

The published information on the reactivity of metribuzin or other aminotriazinone herbicides under purely chemical conditions is rather limited [41-47]. Selected chemical reactions affecting the substituents of the heterocyclic ring of metribuzin occur in a chemical environment and they also appear to participate in the main biotransformation reactions of this herbicide in biological systems. Oxidation and hydrolysis reactions have been reported to affect selected substituents of the metribuzin ring in both chemical and biological systems, while conjugation reactions of metribuzin to natural products (e.g., glucose, glutathione) have been reported primarily in biological systems [41-52]. In biological systems these reactions of metribuzin appear to be enzymatically catalyzed and they will be discussed in detail in a later section of this chapter. A brief summary of the chemical reactions of metribuzin will be presented in this section.

1.   Oxidation Reactions

a. N-Deamination. The amino group in the 4-position of the heterocyclic triazinone ring is very important for the herbicidal activity of metribuzin or other herbicides in this group [3-9]. Pape and Zabik [41] were the first to report that irradiation of metribuzin with ultraviolet (UV) light (300-350 nm) induces the deamination of this herbicide in an aqueous solution. The proposed mechanism for the photoinduced deamination of metribuzin or other aminotriazinones includes an intramolecular hydrogen transfer from the amino group to the carbonyl oxygen in analogy to the γ-hydrogen abstraction of aliphatic ketones [41,45]. The excited carbonyl group is believed to yield a "biradical" intermediate, which then abstracts an amine hydrogen via an intramolecular five-member cyclic transit state

followed by electron shift, with elimination of NH to yield the deaminated product [41] as shown in (3).

[3]

Subsequent studies on the photoinduced deamination of 4-amino-
1,2,5-triazin-5(4H)-one derivatives showed that deamination of these
compounds is dependent on oxygen and the nature of the solvent
used in the reaction [42–45]. Deamination of aminotriazinones under
light conditions requires the combined effect of oxygen and the solvent to produce a cleavage of the amino group since the proton in
the deaminated product is derived from the solvent (usually water)
[44,45]. This evidence lends credence to the intramolecular mechanism proposed for the deamination reaction and shown in (3). Further experiments with a series of 6-*tert*-butyl-3-methylthio-1,2,4-
triazin-5(4H)-one derivatives showed that the formation of the respective deaminated products occurred only when an N—N bond was
present at the 4 position of the heterocyclic ring of these compounds
[44]. However, the deamination reaction was completely independent
of the substituent on the nitrogen atom which is split off [44].

Deamination of aminotriazinone herbicides is also an important
metabolic reaction involved in the degradation of these compounds in
biological systems. In contrast, however, to the photochemical reaction, the biological deamination of aminotriazinones does not require
light and is catalyzed by peroxisomal oxidative enzymes [47–52].

Recently, Nakayama et al. [46] reported that metal-catalyzed
oxidation of selected 4-amino-1,2,4-triazinone herbicides with *tert*-
butyl hydroperoxide resulted in their deamination under dark

conditions. They suggested that hydroperoxide species may play an important role in the reactivity of these herbicides, thus providing a chemical model for their metabolism through a deamination reaction. Cleavage of the 4-amino group of metribuzin has been also obtained as a minor product of the oxidation of this herbicide with m-chloroperbenzoic acid (MCPBA) [47].

b. *Sulfoxidation of the Methylthio Substituent at C-3.* In recent studies on the reactivity of metribuzin under pure chemical conditions, Bleeke and Casida [47] reported that the major products of the peracid oxidation of metribuzin or deaminated metribuzin (DA) with m-chloroperbenzoic acid (MCPBA) were their corresponding sulfoxides. The sequence of reactions involved in the peracid oxidation of metribuzin with MCPBA is summarized in (4) according to Bleeke and Casida [47].

[4]

The sulfoxide of metribuzin was the principal product formed when the ratio of the oxidant (MCPBA) to metribuzin was 1:2. Further oxidation of metribuzin sulfoxide to the diketo derivative (DK) is also dependent on the oxidant:herbicide ratio and with increasing oxidant (4:1, MCPBA:metribuzin) DK was the major final product [47]. The conversion of metribuzin sulfoxide to DK involves a replacement of −S(O)CH$_3$ with OH followed by tautomerization, but the exact mechanism of this reaction is not known [47]. Cleavage of the methylthio group of metribuzin also has been reported to occur under the conditions of the metal-catalyzed oxidation of metribuzin with *tert*-butyl hydroperoxide as described by Nakayama et al. [46]. Under the conditions of (4), the oxidative deamination of metribuzin occurs at a very slow rate compared with sulfoxidation [47]. The deaminated diketo (DADK) compound is the final product of the oxidation of metribuzin with MCPBA at both the 4 and 5 positions of the heterocyclic ring. This product becomes important only when large amounts of the oxidant (MCPBA) are present [47].

The sulfoxides formed as products of the peracid oxidation of metribuzin or DA are intermediate electrophilic compounds with high reactivity. The metribuzin sulfoxide is less stable and harder to isolate than the sulfoxide of DA [47]. Sulfoxidation of thiol-containing herbicides has been considered as an important bioactivation reaction that predisposes these molecules for a subsequent conjugation with glutathione or other thiols in biological systems [53]. In the aforementioned studies of Bleeke and Casida [47] the sulfoxides of metribuzin or DA reacted readily with thiols (e.g., N-acetylcysteine) or protein in a neutral aqueous solution. In vitro studies on the metabolism of metribuzin in mammalian liver microsomal enzymes or in soybean leaf tissues demonstrated further the importance of the intermediate sulfoxide of metribuzin in the conjugation of this herbicide with mercapturic acids and homoglutathione, respectively [39, 47, 48]. Additional details on the conjugation of metribuzin with glutathione in biological systems will be given in a later section.

2. Hydrolysis Reactions

Acid hydrolysis of metribuzin cleaves the methylthio substituent at the 3 position of the heterocyclic ring [39]. Heat causes tautomerization of the hydrolyzed product to the diketo derivative (DK) as shown in (5). Reaction of metribuzin with more concentrated HCl(6N)

[5]

at a higher temperature (105°C) and longer duration (96 hr) yielded the deaminated diketo (DADK) derivative of this herbicide [39,54].

Bartl and Korte [42,43] reported that hydrolysis of metribuzin to the DK metabolite may occur also under thermochemical conditions (52°C) in an aqueous solution following the mechanism illustrated in (6).

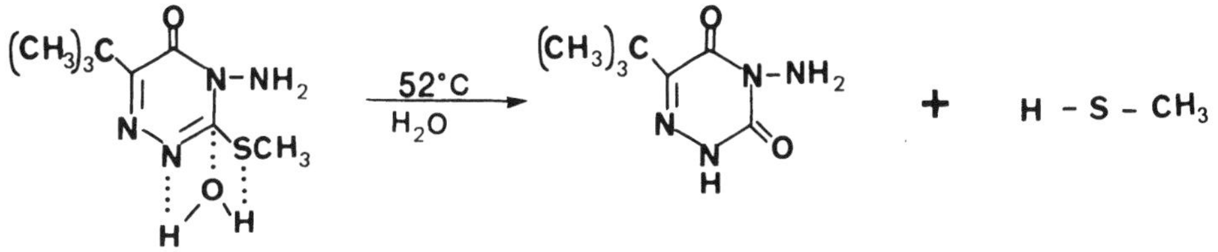

[6]

It is evident therefore that cleavage of the methylthio substituent from the metribuzin ring yielding the DK derivative could proceed either through oxidation with the formation of the metribuzin sulfoxide as an intermediate [see (4)] or through hydrolysis [see (5) and (6)]. Predominance of one reaction over the other would be dependent on the specific conditions of the chemical environment, such as temperature, acidity, and presence of an oxidant as a catalyst [39,43,47].

3.   Other Reactions

Chemical or biochemical reactions affecting the tert-butyl substituent at the 6 position of the triazinone ring have not been reported. Also, direct chemical cleavage of the triazinone ring does not appear to occur; however, hydrolytic cleavage of the triazinone ring has been proposed as a pathway involved in the degradation of the triazinone herbicide metamitron by soil microorganisms (Arthrobacter sp. DSM 20389) or in the soil chemical environment [55]. Dimerization via amino hydrogen abstraction yielding azo compounds and head-head or head-tail dimerization products have been postulated as chemical reactions that may be involved in photochemical transformations of aminotriazinone herbicides [41]. Analytical evidence to support the occurrence of these reactions is not available at present, however.

A chemical reaction having greater relevance to synthesis rather than to degradation of triazinone herbicides is the condensation of metribuzin with 2-methyl-propanal to yield the hydrazone derivative, which is a herbicide known as isomethiozin [31] as shown in (7). As mentioned earlier, isomethiozin is a proherbicide being hydrolyzed readily to the phytotoxic metribuzin in susceptible plants [22].

$$(CH_3)_3C \overset{O}{\underset{N}{\Vert}} \begin{array}{c} N-NH_2 \\ \\ SCH_3 \end{array} + (CH_3)_2CHCHO \longrightarrow$$

$$(CH_3)_3C \overset{O}{\underset{N}{\Vert}} \begin{array}{c} N-N=CH-CH(CH_3)_2 \\ \\ SCH_3 \end{array}$$

[7]

F. Formulations

Similar to other herbicides, formulation of metribuzin for field application is dependent upon certain physical-chemical properties of this herbicide, such as solubility in water and organic solvents, melting point, etc.

The most common formulations of metribuzin are wettable powders prepared by mixing and grinding of the active ingredient with carriers, diluent, wetting, and dispersing agents. Commercial wettable powder formulations contain 50 or 75% of metribuzin as active ingredient and they are marketed under the trade names Sencor 50 or 75 W (trademark of Mobay Chemical Corp.) or Lexone 50 or 75 W (trademark of DuPont) [19,28]. A dry flowable formulation containing 75% of metribuzin is also marketed as Lexone DF by DuPont, while both Mobay and DuPont market a flowable suspension containing 42.1% of metribuzin (480 g/liter) as Lexone 4L or Sencor 4, respectively [19].

Recently, a 75% water-dispersible granular formulation of metribuzin has been introduced and marketed by Mobay under the trade name Sencor Sprayule [19].

Water in volumes of 94 to 374 liters/ha is used as the usual carrier for the spraying of metribuzin in the field. Sufficient agitation during mixing and spraying is imperative to ensure a uniform spray mixture. Metribuzin is compatible with most liquid or wettable powder formulations of other pesticides except in highly concentrated mixtures [28]. It is recommended, however, that each pesticide formulation should be tested for mixing with metribuzin before using such combinations in the field. Marketed formulations of metribuzin are not corrosive and do not damage the spraying equipment. The estimated shelf life of metribuzin formulation exceeds 2 years under normal storage conditions [28].

II. DEGRADATION PATHWAYS

Studies on the fate of metribuzin or other triazinone herbicides have concentrated primarily on higher plants and soils. Degradation studies of metribuzin in animals, microorganisms, and aquatic environments are limited. A survey of the available literature, however, demonstrates that the degradation pathways of metribuzin in plants, animals, and soils have many features in common. Oxidative, hydrolytic, and conjugation reactions appear to be the major reactions involved in the transformation of this herbicide by biological systems. As a result of these reactions both polar and nonpolar metabolites of metribuzin are formed in selected biological systems. In addition, metribuzin or some of its nonpolar metabolites are incorporated into the water-insoluble fraction of biological systems. Differential metabolism or differential rate of metabolism appear to be very important for the crop selectivity of this herbicide under field conditions, and they will be discussed in more detail at a later section.

Techniques used for the separation and characterization of metribuzin metabolites have ranged from simple thin layer chromatography (TLC) separations [38,39,47,48,54,58–70] to more elaborate chromatographic procedures, such as gas chromatography and high performance liquid chromatography (HPLC) [32–40,47,72], as well as spectral procedures including nuclear magnetic resonance (NMR), mass spectrometry, infrared (IR) and ultraviolet (UV) spectroscopy [38,39,47,48].

A. Degradation in Plants

The metabolic fate of metribuzin has been reported in several higher plants including either such crop species as soybeans [39, 49–52, 56–65], sugarcane [66], tomato [38,54,67], potato [68–70], wheat [71], and lentils [72] or weeds, such as barnyardgrass [73], American nightshade [73], pitted morningglory [74], entire-leaf morningglory [74], and hemp sesbania [57].

Phase I metabolic reactions [53] demonstrated in the degradation of metribuzin by higher plants include N-deamination, sulfoxidation, and demethylthiolation. These reactions result in the formation of nonpolar metabolites with reduced or modified phytotoxicity (e.g., deamination and demethylthiolation) or they predispose the metribuzin molecule for subsequent metabolism in the secondary phase (e.g., sulfoxidation). Phase II reactions involved in the metabolism and detoxification of metribuzin in higher plants include N-glucosylation, acylation of the N-glucoside conjugate with malonic acid, and conjugation of the metribuzin sulfoxide with homoglutathione.

A summary of the metabolic reactions and pathways involved in the degradation of metribuzin by higher plants is given in (8). Details of these reactions and pathways are discussed below.

[8]

1. N-Deamination

Deaminated (DA) metribuzin has been reported as a common metabolite of metribuzin in degradation studies with several plants under both in vivo and in vitro conditions [54,56-74]. DA appears to be a major metabolite of metribuzin in soybeans [50-52,57,62,64, 65] but a relatively minor metabolite in other crops, such as tomato [38,54] and sugarcane [66]. In addition, several investigators have failed to detect this metabolite in selected soybean cultivars [39,58, 59]. According to Fedtke [52], the discrepancies in the literature related to the detection and importance of DA as a metabolite of metribuzin in soybean plants could be attributed to two facts. First, experiments on the metabolism of metribuzin using leaf discs could be misleading since these tissues rapidly lose their ability to deaminate metribuzin because of tissue damage. Second, the exact pattern of metribuzin metabolism and detoxification in soybeans is dependent upon the concentration of metribuzin supplied to the plant tissues. Studies on the kinetics of the formation of DA and conjugated metribuzin in soybeans revealed a precursor/product relationship [52]. Data presented by Fedtke and his colleagues [50-52] indicate that at micromolar levels of metribuzin deamination appears to be the predominant metabolic reaction involved in the degradation of this herbicide in soybeans, however, at higher concentrations (>100 μM) conjugation of metribuzin to homoglutathione appears to predominate over the deamination reaction.

The rate of metribuzin deamination was found to be highly correlated with the degree of tolerance of selected soybean cultivars [51]. As discussed in an earlier section of this chapter, deamination of metribuzin is a common photochemical reaction. It is induced by light, requires oxygen, and is dependent on the nature of the solvent used in the reaction [41-45]. In contrast to the photochemical reaction, the biological deamination of metribuzin does not require light and is enzymatically catalyzed. Peroxisomal enzyme preparations from soybean leaves are able actively to deaminate metribuzin under reducing conditions. Thus, enzymatic deamination of this herbicide under in vitro conditions could be obtained in the presence of a reductant, such as reduced glutathione (GSH), ascorbate or cysteine, several redox cofactors, such as FMN and cytochrome c, and under a nitrogen atmosphere [51]. The optimum pH for the reaction mediated by peroxisomal enzymes was 7.5. The released exocyclic amino group of the DA metribuzin is liberated as ammonia during the enzymatic reaction [50].

Pure and active peroxisomal preparations from soybean leaves also were able to deaminate metribuzin under air, although at a lower rate [51]. This observation coupled with the requirement of oxygen for the photochemical deamination of metribuzin suggests that under in vivo conditions the deamination of metribuzin in soybean leaves may be mediated by mixed function oxidase enzymes. Support for such a postulation comes from in vitro studies with mammalian microsomal enzymes, demonstrating the ability of these enzymes to deaminate metribuzin [47,48]. The possible existence of an inhibitor of an enzyme involved in the metabolism of metribuzin in susceptible soybean cultivars has been proposed by Oswald et al. [61] and is further supported by recent work of Fedtke and Schmidt [51]. They showed that peroxisomal enzyme preparations from susceptible soybean cultivars passed through a Sephadex G50 column were able to deaminate metribuzin at a rate comparable to that of peroxisomal preparations from leaves of tolerant soybean cultivars. This observation indicates that the inhibitor that is normally present in the leaves of the susceptible soybean cultivars could be removed by this treatment [51]. Enzymatic deamination has been also reported as an important reaction involved in the selective action of the herbicide metamitron [49].

2. Sulfoxidation of the $-SCH_3$ Group at C-3 Position

Oxidation of the 3-methylthio group of the heterocyclic ring of metribuzin to a sulfoxide intermediate recently has been identified as an important biotransformation involved in the metabolism of this herbicide in soybeans [39]. This reaction appears to be a bioactivation reaction predisposing the molecule of metribuzin for a subsequent conjugation with the tripetide homoglutathione in the leaves of tolerant soybean cultivars [39]. Similar pathways have been reported for the methylthio-s-triazine [75–77] and thiocarbamate [78–80] herbicides undergoing metabolism in biological systems.

Studies on the involvement of enzyme systems in this reaction in plant tissues have not been reported as yet; however, comparative in vitro metabolism studies with rat or mice microsomal preparations suggest that mixed function oxidase enzymes are mediating the oxidation of metribuzin to its sulfoxide [39,47,48], prior to its conjugation with glutathione. Conjugation to glutathione occurred only when the mixed function oxidase cofactor, NADPH, together with GSH and metribuzin were added to washed microsomal preparations [39].

3. Demethylthiolation

Diketo (DK) metribuzin, the product of hydrolytic demethylthiolation of the herbicide, has been reported as a minor metabolite of metribuzin in soybeans [50–52,58,62,65] and other plants [54,66, 68–70,72,73]. It is very likely that, in biological systems, DK is formed by the hydrolysis of the unstable sulfoxide of metribuzin [39]. In soybeans, DK has been found to be an active precursor of insoluble residue formation or it could undergo further metabolism, forming a conjugate with malonic acid [39]. In addition, DK could undergo deamination forming the deaminated diketo (DADK) derivative of metribuzin which has been reported as a possible metabolite of this herbicide in several plant species [50-52,54,58,59,62,65,66, 68-70,72,73]. DADK appears to be a terminal nonphytotoxic metabolite of metribuzin in soybeans or in other plant species [39]. In early studies on the degradation of metribuzin in susceptible soybean cultivars, DADK was reported as a major metabolite of this herbicide [58,59]. Recent and more elaborate studies on the degradation of this herbicide, however, have demonstrated that DADK or DK are minor products in soybeans [39,50-52,62,64,65], tomato [38, 54], potato [68-70], and sugarcane [66]. Conjugates of metribuzin to natural substances, such as glucose or homoglutathione, are currently considered to be the major metabolic pathways involved in the degradation of this herbicide in tolerant plants, such as tomato and soybean, respectively [38,39].

4. Conjugation to Glucose

The formation of polar metabolites of metribuzin in leaf tissues of tolerant plant species had been reported in early studies on the degradation of this herbicide in higher plants [54,59,66]. Speculations on the chemical nature of these polar metabolites of metribuzin suggested glucosides [54,59]. It was only recently, however, that conclusive evidence was presented by Frear and his colleagues [38, 39] to demonstrate that some of the polar metabolites of metribuzin in tolerant tomato and soybean cultivars were indeed N-glucoside conjugates of this herbicide.

The major pathway of metribuzin metabolism and detoxification in tolerant tomato cultivars involves the formation of a β-D-(N-glucoside) conjugate, followed by a rapid acylation of this conjugate with malonic acid to form the malonyl β-D-(N-glucoside) conjugate [38]. The formation of the N-glucoside conjugate of metribuzin in tomato plants is an enzymatic reaction catalyzed by a metribuzin-N-glucosyl transferase requiring UDP-glucose as the sugar substrate [38]. This enzyme has been isolated and partially purified from tomato leaf tissues and has a broad optimum pH ranging from 7.6 to 8.5 for sufficient activity [38]. Differences in the levels of UDP-glucose:metribuzin N-glucosyl transferase activity were detected in leaf tissues of tolerant and sensitive tomato cultivars, indicating that differential metabolism or rate of metabolism play a major role in the intraspecific selectivity of this herbicide [38]. In studies with leaf tissues from young seedlings, specific activities of enzyme preparations from metribuzin-tolerant tomato cultivars (Fireball, Harvestee, and Vision) were consistantly greater (1.5-fold) than those from metribuzin-sensitive cultivars (Ontario 771, Heinz 1706, and Trimson) [38]. Studies with leaf tissues from older seedlings of the same cultivars, however, failed to show such differences, thus confirming earlier reports that tomato tolerance to metribuzin increases with plant age [13,81,85].

Additional support for the importance of the N-glucoside conjugate in the detoxification of metribuzin in tomato comes from field or greenhouse studies showing a reduced tolerance of tomato plants to metribuzin when the herbicide was applied on cloudy days or when the tomato plants were maintained under low light conditions prior to herbicide treatment [13,81-85]. Such conditions are likely to reduce the carbohydrate reserves and(or) UDP-glucose substrate levels needed for the detoxification of metribuzin via the N-glucoside conjugate pathway. Similar conclusions have been reported in studies examining the effect of environmental conditions, such as light, on the N-glucoside conjugate formation of the herbicides chloramben in soybeans [86] and pyrazon in sugar beets [87].

In contrast to its fate in tomato [38], metabolic detoxification of metribuzin via the N-glucoside conjugate pathway is only of minor importance in soybeans [39]. The limited capacity of soybeans to glucosylate metribuzin was further demonstrated by unsuccessful attempts to isolate and purify a UDP-glucose:N-glucosyl transferase that uses metribuzin as a substrate similar to the one reported earlier for tomato [39]. The N-glucoside conjugate of metribuzin in soybeans is further acylated with malonic acid to yield the malonyl N-glucoside of this herbicide [39].

Conjugates of metribuzin with natural products have been also suggested in studies on the fate of this herbicide in plant species such as sugarcane [66], potato [70], and American nightshade [73], however, detailed studies on the conclusive identification of these conjugates as the N-glucoside conjugate of metribuzin are not presently available.

In a recent study, Britton et al. [88] reported that the metabolism of metribuzin via the N-glucoside conjugate pathway in soybeans was significantly reduced when metribuzin was applied to soybean seedlings in combination with the organophosphate insecticide phorate. Earlier studies by other investigators [60,67] had also demonstrated that pretreatment of soybean or tomato plants with carbamate (carbaryl) or organophosphate (malathion, phorate) insecticides significantly inhibited the metabolism of metribuzin in these plants.

N-glucoside conjugates of metribuzin or of other herbicides [89] are resistant to enzymatic hydrolysis and may persist as terminal metabolites in those plant tissues where they are formed [86,87,89].

5. Acylation of Metribuzin Metabolites with Malonic Acid

Acylation reactions of xenobiotics containing carboxyl ($-COOH$), amide ($-CONH_2$), or aniline ($PhNH_2$) groups in their molecules are relatively common in the degradation of these chemicals by higher plants [53,90]. Acetylation, formylation, and malonic acid conjugation are the most common acylation reactions occurring on xenobiotics or their metabolites. These reactions are believed to reduce the phytotoxicity of the parent compounds [53,90].

In the case of metribuzin, the acylation of its N-glucoside conjugate with malonic acid appears to be quite common during the metabolism of this herbicide in soybeans and tomato [38,39]. In addition, conjugation of the DK metabolite of metribuzin with malonic acid has been reported as a minor metabolite of this herbicide in soybeans [39].

6. Conjugation to Homoglutathione

Conjugation of metribuzin with plant peptides such as homoglutathione has been reported so far as a major metabolic pathway of this herbicide only in soybeans [39]. A reactive (electrophilic) sulfoxide intermediate metabolite of metribuzin is the actual form of the herbicide that conjugates with homoglutathione in soybeans. Sulfoxidation, therefore, appears to activate metribuzin and predispose it for a subsequent conjugation with the tripeptide homoglutathione, similar to what happens in the metabolism of methylthios-triazine and thiocarbamate herbicides by biological systems [75–80].

Although detailed information on the isolation and characterization of an enzyme system that mediates the conjugation of metribuzin sulfoxide with homoglutathione is presently limited, it seems likely that this conjugation reaction could be catalyzed by a GSH-S-transferase system similar to the one involved in the conjugation of the herbicide acifluorfen with homoglutathione in soybeans [91]. Indirect support for such a speculation comes from a recent study by Gaul et al. [92], who showed that under field or growth room conditions, tridiphane, a well-known inhibitor of plant GSH-S-transferases, synergized the phytotoxicity of metribuzin on soybeans. Tridiphane failed to synergize the activity of metribuzin on tomato, supporting further the notion that conjugation of metribuzin with glutathione is not an important pathway involved in the detoxification of this herbicide in tomato [92].

B. Degradation in Soil

Similar to the situation with other soil-applied herbicides, loss of metribuzin in soils could be the result of its biological or nonbiological degradation. Soils represent a complex system, and degradation of herbicides in solis may be influenced by many factors, such as concentration of a given herbicide, diffusion, sorption (adsorption/desorption), and climatic factors including temperature, moisture, and duration of sunlight [93]. Since both the biological and the

nonbiological degradation of herbicides in soils are critically dependent upon their sorption behavior, the adsorptive behavior of metribuzin, along with its nonbiological degradation in soils, will be discussed in the present section. Microbial degradation and the desorption behavior of this herbicide will be covered in later sections.

1. Adsorption of Metribuzin to Soil Particles

Early reports on the behavior of metribuzin in soils revealed
the existence of a high negative correlation of metribuzin phytotoxicity with soil organic matter and soil pH [94,95]. It was concluded
that the herbicidal inactivation of metribuzin in soils was due to a
pH-dependent adsorption of this chemical to the organic matter of
the soil [95]. Subsequent studies by other investigators have confirmed the high affinity of this herbicide for soil organic matter
[96-102]. Metribuzin adsorption to soil substrates containing organic
matter is dependent on time and on the concentration of herbicide
[26,99]. Adsorption of metribuzin to soil organic matter increased
with time, whereas it decreased with increasing metribuzin concentration.

The participation and importance of soil clay colloids in the soil
adsorption of metribuzin is less clear. Sharom and Stephenson [96]
failed to find any significant correlation between metribuzin adsorption and clay content of several Canadian soils. The negligible participation of inorganic substrates in the soil adsorption of metribuzin
has been also reported by Lafleur [99] and Kerpen and Schleser
[100]. Other investigators, however, have shown that soil clay colloids are important for the adsorption of this herbicide in selected
soils [26,97,98]. Furthermore, a competitive effect between the combination of various levels of clay and organic matter as related to
metribuzin adsorption has been suggested [97].

The influence of soil pH on the adsorption of metribuzin to soil
particles also has been controversial. Data from Ladlie et al. [24]
presented in Table 2 clearly indicate that adsorption of metribuzin to
soil particles increases with a decrease in soil pH. The dependence
of the soil adsorption of this herbicide on soil pH has been reported
by other investigators [95,98,103,104]. Thus, the adsorption of
metribuzin to soil particles appears to be analogous to that of the
s-triazine herbicides [105,106]. Taking into consideration the surface acidity of clays, which is usually three to four pH units lower
than the pH of the bulk soil solution and the pK_a value of metribuzin ($pK_a = 1.0$), Ladlie et al. [24] suggested that the expected
optimum pH for maximum adsorption of this herbicide to soil colloids
should be between pH 4.0 and 5.0. Data in Table 2 confirm this
speculation; however, the influence of pH on the adsorption of
metribuzin to soil colloids has been disputed by some investigators
[100,104,107].

TABLE 2

[14C] Metribuzin Adsorption by Hillsdale Sandy Clay Loam
at Various pH Levels

	Time after treatment (hr)				
	24	72	144	288	
pH	[14C] metribuzin adsorption[a] (% adsorption)[b]				Average
4.6	15.6 f	17.8 h	21.1 i	27.1 k	20.4 e
5.1	12.4 cd	13.0 d	18.5 h	21.9 j	16.4 d
5.6	10.4 b	11.8 c	16.8 g	20.8 i	14.9 c
6.1	9.9 ab	10.6 b	13.8 e	17.9 h	13.0 b
6.7	9.2 a	10.2 b	11.7 c	15.5 f	11.6 a

[a]Means followed by similar letters are not significantly different at the 5% level by Duncan's multiple range test.
[b]79.4 nm/test tube.
Source: From Ref. 24.

Detailed information on the mechanisms involved in the adsorption of metribuzin to soil particles is limited. The methylthio and amino groups at the C-3 and N-4 positions of the heterocyclic ring of this herbicide could influence the electron density of this molecule and the formation of binding forces between the herbicide and soil particles [96]. Protonation of the weakly basic metribuzin molecule is possible at low soil pH and may be a mechanism involved in the soil adsorption of this herbicide in much the same way that has been described for the adsorption of the s-triazine herbicides [105,106]. In model experiments employing UV and IR spectroscopy, Schmidt [95] proposed hydrogen bonding and binding to aromatic carboxylate ions as potential mechanisms involved in the adsorption of metribuzin to soil organic matter. More recently [99], metribuzin adsorption to peat or soil organic matter was viewed as a multistage process, including the following potential mechanisms: (a) movement (Brownian, mechanical) of the herbicide to the organic matter solution interface, (b) adsorption at accessible external organic matter sites, and (c) gradual invasion of restricted inner surfaces by a dynamic adsorption-desorption–movement replenishment sequence, as the system adjusts to a minimum free energy equilibrium condition.

Under field conditions, adsorption of metribuzin to soil particles is very important since it influences the persistence and herbicidal activity of this herbicide. In any given soil, the adsorption rate of metribuzin will determine the amount of free metribuzin available for weed control or the degree of dissipation by means of leaching and biological or nonbiological degradation.

2. Nonbiological Degradation of Metribuzin in Soils

In the field, herbicides are applied to an environmental complex composed of soil, water, and air. Under these conditions, the environmental degradation of herbicides could involve the attraction of a reagent on a reactive compound resulting in nonbiological transformations of herbicides [108].

Although it is not always easy to distinguish between the biological and nonbiological degradation of a given herbicide in selected soils, early studies on the fate of the herbicide metribuzin in soils emphasized the importance of the nonbiological degradation of this herbicide. Hyzak and Zimdahl [109] failed to find any induction period following application of metribuzin to soils and concluded that nonbiological degradation of this herbicide is equally important or more important than its biological degradation. Since soil components are normally present in vast quantities compared with the herbicide, the rate of the decomposition reaction of each herbicide should follow first-order kinetics [110]. Indeed, the rate of degradation of metribuzin and its isopropyl and cyclohexyl analogues in soils under field and laboratory conditions appears to be best described by first-order kinetics [109]. Later studies by other investigators [24,111,112] confirmed these results by showing that the degradation of metribuzin in soils follows first-order kinetics. In other studies, however, it was found that under different laboratory incubation systems the decomposition of metribuzin in soils could follow kinetics of fractional order (below first order) or orders higher than first [109,113]. To explain these results, Kampson-Jones and Hance [114] suggested that the soil decomposition of metribuzin could involve consecutive or competitive reactions (e.g., competition between the loss of the methylthio and amino groups) or that the artificial conditions in the laboratory could introduce other limiting factors in addition to the concentration of metribuzin. Furthermore, the involvement of diffusion-controlled processes also has been postulated as a factor that could influence the kinetics of the soil degradation of metribuzin under laboratory conditions [113]. Since these explanations are not entirely conclusive, Hance and Haynes [113] proposed that decomposition experiments giving reaction orders greater than one, such as those of the metribuzin degradation under laboratory conditions, should be verified in more than one experimental system.

Several edaphic or environmental parameters are known to in-fluence the degradation of metribuzin in soils. The importance of metribuzin adsorption to soil colloids has been discussed earlier. Amont other factors, temperature appears to be important. Hyzak and Zimdahl [109] showed that the degradation of metribuzin in soils increased linearly with temperature and similar observations have been reported by Savage [111] and Bouchard et al. [101]. A re-verse relationship between metribuzin degradation and soil tempera-ture has been reported by Kempson-Jones and Hance [114]. They found that the rate of metribuzin degradation was slower at 20°C compared with that at 10°C. Degradation of metribuzin also has been documented in soil samples awaiting residue analysis that were stored at −37°C [40]. This report further underlines the importance of the nonbiological degradation of metribuzin in soils, since at these low temperatures (−37°C) little or no biological degradation of this herbi-cide would be expected to take place. The degradation of metribuzin in soils has been shown to be influenced also by soil depth, being slower in soil samples from deeper horizons rather than in surface soils [101,114].

While the degradation of metribuzin in nonsterile soils appears to be dependent on soil moisture [109], degradation of this herbicide in sterilized soils could proceed under relatively dry conditions [115]. The rate of the degradation kinetics of metribuzin under these con-ditions was determined to be less than first order [115].

Although studies on the separation and identification of metribu-zin metabolites formed as a result of the nonbiological degradation of this herbicide in soils are limited, DA, DK, and DADK have been reported as the most common metabolites of this herbicide in soils [34,96,116]. The mechanisms of the reactions that could lead to the formation of these metabolites of metribuzin in a soil environment were described in detail in an earlier section of this chapter.

C. Degradation in Animals

Studies on the degradation of metribuzin in animals are very limited. It was only recently that studies on the metabolic fate of metribuzin in rats, mice, and their microsomal oxidation systems were reported [47,48]. The results of both in vitro and in vivo studies were comparable and demonstrated the importance of oxidative and conjugation reactions in the mammalian metabolism of metribuzin.

N-deamination appeared to be more important in the metabolism of metribuzin in rats than in mice, and it was catalyzed by mixed function oxidase enzymes [47]. Mercapturic acid derivatives were the major metabolites of metribuzin and DA−metribuzin detected in urine of intraperitoneally treated mice and orally treated rats [47]. Protein-bound metribuzin and DA-metribuzin were also detected as metabolic products in both animals [47,48]. Ring cleavage followed by decarboxylation was not reported in plants nor does it appear to be involved in the degradation of this herbicide in animals [47]. Sulfoxidation of the methylthio group of metribuzin or DA-metribuzin, catalyzed by mixed function oxidases, was found to be an important bioactivation reaction predisposing these molecules for their subsequent conjugation to glutathione (GSH) or to mercapturic acid [47, 48]. The conjugation of metribuzin or DA-metribuzin with GSH was catalyzed by microsomal GSH S-transferase enzymes isolated from rat or mouse liver [47,48].

The metabolic reactions involved in the degradation of metribuzin in mammalian systems are summarized in (9), according to reports of Bleeke and Casida [47] and Bleeke et al. [48].

D. Degradation by Microorganisms

The degradation of metribuzin by soil microorganisms is currently considered as an important process involved in the environmental fate of this herbicide. Although specific studies with soil microorganisms degrading metribuzin are limited, a number of reports have demonstrated the importance of microbiological activity in the degradation of metribuzin in soils. Marked reductions in the capacity of selected soils to degrade metribuzin have been shown to result from fumigation [104], sterilization by autoclaving or irradiation with γ-rays [96,111], and treatment with microbial inhibitors [24]. In addition, amendment of natural soils with sugars, such as glucose, resulted in significant increases in the rate of the degradation of metribuzin in these soils [111]. Furthermore, a number of edaphic and environmental factors favoring microbial activity, such as temperature, pH, and soil moisture, were also found to favor the degradation of metribuzin in soils [24,111]. Data from Ladlie et al. [24], presented in Table 3, illustrate the influence of soil pH and of sodium azide (NaN$_3$) on the evolution of $^{14}CO_2$ from soil-applied metribuzin.

Metribuzin metabolites that have been detected in soils and believed to result from microbial degradation of this herbicide include DA, DK, and DADK [96,116]. Although studies with specific microorganisms capable of deaminating metribuzin in soils are not yet available, a number of soil bacteria, including *Arthrobacter* and *Pseudomonas* species, as well as soil fungi, such as *Rhizopus japonicus* and *Cunninghamella echinulata* Thaxter have been reported to rapidly deaminate the triazinone herbicide metamitron [117]. *Aspergillus niger* and other soil fungi did not metabolize metamitron [117].

In studies with ^{14}C-ring-labeled metribuzin, Schumacher [118] reported that the heterocyclic ring of this herbicide could be cleaved by soil microorganisms in a stepwise manner. The release of $^{14}CO_2$

[9]

TABLE 3

Metribuzin Degradation to CO_2 in 12 Weeks,
as Affected by Soil pH and Sodium Azide

pH	[14]C evolved as [14]CO$_2$[b] (%)
4.6	4.6 c
4.6 NaN$_3$	0.4 a
5.1	8.6 d
5.6	10.2 e
6.1	13.5 f
6.7	17.9 g
6.7 NaN$_3$	1.1 b

[a]Means within columns followed by similar
letters are not significantly different at the
5% level by Duncan's multiple range test.
The sodium azide treatment approximates
sterilization.
[b]0.0256 μCi/flask.
Source: From Ref. 24.

from [14]C-3 labeled metribuzin was more rapid than that released
from [14]C-5 labeled metribuzin. Cleavage of the metribuzin ring
under soil conditions has been also reported by Prestel et al. [69].
Thus, in contrast to s-triazine herbicides, whose ring structure
appears to be fairly resistant to microbial attack [119], the ring of
metribuzin or other triazinone herbicides could be cleaved by soil
microbes [69,118,120]. Hawck and Stephenson [121] suggested that
the symmetrical resonating structure of the s-triazine molecule contributed to differences in degradation rates. Asymmetrical chloro-
triazines were degraded more quickly than their corresponding symmetrical triazines. It was suggested that polymerization resulting
in an increased symmetry would lead to stability, whereas labile
products of biochemical transformation tend toward asymmetry [121].
    In a recent study, Engelhardt et al. [55] reported that cultures
of the soil bacterium Arthrobacter sp. DSM 20389 were capable of
degrading the triazinone ring of the herbicide metamitron by hydrolytic cleavage of the amide bond (N-4 and C-5 positions on the

heterocyclic ring). Benzoylformic acid acetylhydrazone and benzoyl-formic acid were identified as the major products of the ring cleavage of metamitron by *Arthrobacter*. In addition, 3-methyl-6-phenyl-1,2, 4,5-tetrazine was identified as a major product formed as a result of decarboxylation and oxidation of metamitron by *Arthrobacter* in the presence of light [55]. Deaminated metamitron was only a minor metabolic product of metamitron in this study. Although it is quite possible that metribuzin would be degraded by soil microorganisms following the same reaction sequences reported for metamitron [55], further experiments with specific soil microorganisms would be needed for confirmation.

E. Degradation by Phototransformation

Studies on the photochemical degradation of metribuzin have been conducted primarily under laboratory rather than actual field conditions. Therefore, the potential contribution of this process in the environmental fate of this herbicide under practical conditions is not well understood at the present time.

The major reaction involved in the phototransformation of metri-buzin is deamination yielding DA-metribuzin as the major metabolic product [41–45]. Deamination of aminotriazinone herbicides in the laboratory under UV light (300–350 nm) is dependent upon oxygen and the solvent used in the reaction mixture [44,45]. The mechan-ism of the photochemical deamination of metribuzin was discussed in an earlier section of this chapter and is illustrated in (3). Apart from DA-metribuzin, azo compounds and head–head or head–tail dimer-ization compounds have been postulated as possible products formed during the photochemical transformation of metribuzin under laboratory conditions [41].

F. Dissipation and Persistence Under
Field Conditions

The amount of metribuzin in the soil available for plant uptake is dependent upon a number of transfer and transformation processes that may attack the molecule of this herbicide. Many of the degrada-tive processes of metribuzin discussed in earlier parts of this chap-ter, such as chemical or microbial degradation in soils, phototrans-formation, and adsorption to soil particles, will influence the residual life and activity of this herbicide. In addition, movement of metribu-zin by leaching in the soil profile will greatly influence the persist-ence and phytotoxicity of this herbicide under field conditions.

The mobility of metribuzin in soils is inversely related to the soil adsorptive capacity. In general, metribuzin is relatively mobile in sandy and mineral soils but very immobile in soils with high or-ganic matter [96]. Other studies, however, have shown a significant

correlation between metribuzin mobility and soil clay content [97].
Ladlie et al. [122] showed that metribuzin leaching in soils increased
with increasing soil pH. Residue analysis of soil samples in a Hillsdale sandy clay loam soil showed that greater amounts of metribuzin
residues were extractable at soil pH 6.7 than 4.6 [122]. Ladlie et al.
[122] concluded that the effect of soil pH on metribuzin mobility was
indirect, resulting from the influence of soil pH on the adsorption
of metribuzin to soil clay colloids. Metribuzin leaching from the zone
of soil application is also dependent on the amount of rainfall or irrigation that occurs under field conditions [96]. Reduced efficacy
of metribuzin for preemergence weed control could result under high
rainfall conditions in soils with low organic matter [96]. Loss of
metribuzin by leaching was negligible in muck (organic) soils of
Ontario regardless of rainfall rate [96].

The combined effect of all transfer and transformation processes
on the molecule of metribuzin eventually will determine its persistence
and residual activity in soils under field conditions. Most of the research on the residual life of metribuzin has shown it to be nonper-
sistent. Bioassay results indicated that metribuzin phytotoxicity to
sensitive plants such as cucumber or oats, was lost in several soil
types within a few weeks following applications [96,97,101,122-126].
Depending on the soil type and variations in edaphic or environmental
conditions, such as pH, temperature, and soil moisture, the half-life
of metribuzin dissipation in soils could range from a few days to 4 or
more months [96,97,101,122-126]. The rapid loss of metribuzin from
soils indicates that the residual activity of metribuzin on susceptible
crops will not be a problem under warm and humid conditions during
the growing season; however, due to variations in soil conditions,
persistence problems with metribuzin have been reported and several
approaches have been attempted to minimize their impact. Changes
in the chemistry of the metribuzin molecule can reduce the residual
life of this herbicide. Thus, substitution of the methylthio group in
the heterocyclic ring of metribuzin by an ethylthio group results in
a herbicidal molecule that is less persistent than metribuzin [127].
Other approaches to overcome the residual problems of metribuzin
include the use of tolerant crop cultivars or chemical protection of
the activity of metribuzin on susceptible crops. These approaches
will be discussed in detail later in this chapter.

III. MODE OF ACTION

A. Uptake and Translocation

In analogy to the situation with other photosynthesis-inhibiting
herbicides, metribuzin or other triazinone herbicides are readily
taken up by the roots and translocated to the shoots and leaves of

treated plants in the apoplast [38,39,57,58,61,64,65,128–130]. When plants are grown in nutrient solutions containing radiolabeled metribuzin, rapid uptake occurs readily in all plants regardless of their degree of susceptibility or tolerance to this herbicide [38,39,51,54,58, 59,62,63,65,70]. Following a short, initial period of rapid uptake associated with the saturation of the adsorptive sites of the roots, uptake and translocation of metribuzin are directly proportional to the transpiration rate [131,132]. Root uptake of metribuzin is passive and it is not influenced by metabolic inhibitors as long as these do not affect transpiration [132]. Factors affecting the rate of transpiration, such as temperature, humidity, light intensity, and stomatal aperture, would also affect the root uptake of metribuzin [131,132]. Exposure of soybean seeds to preemergence cold stress reduced the uptake of [14C]metribuzin by the germinated seedlings [88]. Increased uptake of metribuzin is observed with increasing concentrations of the herbicide and time of exposure to this herbicide [58,59, 63,74]. Radiolabel from 5×10^{-7} M [14C]metribuzin accumulated only in the leaf veins while that from 2.5×10^{-6} M of [14C]metribuzin moved throughout the leaves of tolerant (Tracy M) or susceptible (Tracy) soybean cultivars [63]. A decrease in the uptake of [14C]-metribuzin associated with increases in exposure time has been reported so far only in studies with the metribuzin-sensitive weed entireleaf morningglory [74].

Following postemergence treatment, metribuzin can be absorbed by penetrating the leaf surface via an "aqueous route" of the hydrated cuticle [131]. However, very little foliarly applied metribuzin is transported basipetally out of the treated plant parts [128]. In a recent study, Frank and Beste [129] reported that leaf absorption and basipetal movement appeared to be appreciable following postemergence applications of this herbicide to tomato and jimsonweed. Gentle abrasion of the leaf surface with carborundum was shown to increase the foliar penetration of metribuzin in tomato but not in jimsonweed [129]. Use of surfactants did not enhance the foliar uptake of metribuzin in wheat [133].

Studies employing autoradiography and quantitative determinations to monitor the root uptake and distribution of [14C]metribuzin in susceptible dicotyledonous species have demonstrated that detectable radiolabel is first observed in the veins, but with time it becomes interveinal and accumulates at the leaf margins and tips where transpiration is the greatest [54,58,59,62,65]. Retention of 14C label in the roots and stems of these plants was very limited. In tolerant plants or cultivars, however, retention of radioactivity in roots, stems, and major veins of the more mature leaves was very evident [58,59,65,70,74]. This suggests that metribuzin or its metabolites may not be as readily translocated within the tissues of tolerant plants or crop cultivars [59,65]. Extensive vascular localization or compartmentalization of metribuzin, perhaps by binding to lignin in tolerant cultivars, appears to be important in the selectivity of this herbicide toward soybeans or potatoes [39,54,59,70]. Recently, Falb and Smith [65] suggested that the restriction of radiolabel from [^{14}C]metribuzin to the vascular tissues could be the result of metribuzin metabolism yielding a product that cannot penetrate through membranes and is trapped in the veins, which have low levels of chlorophyll and, therefore, limited capacity to photosynthesize. This suggestion presupposes that the metabolism occurs in the root or xylem. A time-dependent incorporation of metribuzin to water-insoluble residues has been reported in metabolic studies of this herbicide with tomato and soybean plants [38,39]. In soybeans, 24 hr after treatment, 10-11% of the radiolabel was associated with incorporated residues of metribuzin; however, residue incorporation does not appear to be an important cause of soybean tolerance to this herbicide [54]. Treatment of soybeans with the insecticide phorate has been reported to increase the distribution of [^{14}C]metribuzin in the shoots of soybean [88].

Although predominant, the translocation of metribuzin in the apoplast is not exclusive. Like the s-triazines or the urea herbicides, metribuzin has been characterized as a "pseudoapoplastic" herbicide [134]. Metribuzin readily penetrates the symplasm of roots or leaves, but because of the inability of the symplasm to retain it for long time, metribuzin is leached into the apoplast and carried away with the transpiration stream [134].

B. Enzymatic Studies

Studies on the effects of metribuzin on selected plant enzymes are very limited. Tested at concentrations ranging from 10^{-8} to 10^{-5} M, metribuzin had no effect on ATPase activity of bean chloroplasts [135]. At 0.5 mM, metribuzin has been reported to lower the extractable activity of phenylalanine-ammonia lyase (PAL) in hypocotyls of light-grown soybean seedlings, 48 hr after treatment [136, 137]. A minor increase in the activity of nitrate reductase of primary leaves of soybean plants by 3.7 μM of metribuzin has been reported also [138].

C. Physiological and Biochemical Effects

Metribuzin is a very active herbicide requiring an application of only 350-700 g/ha of the active ingredient under field conditions [28]. Symptoms of metribuzin toxicity on treated plants are expressed visually as necrosis and bleaching of leaf tissue along the margins, leaf veins, and petioles [59,139,140]. Interveinal chlorosis or necrosis are usually observed in leaves of susceptible crop cultivars treated with this herbicide [59,140]. Studies with metribuzin-sensitive tomato cultivars showed that spraying with this herbicide at blooming time resulted in significant loss of flowers of these plants [140]. It is evident that the symptoms of metribuzin toxicity are typical of those caused by the slow-acting photosynthesis-inhibiting herbicides. The visual injury symptoms of metribuzin appear late, following an initial inhibition of photosynthesis [22]. Apart from its primary action on photosynthesis, metribuzin has been reported to interfere with several other processes of plant metabolism. These physiological and biochemical effects of metribuzin on plants will be discussed in the next sections.

1.   Inhibition of Photosynthesis

a.   *Effects on Photosynthetic Electron Transport.*   The inhibitory effect of metribuzin on photosynthesis has been demonstrated on numerous studies with isolated chloroplasts [4,5,7,8,141–143], isolated plant cells [135,149], algal cells [144–148], and intact plants [150]. Early or more recent studies employing polarographic or spectrophotometric measurements of photoreductions of Photosystem II (Hill reaction) or fluorescence detections of photosynthetically active chlorophyll have demonstrated that metribuzin inhibits photosynthetic electron transport between the primary and secondary electron acceptor of Photosystem II [4,8,142,143,145,151,152]. The pI_{50} value for the inhibition of photosynthetic electron transport by metribuzin has been calculated at 6.5–6.7 [22]. Competitive binding experiments have shown that metribuzin inhibits photosynthetic electron transport at the same site as atrazine and diuron and also binds to the same target site of photosystem II complex of the chloroplasts [8,22,153,154]. Subsequent studies with trypsin-digested chloroplasts indicated that the inhibition of photosynthetic electron transport by metribuzin could be reversed by treatment with trypsin [155–157], confirming further the existence of a common site involved in the inhibition of photosynthesis by metribuzin, atrazine, and diuron [22]. In addition, the experiments with trypsin digestion revealed the proteinaceous nature of the binding component involved in the action of Photosystem II photosynthetic inhibitors. Although trypsin treatment removes the receptor site for the diuron-type inhibitors, it did not affect significantly the affinity of the receptor site to the inhibitor [152,153]. Tischer and Strotmann [153] suggested recently that apart from the protein component, the lipid phase of thylakoid membranes may contribute to the binding of the diuron-type photosynthetic inhibitors, such as metribuzin to the chloroplast. Maximum binding was achieved with 1 mole of the inhibitor per 300–500 moles of chlorophyll, suggesting that the receptor was an electron carrier or a proteinaceous membrane component entrapping the redox components (plastoquinones) of the binding site [153,154]. Extensive research on the identity of this component has shown it to be a 34 kDa protein which is commonly being referred to in the literature as the Q_B protein [158]. Binding of the diuron-type inhibitors to this protein is reversible and it is thought to prevent the proper binding of plastoquinone, bringing about the inhibition of the photosynthetic electron transport in chloroplast thylakoids [159,160]. The characterization of the 34 kDa protein as a receptor target site for the diuron-type herbicides was accomplished with the use of azido-atrazine, which is a specific photoaffinity label [151,158]. Azido-atrazine binds exclusively to the 34 kDa protein. Recent studies by Oettmeier et al. [159,160] using a new azido-triazinone photoaffinity label showed that the protein labeled by azido-triazinone was identical to the 34 kDa herbicide binding protein which is tagged by the azido-atrazine photoaffinity label. This further supports the existence of a common primary receptor for all the diuron-type herbicides that inhibit photosynthesis and answers to some extent the criticism expressed by Gressel [161] as to the involvement of the 34 KDa protein in the binding and action of these herbicides. The existence of subreceptors specific for the binding of various classes of the diuron-type inhibitors has been suggested by several investigators and theoretical models have been proposed [151,158,162]. Additional information on the binding site and mode of action of Photosystem II herbicidal inhibitors can be found in the literature [22,151,158,162].

b. *Effects on the Photosynthetic Apparatus.* The specific inhibition of light reactions in chloroplast membranes by metribuzin will disrupt the existing photosynthetic apparatus and seriously affect the intact plant, causing the typical symptoms of metribuzin injury that were discussed earlier.

Following the initial and rapid inhibition of photosynthetic electron transport, subsequent effects caused by metribuzin include changes in the chloroplast anatomy and morphology. These changes are observed after a few days or one week following plant treatment with sublethal or lethal dosages of metribuzin [22,163,164]. Although research on these effects of metribuzin has not been very extensive, a number of investigators have demonstrated their occurrence and these reports are summarized in Table 4.

Most of the morphological, anatomical, and biochemical changes in the chloroplast induced by metribuzin also have been observed following treatments of susceptible plants with other photosynthesis-inhibiting herbicides [22]. Since similar effects are also induced by light of low intensity, Fedtke [163,165] has suggested that plants treated with sublethal dosages of selected photosynthesis-inhibiting herbicides resemble "shade-adapted" or "low light-adapted" plants. The similar response of plants exposed to low light intensities or to chemicals inhibiting photosynthesis is thought to be induced by a carbohydrate stress related to a decreased rate of photosynthetic electron transport [163].

TABLE 4

Changes in Chloroplast Morphology and Biochemical Composition Induced by Metribuzin

Response	Plant species	Reference
Dilated thylakoids	*Beta vulgaris* L.	164
Decrease of starch grains	*Beta vulgaris*	164
Decrease of plastoglobuli	*Beta vulgaris*	164
Decrease of chlorophyll a/b	*Glycine max* (L.) Merr.	163
Decrease of chlorophyll (a + b)	*Glycine max*	136,137
	Pisum sativum L.	143
	Lycopersicon esculentum Mill	140
	Euglaena gracilis	146
Increase of chlorophyll (a + b)	*Bumilleriopsis filliformis*	145
Decrease of carotenoids	*Psium sativum*	143

c. *Effects on the Photosynthesis of Intact Plants*. The chemical energy accumulated in nicotinamide adenine dinucleotide phosphate (NADPH) and adenosine triphosphate (ATP) as a result of the photoelectron flow in grana of plant chloroplasts is used for the fixation of carbon dioxide in the stroma that contains the enzymes for the carbon dioxide fixation pathways. Inhibition of reactions in the chloroplast by herbicidal inhibitors causes a suppression of the carbon dioxide reduction. This raises the intercellular CO_2 concentration in the mesophyll and diminishes the influx of CO_2 into the leaves [166]. The CO_2 exchange in the leaves of herbicide-treated plants can be measured quantitatively with the use of an infrared gas analyzer [166]. Inhibitions of CO_2 assimilation in intact plants by metribuzin or other triazinone herbicides have been reported [150, 166,167]. A decrease in CO_2 assimilation is usually observed within a few hours following plant treatment with postemergence applications of the herbicide or treatment of plant roots in nutrient solution under conditions that favor transpiration [150]. The inhibition of CO_2 assimilation and of carbohydrate synthesis leads to starvation of the

plants, since energy-requiring processes continue to operate within
the treated plant [163,166]. Starvation alone, however, does not
completely explain the phytotoxicity of photosynthesis-inhibiting
herbicides. Reactive free radicals are formed in plant tissues treated
with selected herbicides, and their involvement in the phytotoxicity
of these chemicals has been demonstrated [22]. Specific studies on
the possible generation and involvement of free radicals in metribuzin
phytotoxicity have not been conducted, however. Photooxidation of
chloroplast pigments also has been suggested as a possible explanation for the observed disintegration of chloroplasts in the light and
the light-dependent phytotoxicity of photosynthetic inhibitors [22].
Inhibition of carotenoids by metribuzin was reported recently by
Singh et al. [143].

Although metribuzin is a potent inhibitor of photosynthesis, at
concentrations of 0.0625 and 0.125 ppm, it slightly increased the
photosynthetic capacity of tomato plants 21 to 42 days after treatment
[168]. The reasons for the increase of photosynthetic capacity of
tomato cultivars by metribuzin are not understood at present. It
was suggested that the role of metribuzin in the alteration of the
photosynthetic apparatus may be of economic importance and thus
needs to be investigated in the future [168].

2. Structure–Activity Correlations

Following the discovery of the herbicidal properties of the
1,2,4-triazinone compounds, extensive studies on the structure-
activity correlations of these herbicides have been conducted primarily by a Bayer AG group headed by Draber [3-7,9] or by Trebst
and his colleagues [8]. The objectives of these studies were to (a)
identify substituents required for maximum inhibitory effectiveness,
(b) relate physicochemical properties of the herbicides to the inhibitory action, and (c) identify the interactions between substituents of inhibitors and postulated receptors in the chloroplast membranes.

The results of these studies showed that compounds with alkoxy
(e.g., $-OCH_3$), alkythio (e.g., $-SCH_3$), or alkylamino (e.g.,
$-NHCH_3$) substituents at the C-3 position, an amino ($-NH_2$) group
at the N-4 position and an alkyl, cyclohexyl, or phenyl group at the
C-6 position of the heterocyclic 1,2,4-triazinone ring were particularly effective as inhibitors of the Hill reaction of chloroplast membranes [4,5,9]. In many cases, the potential of each compound for
inhibition of the Hill reaction was found to correlate well with its
phytotoxicity under field conditions [9]. Empirical equations de-
scribing the best fit of such data to a regression curve have been
calculated and are available in the literature [4-6]. From such
equations it was concluded that a good quantitative correlation between the activity of a compound on the Hill reactions and its hydrophobic properties exists and is dependent upon variations of the

substituent at the C-6 position of the heterocyclic ring [6]. Thus, substituents at the C-6 position are important for the lipophilicity of the whole molecule and influence its penetration properties. Some degree of lipophilicity has been also considered essential for a good fit of the inhibitor to its receptor site [6]. The activity of triazinone herbicides was also found to be sensitive to steric effects exerted by substituents at the C-3 position of the heterocyclic ring. In contrast, electronic factors appeared to play a minor role in the activity and binding of these herbicides [6]. The amino substituent at the N-4 position was found to be very important for the binding of the molecule to its receptor, possibly through inflicting a favorable electronic distribution on the heterocyclic system or by direct interaction with some part of the receptor [6].

Recently, Draber and Fedtke [169] introduced the concept of a binding area rather than a binding site and described a theoretical model for the binding of amino-triazinone herbicides, such as metribuzin and metamitron, to chloroplast membranes. They proposed that the binding area consists of an ensemble of few obligatory and many facultative subreceptors from which only a selection has to be occupied to ensure efficient binding of these molecules to chloroplast and inhibition of photosynthetic electron transport. According to this model, the carbonyl group at the C-5 position and the amino group at the N-4 position of the triazinone ring provide the obligatory subreceptor, which may influence the hydrogen bonding properties of the molecule to the site. The substituent at the C-6 position was described as a hydrophobic pocket, which does not impose any strict dimentional requirements for the binding of the molecule to its site. Finally, the substituent at the C-3 position of the heterocyclic ring was described as a facultative subreceptor, which imposes strict dimensional requirements (e.g., length, width) for the binding of the molecule to the chloroplast site.

3.   Other Biochemical and Physiological Effects
     Induced by Metribuzin

Apart from its strong effect on photosynthesis, metribuzin has been shown to interfere with a number of other biochemical and physiological responses of treated plants. In most cases, these effects of metribuzin have been characterized as secondary or indirect effects deriving from its primary or direct effect on photosynthesis.

Thus, sublethal concentrations of metribuzin or of other photosynthesis-inhibiting herbicides are known to affect the nitrogen metabolism of treated plants. A strong increase of soluble amino acids, soluble protein, and total nitrogen content in plants treated with photosynthetic inhibitors, including metribuzin, has been reported [22]. These effects currently are viewed as a consequence of an increased supply of reduced nitrogen observed after treatment with photosynthetic inhibitors and emphasize further the similarity of responses elicited by either light of low intensity or treatment with sublethal concentrations of photosynthetic inhibitors [163,165]. The increased in vitro nitrate reductase activity in soybean plants treated with sublethal doses of metribuzin [163] is believed to result from the increased levels of nitrate in the cells of these plants [22]. An increase in nitrite accumulation in the cells of plants treated with metribuzin or other photosynthetic inhibitors has been suggested as a secondary mode of action of these herbicides by Klepper [170]. This accumulation of nitrite is a consequence of the primary effect of photosynthetic inhibitors on photosynthetic electron transport, which causes a shortage of reducing equivalents needed for the reduction of ferredoxin which in turn mediates the reduction of nitrate and sulfate in higher plants [170-172].

A significant inhibitory effect of metribuzin on the leaf protein content of soybeans has been reported [143]; however, protein synthesis measured by the incorporation of $[^{14}C]$leucine into isolated mesophyll cells of velvetleaf was not affected by concentrations of metribuzin as high as 100 μM [149].

Recently, Vavrina and Phatak [173] described the effects of metribuzin on dark respiration of tomato seedlings. They suggested that a metribuzin-induced increase of dark respiration of tomato diverts photosynthates that could otherwise be used for growth, causing a growth retardation effect on tomato. Other investigators [174, 175], however, reported that metribuzin inhibited the respiration of selected plants. An effect of metribuzin on the patterns of cumulative water use of tomato has been reported recently and was found to be correlated with the differential tolerance of tomato cultivars to this herbicide [176]. Lower rates of metribuzin stimulated while high rates of metribuzin reduced the daily water use in the susceptible cultivar Heinz 1706. The observed promotion of growth of this cultivar following treatment with sublethal doses of metribuzin, coupled with the slight stimulation of photosynthesis of tomato plants described in earlier sections, supports the view that stimulatory effects of metribuzin on tomato growth may be of economic value [168,176].

Significant increases in total fatty acid concentration of soybean oil induced by metribuzin on several soybean cultivars have been reported [177]; however, the effects of metribuzin on soybean oil quality were insignificant [177,178]. The insensitivity of lipid synthesis to this herbicide also has been demonstrated in metabolic studies examining the effects of metribuzin on the incorporation of $[^{14}C]$acetate into lipids of isolated mesophyll cells of velvetleaf [149]. Recently, Hoagland and Duke [136,137] reported a decrease in the anthocyanin content of hypocotyls of light-grown soybean seedlings by 0.05 μM of metribuzin, suggesting a potential secondary effect of this herbicide on flavanoid biosynthesis.

Studies on the potential effects of metribuzin on nucleic acid synthesis and metabolism in higher plants are limited. Interference of s-triazine herbicides with nucleic acid synthesis and metabolism of treated plants has been documented [22], and similar effects might be expected by the structurally related triazinone herbicides. A strong effect of metribuzin on RNA synthesis of isolated leaf cells of velvetleaf has been reported recently [149]. Although this effect of metribuzin on RNA synthesis could be indirect, resulting from its strong interference with photosynthesis, further research will be needed to exclude the possibility of a direct effect of aminotriazinone herbicides on nucleic acid metabolism and synthesis of higher plants. The structural similarity of aminotriazinone herbicides and of their degradation products in plants to precursor bases of nucleic acids would seem to support the notion of a direct effect of these chemicals on plant nucleic acids. In particular, the great similarity of the deaminated diketo (DADK) metabolite of metribuzin to thymine (2-methyl-7,4-dioxopyrimidine) indicates that DADK may act as an antimetabolite disturbing the nucleic acid synthesis of susceptible plants treated with metribuzin and needs to be examined further. Studies based on the structural similarity of s-triazine herbicides or their metabolites to precursor bases of nucleic acids have demonstrated the potential interference of these chemicals with plant nucleic acid metabolism [179-181].

Sublethal doses of metribuzin did not provoke any increases in glutathione biosynthesis as a "stress response" in corn or soybeans treated with this herbicide [182].

Finally, studies with nonchlorophyllous, dark-grown suspension cultures of soybean [61] or tomato [183] showed that the phytotoxicity of metribuzin is not restricted to photosynthesis. Growth inhibitions of these nonphotosynthetic tissue cultures by metribuzin were evident in these studies. The magnitude of metribuzin concentrations needed for eliciting these effects, however, was much greater than that of concentrations that were lethal to plant seedlings [61,183]. Specific sites or processes that may be affected by metribuzin in nonphotosynthetic heterotrophic suspension cultures were not examined.

D. Selectivity

Although metribuzin is an effective herbicide, it has a relatively narrow margin of safety in several crops that restricts the range of its uses. Nevertheless, inter- and intraspecific differential responses of many plants to metribuzin have been reported by several investigators. Solanaceous crops or weeds including tomato, potato, black nightshade, and American nightshade are tolerant to metribuzin [12-15,73]. In contrast, other solanaceous crops or weeds, such as

eggplant, tobacco, peppers, and jimsonweed, are susceptible to
metribuzin [13,129,184]. Legume crops, such as soybeans, alfalfa,
and lentils, as well as grass crops, such as wheat, barley, and
sugarcane, are also tolerant to this herbicide [11,16–19,72]. Intraspecific differential tolerance to metribuzin has been demonstrated
with cultivars of soybean [39,51,52,61–65,185–194], tomato, [12,38,
54,128,140,142,183,195–199], potato [15,68,70,139,197,200–206],
sweetpotato [207,208], winter wheat [71,209–211], barley [212,213],
bermudagrass [150], asparagus [197], carrots [197], cowpeas [197,
214], and lentils [197].

Decreased effectiveness of a specific herbicide in a particular
plant species may result from a number of different causes including:
(a) reduced or lack of herbicide uptake; (b) restriction of translocation or compartmentalization of the herbicide; (c) inactivation of
the herbicide by metabolic detoxification; (d) modification of the
properties of the herbicidal target site; (e) overproduction of target
enzymes; and (f) overproduction of substrates able to reverse the
herbicide-induced inhibition of plant growth [53,215]. Many of these
factors have been shown to be important for the observed inter- or
intraspecific differential plant tolerance to metribuzin. Selected examples illustrating the involvement of these factors in the interspecific differential plant tolerance to metribuzin are shown in Table
5. A number of bioassay methods for screening and selecting plants
for intraspecific differential tolerance to metribuzin have been de-
vised and used with varying degrees of success. Such methods include the sinking leaf disk test [202], nutrient culture methods [187,
196,205], or appropriately manipulated plant cell cultures [61,183,
199,216].

Differential retention or uptake of metribuzin by tolerant and
susceptible species has been reported to be of minor significance in
inter- or intraspecific differential responses to metribuzin [54,58,
59,62–74,128,129,130,133]. Differential translocation appears to be
of greater importance. Extensive vascular localization or compartmentalization of absorbed metribuzin seems to play a key role in the
observed differential tolerance of selected species or cultivars to
metribuzin. Thus, tolerant soybean [39,58,59,65], potato [70], or
tomato [38,140] cultivars or tolerant weed species, such as entire-
leaf morningglory [74] have been shown to restrict the distribution
of absorbed metribuzin to their roots, stems, or major veins of their
lower leaves. In contrast, susceptible species or cultivars accumulate
greater amounts of unmetabolized metribuzin to interveinal areas of
their leaves where the target site of this herbicide is localized [38,
39,58,59,65,70,74,140].

In most studies, however, differential metabolism or differential
rate of metabolism have been described as the main mechanism responsible for the inter- or intraspecific differential plant tolerance

TABLE 5

Interspecific Differential Tolerance to Metribuzin

Plant species		Mechanism of	
Tolerant	Susceptible	differential response	Reference
Soybean	Hemp sesbania	Differential absorption	58
		Differential translocation	
		Differential metabolism	
Tomato	Jimsonweed	Differential translocation	129
		Differential metabolism	
Winter wheat	Downy brome	Differential retention	133
		Differential absorption	
		Differential metabolism	
Lentils	Tumble mustard	Differential retention	72
		Differential metabolism	
American nightshade	Barnyardgrass	Differential metabolism	73
Entireleaf morningglory	Pitted morningglory	Differential translocation	74
		Differential metabolism	

to metribuzin [38,39,54,59,61-65,70,71, and examples of Table 5].
Differential responses of plant species or cultivars to this herbicide
result from differential capacity for the metabolic detoxification of
metribuzin by the reactions and pathways discussed in Section II.A.
of this chapter. Enzymatic detoxification of metribuzin by conjugation to homoglutathione appears to be the major mechanism conferring
tolerance to soybeans against this herbicide [39]. This mechanism
appears to be independent of light and photosynthesis since it remains present in dark-grown heterotrophic suspension cultures of
soybean [61]. The sensitivity of susceptible soybean cultivars to
metribuzin was attributed not to their inability to metabolize this
herbicide effectively, but to their limited ability to metabolize a substrate that accumulates in their cells and inhibits the enzyme responsible for the detoxification of metribuzin [61]. On the other
hand, the detoxification mechanism contributing to the tolerance of
tomato cell cultures to metribuzin appeared to be dependent on photosynthesis since heterotrophic suspension cultures from tolerant or
susceptible tomato cultivars were uniformly sensitive to this herbicide
[183]. These results appear to support the findings of Frear et al.
[38,39], who showed that the main detoxification pathway of metribuzin in tomato is different from that in soybean and includes conjugation of this herbicide to glucose. It is evident, then, that retention
of the mechanism of metribuzin detoxification in tomato cell cultures
would be dependent on light and photosynthesis, which are required
for the production of carbohydrates like glucose. Early development
of enzymatic activity mediating the conjugation of metribuzin to glucose was found to be limited in young leaves of susceptible tomato
cultivars [38]. Differential metabolism by conjugation has been proposed as the reason for the intraspecific differential tolerance of
winter wheat cultivars to metribuzin [71].

The intraspecific differential tolerance of soybean, tomato, and
potato to metribuzin is influenced by genotype. Genetic analyses
have shown that the sensitivity of soybean [187,188,192], tomato
[198], and potato [205] to this herbicide is controlled by a single
recessive gene. The symbol *hm* has been proposed for the gene
controlling metribuzin sensitive in soybean [188] whereas the symbol
me has been proposed for the gene controlling metribuzin sensitivity
in potato [205]. The potential use of these genes (*hm* and *me*) as
markers in soybean and potato genetics and for cultivar improvement
has been suggested [188,198,205]. In fact, a selection from the
metribuzin-susceptible cultivar Tracy released as "Tracy M" has been
developed through plant breeding efforts of Hartwig et al. [217].
Tracy M supposedly is tolerant to 10 times the rate of metribuzin
required to kill "Tracy." Differential rate of metabolism was identified as the mechanism of metribuzin selectivity in Tracy M and Tracy
[39]; however, successes in plant breeding programs aimed at the

selection of crop cultivar lines tolerant to metribuzin or other herbicides are not as simple as they seem. A recent report by Kilen and
Barrentine [218] demonstrated a close linkage between the gene *hm*
that carries the sensitivity of Tracy soybeans to metribuzin and the
genes Rps_1 that carry the resistance of Tracy soybeans to phytoph-
thora rot. These results indicate that linkages between desirable
and undesirable genes may hinder the progress of breeding programs
by increasing tremendously the population required to recover prefer-
able gene combinations [218].

From the review of the available literature, it is evident that
the selectivity of metribuzin or of other triazinone herbicides is not
related to their mode of action. Studies with isolated chloroplasts
[142], leaf disk [209], or whole seedlings [150] from metribuzin-
tolerant or susceptible plants showed that the Hill reaction or the
fixation of CO_2 were inhibited equally by this herbicide regardless
of their tolerance or susceptibility. The inhibition of net photosynthesis by metribuzin or metamitron in tolerant plants was transitory,
and recovery of the photosynthetic rate of tolerant plants was observed within a relatively short time [150,167]. These results further
emphasize the significance of differential translocation and metabolism
in the selectivity of this herbicide.

Although the discovery of higher plant mutants acquiring resistance to metribuzin at the chloroplast level has not yet been reported, the existence of a high degree of cross-resistance to metribuzin in weed biotypes resistant to *s*-triazine herbicides has been demonstrated [219]. The isolation of metribuzin-resistant mutants from
Chlamydomonas reinhardii cells mutagenized with 5-fluorodeoxyuridine
and ethylmethanesulfonate has been reported recently [220,221].
Use of ^{14}C-labeled metribuzin and of photoaffinity labels showed that
the basis of resistance of the *Chlamydomonas* mutants to metribuzin
was associated with a modification of the metribuzin binding site in
the photosystem II complex of thylakoid membranes [220]. Thus, the
mechanism of this resistance of algae to metribuzin is similar to that
described for the development of resistance in weed biotypes resistant to *s*-triazine herbicides [219]. These results indicate that
although differential translocation and metabolic detoxification appear
to be the main mechanism for inter- or intraspecific differential plant
responses to metribuzin, the development of plant biotypes resistant
to this herbicide resulting from modifications at its target site would
be possible in the future if the required conditions for selective
pressure are met. If this happens, the manipulation of crop tolerance to metribuzin by means of genetic engineering procedures may
be feasible in the future.

In addition, an understanding of the antagonistic interactions
of metribuzin with other agrochemicals in higher plants may lead to
the development of safening or protective chemicals that could

increase the narrow margin of selectivity of this herbicide in selected crops. These subjects will be discussed in more detail later.

E. Factors Affecting Metribuzin Phototoxicity

The formation of the N-glucoside conjugate of metribuzin is very important in the tolerance of tomato to metribuzin. Thus such conditions as cloudy weather or low light intensity unfavorable to the synthesis and accumulation of glucose, making it unavailable for metribuzin detoxification, can result in tomato injury [13,81-85].

Crop varieties may differ in their sensitivity to metribuzin. Examples of sensitive tomato varieties are Ontario 771, Heinz 1706, and Trimison [38]. Difference in intraspecific tolerance to metribuzin has also been observed in potato, sweetpotato, winter wheat, barley, bermudagrass, asparagus, carrots, cowpeas, and soybeans. These have been previously discussed in the Selectivity section. It is interesting to note that tetraploid soybeans appear relatively more tolerant to metribuzin than diploid soybeans [222].

Heavy rainfall or irrigation may result in leaching of metribuzin in sandy soils with low organic matter [96]. High soil clay content may retard leaching [97], whereas, high soil pH may increase it [122]. The leaching of metribuzin can be great enough to affect weed control and, in some instances, increase injury to sensitive cultivars.

Soil pH has a marked effect on metribuzin availability [24]. Fields highly variable in pH may show weed control failure in areas of low pH to crop injury in areas of high pH.

The carryover of atrazine residues from previous years can result in a synergistic interaction with metribuzin, increasing metribuzin uptake and consequently increasing crop injury [223-225].

F. Interactions and Protectants

Following the introduction of metribuzin as a herbicide for soybean, there was concern that metribuzin might increase soybean injury from atrazine residue carryover, especially in alkaline areas. Synergistic interactions were reported and are discussed in the previous section. They appear to be dependent on the extent of atrazine carryover.

Metribuzin is generally used in combination with other herbicides to broaden the spectrum of weed control. Of the herbicides with which it may be in combination, trifluralin appears to interact antagonistically with metribuzin. Trifluralin reduced phytotoxicity of metribuzin to soybean [226,227], fababean [228], and tomato [229]. Metribuzin residues on the other hand antagonized glyphosate activity [230].

Tridiphane inhibits glutathione-s-transferases and is under development for use with the s-triazines to increase control of several annual grasses in corn by inhibiting the s-triazine metabolism in these grasses. Corn remains tolerant to the s-triazine due to the operation of alternative metabolism routes. Soybean tolerance to metribuzin, however, is dependent on the activity of glutathione-s-transferase and the expected increase in metribuzin phytotoxicity to soybean in the presence of tridiphane has been observed [92]. Since tomato tolerance to metribuzin is based on metabolism via glycoside formation, the presence of tridiphane should be, and has been, reported to be without effect [92].

Although trifluralin antagonizes metribuzin activity and the organophosphate insecticides antagonize trifluralin activity [231], the organophosphate insecticides increase metribuzin injury. Synergistic interactions of metribuzin and aldicarb [231,232], carbofuran [233], disulfoton [60,234], ethoprop [235], fensulfothion [233], phenamiphos [232], phorate [60,232], or terbufos [232] have been reported in soybean. Metribuzin and azinphosmethyl [236], carbaryl [67], endosulfan [236], and malathion [67] interacted on tomato. These insecticides may exert their synergistic action by inhibiting the N-glucoside conjugating pathway of metribuzin metabolism [88].

Soybean injury from metribuzin is usually related to the seeding of sensitive varieties, high pH soils, high rainfall after metribuzin application moving the herbicide to the soybean root zone, or a combination of these factors. Broadening the margin of selectivity, or protection of crops from metribuzin, has been approached genetically [188], chemically [237], and through formulation [238,239]. In the latter approach, metribuzin lignin complexes have been evaluated for reduced leaching in soil [240]. The plant growth regulator, daminozide [butanedioic acid mono(2,2-dimethyl hydrazide)], appears to protect potatoes from metribuzin injury by increasing the concentration of soluble sugars, making them more available for conjugation with metribuzin [237].

IV. CONCLUSIONS

Metribuzin is an effective herbicide for weed control in soybean, potato, and tomatoes. Since its selectivity is not very narrow, it is commonly applied in combination with other herbicides to broaden the spectrum of weeds controlled. This use of herbicide combinations appears to have forestalled resistant weed problems with metribuzin. Although the action of metribuzin is very similar to that of the s-triazine, atrazine, crop tolerance and fate in soils are markedly different. Rapid metabolism of metribuzin provides the basis for tolerance in the principal crop in which it is used. Furthermore, the metabolism in soybean differs from that of tomato.

In the soil, leaching of metribuzin is greater than that observed for atrazine. Similarly, the availability of metribuzin to plants in high pH soils is greater than for atrazine. Finally, the dissipation of metribuzin is faster than for atrazine. Thus, carryover problems with metribuzin residues are not as great as those associated with the more persistent s-triazines.

Prospects for altering or broadening the selectivity of metribuzin are excellent through both genetic manipulation and the use of chemical protectants or safeners.

REFERENCES

1. A Dornow, H. Menzel, and P. Marx, *Chem. Ber.*, 97, 2173 (1964).
2. K. Westphal, W. Meiser, L. Eue, and H. Hack (to Bayer AG), German Pat. 1,519,180 (1966).
3. W. Draber, K. Dickore, K. H. Buchel, A. Trebst, and E. Pistorius, *Naturwissenchaften.*, 55, 446 (1968).
4. W. Draber, K. H. Buchel, K. Dickore, A. Trebst, and E. Pistorius, *Prog. Photosyn. Res.*, 3, 1789 (1969).
5. W. Draber, K. H. Buchel, and K. Dickore, in *Proc. 2nd IUPAC Congress of Pesticide Chemistry Tel Aviv, 1971* (A. S. Tahori, ed.), Butterworths, London, 1972, p. 153.
6. W. Draber, K. H. Buchel, and H. Timmler, in *Mechanisms of Pesticide Action* (G. K. Kohn, ed.), ACS Symposium Series 2 Am. Chem. Soc., Washington, D.C., 1974, p. 100.
7. W. Draber, *Z. Naturforsch.*, 34c, 973 (1979).
8. A. Trebst and H. Wietoska, *Z. Naturforsch.*, 30c, 499 (1975).
9. R. R. Schmidt, W. Draber, L. Eue, and H. Timmler, *Pestic. Sci.*, 6, 239 (1975).
10. L. Eue, *Pflanzenschutz Nachr.*, 25, 175 (1972).
11. L. Eue and L. Lembrick, *Pest. Articles & New Summaries*, 19, 254 (1973).
12. J. Fortino and W. E. Splittstoesser, *Weed Sci.*, 22, 615 (1974).
13. J. F. da Silva and G. F. Warren, *Weed Sci.*, 24, 612 (1976).
14. L. Eue and H. Tietz, *Pflanzenschutz Nachr.*, 26, 208 (1973).
15. A. D. Cohick, *Pflanzenschutz Nachr.*, 26, 23 (1973).
16. L. R. Hawf and T. B. Waggoner, *Pflanzenschutz Nachr.*, 26, 35 (1973).
17. H. D. Coble and J. W. Schrader, *Weed Sci.*, 21, 308 (1973).
18. D. W. Young and T. B. Waggoner, *Pflanzenschutz Nachr.*, 26, 52 (1973).
19. D. M. Miller, *Crop Protection Chemicals Reference*, Chemical and Pharmaceutical Publishing Co., New York, 1st edition, 1985.
20. H. Lembrich, *Pflanzenschutz Nachr.*, 31, 197 (1978).

21. R. R. Schmidt and C. Fedtke, *Pestic. Sci.*, *8*, 611 (1977).

22. C. Fedtke, *Biochemistry and Physiology of Herbicide Action*, Springer Verlag, Berlin, 1982.

23. W. Kolbe, *Pflanzenschutz Nachr.*, *28*, 257 (1975).

24. J. S. Ladlie, W. F. Meggitt, and D. Penner, *Weed Sci.*, *24*, 477 (1976).

25. J. Weber, *Weed Sci.*, *28*, 467 (1980).

26. P. W. Albro, C. E. Parker, E. O. Abusteit, T. C. Mester, J. R. Hass, Y. S. Sheldon, and F. T. Corbin, *J. Agric. Food Chem.*, *32*, 212 (1984).

27. C. S. Hartley and L. J. Graham-Bryce, *Physical Principles of Pesticide Behaviour: The Dynamics of Applied Pesticides in the Local Environment in Relation to Biological Response*, Academic Press, London, 1980.

28. Weed Science Society of America, *Herbicide Handbook*, 5th ed., 1983, p. 317.

29. E. Loser and G. Kimmerle, *Pflanzenschutz Nachr.*, *25*, 186 (1972).

30. F. Heimbach, *Pflanzenschutz Nachr.*, *36*, 177 (1983).

31. G. Jager, in *Chemistry of Pesticides* (K. H. Buchel, ed.), Wiley (Interscience), New York, 1983.

32. F. G. von Stryck, *J. Chromatogr.*, *56*, 345 (1971).

33. G. R. B. Webster, S. R. MacDonald, and L. P. Sarna, *J. Agric. Food Chem.*, *23*, 74 (1975).

34. J. S. Thornton and C. W. Stanley, *J. Agric. Food Chem.*, *25*, 380 (1977).

35. H. J. Jarczyk, *Pflanzenschutz Nachr.*, *31*, 84 (1978).

36. H. J. Jarczyk, *Pflanzenschutz Nachr.*, *36*, 63 (1983).

37. D. F. Brown, L. M. McDonough, D. K. McCool, and R. I. Papendick, *J. Agric. Food Chem.*, *32*, 195 (1984).

38. D. S. Frear, E. R. Mansager, H. R. Swanson, and F. S. Tanaka, *Pestic. Biochem. Physiol.*, *19*, 270 (1983).

39. D. S. Frear, H. R. Swanson, and E. R. Mansager, *Pestic. Biochem. Physiol.*, *23*, 56 (1985).

40. G. R. B. Webster and G. J. Reimer, *Pestic. Sci.*, *7*, 292 (1976).

41. B. E. Pape and M. J. Zabik, *J. Agric. Food Chem.*, *20*, 72 (1972).

42. P. Bartl and F. Korte, *Chemosphere*, *4*, 169 (1975).

43. P. Bartl and F. Korte, *Chemosphere*, *4*, 173 (1975).

44. P. Bartl, H. Parlar, and F. Korte, *Z. Naturforsch.*, *31b*, 1122 (1976).

45. H. Parlar and F. Korte, *Chemosphere*, *8*, 797 (1979).

46. Y. Nakayama, Y. Sanemitsu, H. Yoshioka, and A. Nishinaga, *Tetrahedron Lett.*, *23*, 2499 (1982).

47. M. S. Bleeke and J. E. Casida, *J. Agric. Food Chem.*, *32*, 749 (1984).

48. M. S. Bleeke, M. T. Smith, and J. E. Casida, *Pestic. Biochem. Physiol.*, *23*, 123 (1985).

49. C. Fedtke and R. R. Schmidt, *Z. Naturforsch.*, *34c*, 948 (1979).

50. C. Fedtke, *Naturwissenschaften.*, *70*, 199 (1983).

51. C. Fedtke and R. R. Schmidt, in *IUPAC Pesticide Chemistry: Human Welfare and the Environment* (J. Miyamoto, ed.), Pergamon Press, Oxford, 1983, p. 177.

52. C. Fedtke, *Pestic. Sci.*, *17*, 65-66 (1986).

53. K. K. Hatzios and D. Penner, *Metabolism of Herbicides in Higher Plants*, Burgess Publishing Co., Minneapolis, 1982.

54. C. R. Stephenson, J. E. McLeod, and S. C. Phatak, *Weed Sci.*, *24*, 161 (1976).

55. G. Engelhardt, W. Ziegler, P. R. Wallnofer, H. J. Jarczyk, and L. Oehlmann, *J. Agric. Food Chem.*, *30*, 278 (1982).

56. R. W. Schumacher, L. Thompson, Jr., and C. E. Rieck, *Abstr. Weed Sci. Soc. Am.*, *13*, 60 (1973).

57. R. W. Schumacher, L. Thompson, Jr., and C. E. Rieck, *Abstr. Weed Sci. Soc. Am.*, *14*, 123 (1974).

58. T. G. Hargroder and R. L. Rogers, *Weed Sci.*, *22*, 238 (1974).

59. A. E. Smith and R. E. Wilkinson, *Physiol. Plant.*, *32*, 253 (1974).

60. R. M. Hayes, W. W. Witt, K. V. Yeargan, and H. G. Raney, *Proc. South. Weed Sci. Soc.*, *29*, 95 (1976).

61. T. H. Oswald, A. E. Smith, and D. V. Phillips, *Pestic. Biochem. Physiol.*, *8*, 73 (1978).

62. B. L. Mangeot, F. E. Slife, and C. E. Rieck, *Weed Sci.*, *22*, 267 (1979).

63. E. F. Eastin, *Proc. South. Weed Sci. Soc.*, *34*, 263 (1981).

64. M. W. Bugg, R. K. Mann, and W. W. Witt, *Proc. South. Weed Sci. Soc.*, *34*, 244 (1981).

65. L. N. Falb and A. E. Smith, *J. Agric. Food Chem.*, *32*, 1425 (1984).

66. H. W. Hilton, N. S. Nomura, W. L. Yauger, and S. S. Kameda, *J. Agric. Food Chem.*, *22*, 578 (1974).

67. G. R. Stephenson, S. C. Phatak, R. I. Makowski, and W. J. Bouw, *Can. J. Plant Sci.*, *60*, 167 (1980).

68. R. H. Callihan, G. F. Stallknecht, R. B. Dwelle, and M. Blicharczyck, *Am. Potato J.*, *53*, 253 (1976).

69. D. Prestel, I. Weisgerber, W. Klein, and F. Korte, *Chemosphere*, *5*, 137 (1976).

70. S. W. Gawronski, L. C. Haderlie, and R. H. Callihan, *Abstr. Weed Sci. Soc. Am.*, *23*, 80 (1983).

71. R. L. Ratliff, T. F. Peeper, T. G. Wheless, E. Basler, and H. Nguyen, *Proc. South. Weed Sci. Soc.*, *37*, 365 (1984).

72. E. E. Hassanein, D. R. Gealy, and L. A. Morrow, *Abstr. Weed Sci. Soc. Am.*, *24*, 78 (1984).

73. M. A. Maun and W. J. McLeod, *Can. J. Plant Sci.*, *58*, 485 (1978).

74. T. L. Kelley and L. R. Oliver, *Proc. South. Weed Sci. Soc.*, *38*, 481 (1985).

75. C. T. Bedford, M. J. Crawford, and D. H. Hutson, *Chemosphere*, *4*, 311 (1975).

76. M. J. Crawford, D. H. Hutson, and G. Stoydin, *Xenobiotica*, *10*, 169 (1980).

77. D. H. Hutson, in *Sulfur in Pesticide Metabolism and Action* (J. D. Rosen, P. S. Magee, and J. E. Casida, eds.), ACS Symposium Series 158 Am. Chem. Soc., Washington, D.C., 1981, p. 53.

78. J. E. Casida, E. C. Kimmel, H. Ohkawa, and R. Ohkawa, *Pestic. Biochem. Physiol.*, *5*, 1 (1975).

79. J. P. Hubbell and J. E. Casida, *J. Agric. Food Chem.*, *25*, 404 (1977).

80. Y. S. Chen, I. Schupan, and J. E. Casida, *J. Agric. Food Chem.*, *27*, 709 (1979).

81. S. C. Phatak and G. R. Stephenson, *Can. J. Plant Sci.*, *53*, 843 (1973).

82. S. C. Phatak, *Abstr. Weed Sci. Soc. Am.*, *14*, 121 (1974).

83. G. H. Friesen and A. S. Hamill, *Can. J. Plant Sci.*, *58*, 1115 (1978).

84. C. E. Beste, *Proc. Northeast. Weed Sci. Soc.*, *30*, 166 (1979).

85. M. K. Pritchard and G. F. Warren, *Weed Sci.*, *28*, 186 (1980).

86. C. R. Swanson, R. H. Hodgson, R. E. Kadunce, and H. R. Swanson, *Weeds*, *14*, 323 (1966).

87. G. R. Stephenson, D. R. Diley, and S. K. Ries, *Weed Sci.*, *19*, 406 (1971).

88. D. W. Britton, F. T. Corbin, D. P. Schmitt, J. R. Breadley, Jr., J. W. van Duyn, and H. D. Coble, *Proc. South. Weed Sci. Soc.*, *35*, 367 (1982).

89. D. S. Frear, in *Bound and Conjugated Pesticide Residues* (D. D. Kaufman, G. G. Still, G. D. Paulson, and S. K. Bandal, eds.), ACS Symposium Series 29, Am. Chem. Soc., Washington, D. C., 1976, p. 35.

90. J. Iwan, in *Bound and Conjugated Pesticide Residues* (D. D. Kaufman, G. G. Still, G. D. Paulson, and S. K. Bandal, eds.), ACS Symposium Series 29, Am. Chem. Soc., Washington, D.C., 1976, p. 132.

91. D. S. Frear, H. R. Swanson, and E. R. Mansager, *Pestic. Biochem. Physiol.*, *20*, 299 (1983).

92. S. Gaul, G. Ezra, G. R. Stephenson, and K. R. Solomon, *Abstr. Weed Sci. Soc. Am.*, *25*, 75 (1985).

93. J. W. Hamaker and C. A. I. Goring, in *Bound and Conjugated Pesticide Residues* (D. D. Kaufman, G. G. Still, G. D. Paulson,

and S. K. Bandal, eds.), ACS Symposium Series 29, Am. Chem. Soc., Washington, D.C., 1976, p. 219.

94. R. R. Schmidt, *Proc. Eur. Weed Res. Symp. Herbicides–Soil*, 1973, p. 24.

95. R. R. Schmidt, *Int. Plant Prot. Congr.*, *8*, 640 (1975).

96. M. S. Sharom and G. R. Stephenson, *Weed Sci.*, *24*, 153 (1976).

97. K. E. Savage, *Weed Sci.*, *24*, 525 (1976).

98. J. R. Holowid, L. S. Jeffery, and T. C. McCutchen, *Proc. South. Weed Sci. Soc.*, *30*, 51 (1977).

99. K. S. Lafleur, *Soil Sci.*, *127*, 51 (1979).

100. W. Kerpen and G. Schleser, in *Agrochemicals in Soils* (A. Banin and U. Kafkari, eds.), Pergamon Press, Oxford, 1980, p. 141.

101. D. C. Bouchard, T. L. Lavy, and D. B. Marx, *Weed Sci.*, *30*, 629 (1982).

102. I. D. Black, *Aust. Weeds*, *3*, 74 (1984).

103. W. G. Richardson and J. D. Banting, *Weed Res.*, *17*, 203 (1977).

104. P. J. Ballerstedt and P. A. Banks, *Proc. South. Weed Sci. Soc.*, *35*, 331 (1982).

105. J. B. Weber, *Am. Mineral.*, *51*, 1657 (1966).

106. J. B. Weber, S. B. Weed, and T. M. Ward, *Weed Sci.*, *17*, 417 (1969).

107. P. J. Ballerstedt and P. A. Banks, *Proc. South Weed Sci. Soc.*, *36*, 392 (1983).

108. D. G. Crosby, in *Herbicides: Physiology, Biochemistry, Ecology* (L. J. Audus, ed.), Academic Press, London, 1976, p. 65.

109. D. L. Hyzak and R. L. Zimdahl, *Weed Sci.*, *22*, 75 (1974).

110. R. L. Zimdahl, V. H. Freed, M. L. Montgomery, and W. R. Furtick, *Weed Res.*, *10*, 18 (1970).

111. K. E. Savage, *Weed Sci.*, *25*, 55 (1977).

112. J. A. Ivany, J. M. Sadler, and E. R. Kimball, *Can. J. Plant Sci.*, *63*, 481 (1983).

113. R. J. Hance and R. A. Maynes, *Weed Res.*, *21*, 87 (1981).

114. G. F. Kempson-Jones and R. J. Hance, *Pestic. Sci.*, *10*, 449 (1979).

115. G. R. B. Webster, L. P. Sarna, and S. R. MacDonald, *Bull. Environ. Contam. Toxicol.*, *20*, 401 (1978).

116. G. R. B. Webster and G. J. Reimer, *Weed Res.*, *16*, 191 (1976).

117. G. Engelhardt and P. R. Wallnofer, *Chemosphere*, *7*, 463 (1978).

118. R. W. Schumacher, *Metabolism of Metribuzin in Soybean and Soils*, Ph.D. dissertation, University of Kentucky, Lexington, 176 pp.

119. D. D. Kaufman and P. C. Kearney, *Residue Rev.*, *32*, 235 (1970).

120. H. J. Jarczyk, *Proc. Br. Crop Prot. Conf.–Weeds*, *6*, 619 (1976).

121. R. D. Hawck and H. F. Stephenson, *J. Agric. Food Chem.*, *12*, 147 (1964).

122. J. S. Ladlie, W. F. Meggitt, and D. Penner, *Weed Sci.*, *24*, 508 (1976).

123. J. S. Ladlie, W. F. Meggitt, and D. Penner, *Weed Sci.*, *24*, 505 (1976).

124. M. M. Lay and R. D. Ilnicki, *Weed Res.*, *14*, 189 (1974).

125. A. Walker, *Weed Res.*, *18*, 305 (1978).

126. A. E. Smith and B. J. Hayden, *Bull. Environ. Contam. Toxicol.*, *29*, 243 (1982).

127. D. R. Shaw and T. F. Peeper, *Abstr. Weed Sci. Soc. Am.*, *25*, 87 (1985).

128. J. Fortino, Jr. and W. E. Splittstoesser, *Weed Sci.*, *22*, 460 (1974).

129. J. R. Frank and C. E. Beste, *Weed Sci.*, *31*, 445 (1983).

130. H. W. Hilton, N. S. Nomura, S. S. Kameda, and W. L. Yauger, *Arch. Environ. Contam. Toxicol.*, *4*, 385 (1976).

131. M. J. Bukovac, in *Herbicides: Physiology, Biochemistry, Ecology* (L. J. Audus, ed.), Academic Press, London, Vol. 1, 1976, p. 335.

132. K. I. N. Jensen, in *Herbicide Resistance in Plants* (H. M. LeBaron and J. Gressel, eds.), Wiley (Interscience), New York, 1982, p. 133.

133. D. L. Devlin, L. A. Morrow, and D. R. Gealy, *Abstr. Weed Sci. Soc. Am.*, *25*, 6 (1985).

134. C. A. Peterson and L. V. Edgington, *Pestic. Sci.*, *7*, 483 (1976).

135. O. T. deVilliers, M. J. Van der Merwe, and H. M. Koch, *S. Afr. J. Sci.*, *75*, 315 (1979).

136. R. E. Hoagland and S. O. Duke, *Weed Sci.*, *29*, 433 (1981).

137. R. E. Hoagland and S. O. Duke, *Weed Sci.*, *31*, 845 (1983).

138. C. Fedtke, *Weed Sci.*, *27*, 192 (1979).

139. G. T. Graf and A. G. Ogg, *Weed Sci.*, *24*, 137 (1976).

140. S. W. Gawronski, *Weed Sci.*, *31*, 525 (1983).

141. K. H. Buchel, *Pestic. Sci.*, *3*, 89 (1972).

142. V. Souza Machado and C. Ditto, *Sci. Hortic.*, *17*, 9 (1982).

143. K. N. Singh, J. Prakash, A. K. Agrawal, and G. S. Singhal, *Z. Naturforsch.*, *39c*, 464 (1984).

144. J. H. Avrik, D. L. Hyzak, and R. L. Zimdhal, *Weed Sci.*, *21*, 173 (1973).

145. P. Boger and U. Schule, *Weed Res.*, *16*, 149 (1976).

146. J. T. Richardson, R. E. Frans, and R. E. Talbert, *Weed Sci.*, *27*, 619 (1979).

147. C. Fedtke, in *Biochemical Responses Induced by Herbicides* (D. E. Moreland, J. B. St. John, and D. E. Hess, eds.), ACS Symposium Series No. 181, Am. Chem. Soc., Washington, D.C., 1982, p. 231.

148. J. H. Eley, J. F. McConnel, and R. H. Catlett, *Environ. Exp. Bot.*, *23*, 365 (1983).

149. K. K. Hatzios, *Z. Naturforsch.*, *37c*, 276 (1982).

150. Y.-S. Yang and S. W. Bingham, *Weed Sci.*, *32*, 247 (1984).

151. K. Pfister and C. J. Arntzen, *Z. Naturforsch.*, *34c*, 996 (1979).

152. A. V. Volvodin, V. I. Kondatenko, and A. M. Korolev, *Byull. Vsesoyunzyi Referativnyi Zhurnal*, *10*, 55.950 (1982).

153. W. Tischer and H. Strotmann, *Biochim. Biophys. Acta*, *460*, 113 (1977).

154. W. Tischer and H. Strotmann, *Z. Naturforsch.*, *34c*, 992 (1979).

155. P. Gober and K-J. Kunert, *Z. Naturforsch.*, *34c*, 1015 (1979).

156. K. E. Pallett and A. D. Dodge, *Pestic. Sci.*, *10*, 216 (1979).

157. K. E. Pallett and A. D. Dodge, *J. Exp. Bot.*, *31*, 1051 (1980).

158. C. J. Arntzen, K. Pfister, and K. E. Steinback, in *Herbicide Resistance in Plants* (H. M. LeBaron and J. Gressel, eds.), Wiley (Interscience), New York, 1982, p. 133.

159. W. Oettmeier, H-J. Soll, and E. Neumann, *Z. Naturforsch.*, *39c*, 393 (1984).

160. W. Oettmeier, K. Masson, H-J. Soll, and W. Draber, *Biochim. Biophys. Acta*, *767*, 590 (1984).

161. J. Gressel, *Plant Sci. Lett.*, *25*, 99 (1982).

162. D. E. Moreland, J. B. St. John, and F. D. Hess, *Biochemical Responses Induced by Herbicides*, ACS Symposium Series No. 181, Am. Chem. Soc., Washington, D.C., 1982.

163. C. Fedtke, *Weed Sci.*, *27*, 192 (1979).

164. A. Srobarova, V. Kralo'va, and F. Kral'ovic, *Biologia Czechoslovakia*, *38*, 873 (1983).

165. C. Fedtke, *Z. Naturforsch.*, *34c*, 932 (1979).

166. J. L. P. van Ooorschot, in *Herbicides: Physiology, Biochemistry, Ecology*, (L. J. Audus, ed.), Academic Press, London, 1976, Vol. 1, p. 305.

167. J. L. P. van Oorschot and P. H. van Leeuwen, *Weed Res.*, *19*, 63 (1979).

168. C. S. Vavrina, N. C. Glaze, S. C. Phatak, and B. G. Mullinix, *Proc. South. Weed Sci. Soc.*, *36*, 366 (1982).

169. W. Draber and C. Fedtke, in *Advances in Pesticide Science*, Part 2 (H. Geissbuhler, ed.), Pergamon Press, Oxford, 1979, p. 475.

170. L. Klepper, *Nebr. Agric. Exp. Stn. Res. Bull. No. 259*, 42 pp. (1974).

171. D. E. Moreland, *Annu. Rev. Plant Physiol.*, *31*, 597 (1980).

172. C. Fedtke, *Pestic. Sci.*, *8*, 152 (1977).

173. C. S. Vavrina and S. C. Phatak, *Abstr. Weed Sci. Soc. Am.*, *23*, 71 (1983).

174. C. Fedtke, *Pestic. Biochem. Physiol.*, *2*, 312 (1972).

175. H-J. Yeh, *Rpt. Taiwan Sugar Res. Inst. No. 73*, 17 (1976).

176. C. S. Vavrina and S. C. Phatak, *Proc. South. Weed Sci. Soc.*, *37*, 345 (1984).

177. W. S. Hardcastle, R. E. Wilkinson, and C. T. Young, *Weed Sci.*, *22*, 575 (1974).

178. O. I. Prokopenko and G. V. Musina, *Sibiriski Vestnik Sel'skokhozyaistvennoi Nauk*, *15–17*, 12 (1983).

179. H. Graser, *Biol. Zentralbl.*, *88*, 191 (1969).

180. V. F. Ladovin and L. G. Spesivtsev, *Environ. Qual. Saf. Suppl.*, *Vol. III*, 517 (1975).

181. A. Tamperli, H. Turlier, and C. D. Ercegovich, *Z. Naturforsch.*, *21b*, 903 (1966).

182. C. Fedtke, *Z. Pflanzenkr. Pflanzenschutz, Sonder.*, *IX*, 141 (1981).

183. B. E. Ellis, *Can. J. Plant Sci.*, *58*, 775 (1978).

184. W. E. Chappell and L. A. Link, *Weed Sci.*, *25*, 511 (1977).

185. W. S. Hardcastle, *Weed Res.*, *14*, 181 (1974).

186. W. S. Hardcastle, *Pestic. Sci.*, *6*, 589 (1975).

187. W. L. Barrentine, C. J. Edwards, and E. E. Hartwig, *Agron. J.*, *68*, 351 (1976).

188. C. J. Edwards, W. L. Barrentine, and T. C. Kilen, *Crop Sci.*, *16*, 119 (1976).

189. L. M. Wax, E. W. Stoller, and R. L. Bernard, *Agron. J.*, *68*, 484 (1976).

190. W. S. Hardcastle, *Weed Sci.*, *27*, 278 (1979).

191. E. F. Eastin, J. W. Sij, and J. P. Craigmiles, *Agron. J.*, *72*, 167 (1980).

192. W. L. Barrentine, E. E. Hartwig, C. J. Edwards, Jr., and T. C. Kilen, *Weed Sci.*, *30*, 344 (1982).

193. E. O. Abusteit, F. T. Corbin, and J. W. Burton, *Proc. South. Weed Sci. Soc.*, *36*, 378 (1983).

194. C. B. Guy, L. S. Jeffery, and C. R. Graves, *Proc. South. Weed Sci. Soc.*, *36*, 79 (1983).

195. R. C. Henne and R. T. Guest, *Proc. Northeast. Weed Sci. Soc.*, *28*, 253 (1974).

196. V. Souza Machado, J. L. Nonnecke, and S. C. Phatak, *Can. J. Plant Sci.*, *58* (1978).

197. S. C. Phatak, V. Souza Machado, J. Fortino, and N. C. Glaze, *Abstr. Weed Sci. Soc. Am.*, *19*, 115 (1979).

198. V. Souza Machado, S. C. Phatak, and I. L. Nonnecke, *Euphytica*, *31*, 129 (1982).

199. H. F. Harrison, P. Bhatt, and G. Fassuliotis, *Weed Sci.*, *31*, 533 (1983).

200. W. Kolbe and K. Zimmer, *Pflanzeschutz Nachr.*, *25*, 210 (1972).

201. R. L. Zimdahl, *Am. Potato J.*, *53*, 211 (1976).

202. S. W. Gawronski, R. H. Callihan, and J. J. Pavek, *Weed Sci.*, *25*, 122 (1977).

203. J. A. Freeman, *Am. Potato J.*, *56*, 461 (1979).

204. J. A. Ivany, *Can. J. Plant Sci.*, *59*, 417 (1979).

205. H. DeJong, *Euphytica*, *32*, 41 (1983).

206. G. H. Friesen and D. A. Hall, *Weed Sci.*, *32*, 442 (1984).

207. S. C. Phatak, M. Singh, S. A. Harmon, and N. C. Glaze, *Proc. South. Weed Sci. Soc.*, *34*, 107 (1981).

208. H. F. Harrison, Jr., A. Jones, and P. D. Dukes, *Proc. South. Weed Sci. Soc.*, *38*, 118 (1985).

209. M. L. Fischer and T. F. Peeper, *Proc. South. Weed Sci. Soc.*, *35*, 359 (1982).

210. T. J. Runyan, W. K. McNeil, and T. F. Peeper, *Weed Sci.*, *30*, 94 (1982).

211. J. Schroeder Kvien and P. A. Banks, *Abstr. Weed Sci. Soc. Am.*, *23*, 14 (1983).

212. L. C. Haderlie, J. C. Stark, S. W. Gawronski, and D. M. Wesenberg, *Abstr. Weed Sci. Soc. Am.*, *23*, 15 (1983).

213. C. D. Caldwell and P. A. O'Sullivan, *Can. J. Plant Sci.*, *65*, 415 (1985).

214. M. Singh, S. C. Phatak, and N. C. Glaze, *Proc. Fl. State Hortic. Soc.*, *95*, 336 (1982).

215. E. Nielsen, F. Rollo, B. Parisi, R. Cella, and F. Sala, *Plant Sci. Lett.*, *15*, 113 (1979).

216. L. M. Deal and M. L. Christianson, *Abstr. Weed Sci. Soc. Am.*, *25*, 85 (1985).

217. E. E. Hartwig, W. L. Barrentine, and C. J. Edwards, Jr., *Crop Sci.*, *20*, 825 (1980).

218. T. C. Kilen and W. L. Barrentine, *Crop Sci.*, *23*, 894 (1983).

219. H. M. LeBaron and J. Gressel, *Herbicide Resistance in Plants*, Wiley (Interscience), New York, 1982.

220. H. Janatkova and G. F. Wilder, *Biochim. Biophys. Acta*, *682*, 227 (1982).

221. N. Pucheu, W. Oettmeier, U. Heisterkamp, K. Mason, and G. F. Wildner, *Z. Naturforsch.*, *39c*, 437 (1984).

222. E. O. Abusteit, F. T. Corbin, D. P. Schmitt, J. W. Burton, A. D. Worsham, and L. Thompson, Jr., *Weed Sci.*, *33*, 618 (1985).

223. J. S. Ladlie, W. F. Meggitt, and D. Penner, *Weed Sci.*, *25*, 115 (1977).

224. M. D. McGlamery and L. M. Wax, *Proc. North Central Weed Contr. Conf.*, *29*, 83 (1974).

225. C. R. Salhoff and A. R. Martin, *Proc. North Central Weed Contr. Conf.*, *32*, 31 (1977).

226. J. S. Ladlie, W. F. Meggitt, and D. Penner, *Weed Sci.*, *25*, 88 (1977).

227. R. S. Moomaw and A. R. Martin, *Weed Sci.*, *26*, 327 (1978).

228. M. F. Betts and I. N. Morrison, *Weed Sci.*, *27*, 691 (1979).

229. B. B. Messier and R. A. Ashley, *Proc. Northeast. Weed Sci. Soc.*, *34*, 110 (1980).

230. G. W. Selleck and D. D. Baird, *Weed Sci.*, *29*, 185 (1981).

231. L. W. Smith, *Weed Sci. Soc. Am. Abstr.*, 21 (1970).

232. D. D. Waldrop and P. A. Banks, *Weed Sci.*, *31*, 730 (1983).

233. D. P. Schmitt, F. T. Corbin, J. R. Bradley, Jr., and J. W. Van Duyn, *Agrichm. Age*, *25(1)*, 54 (1981).

234. R. B. Hammond, *Ohio Report*, *68*, 11 (1983).

235. S. Irons and O. G. Russ, *Proc. North Central Weed Contr. Conf.*, *32*, 32 (1977).

236. D. T. Warholic, E. S. Mohamed, G. W. Selleck, and R. D. Sweet, *Proc. Northeast Weed Sci. Soc.*, *31*, 215 (1977).

237. S. C. Phatak, C. A. Jaworski, and S. R. Ghate, *Hort. Sci.*, *20*, 690 (1985).

238. M. D. Mahoney and D. Penner, *Proc. North Central Weed Contr. Conf.*, *36*, 103 (1981).

239. B. D. Riggle and D. Penner, *Proc. North Central Weed Contr. Conf.*, *39*, 32 (1984).

240. B. D. Riggle and D. Penner, *Proc. North Central Weed Contr. Conf.*, *39*, 77 (1984).

Chapter 5
CARBAMOTHIOATES

ROBERT E. WILKINSON

Department of Agronomy
University of Georgia Agricultural Experiment Station
Experiment, Georgia

I. INTRODUCTION

Research reports on carbamothioates are numerous, contrasting, and change rapidly. Thus, no claim of a complete literature review is intended. Plant (Table 1) and chemical (Table 2) terminology follow those listed in *Weed Science* Vol. 32 (Suppl. 2) and Vol. 33 (Suppl. 1), respectively. Common and chemical names of the carbamothioates marketed in 1985 are listed in Table 2. Chemical structures, company, year of introduction, and AWLN nomenclature are presented in Table 3. Two carbamodithioate herbicides (CDEC and metham-sodium) are included in this chapter because of similarities in chemistry, activities, degradation, and other properties.

TABLE 1

Plant Nomenclature Species Listed in Text

Common	Latin	Bayer code
Alfalfa	*Medicago sativa* L.	MEDSA
Barley	*Hordeum vulgare* L.	HORVX
Barnyardgrass	*Echinochloa crus-galli* (L.) Beauv.	ECHCG
Birdsfoot trefoil	*Lotus corniculatus* L.	LOTCO
Cabbage	*Brassica oleracea* L.	
Citrus	*Citrus* spp	
Corn	*Zea mays* L.	
Cotton	*Gossypium* spp	
Curlydock	*Rumex crispus* L.	RUMCR
Flax	*Linum usitatissimum* L.	

TABLE 1 (continued)

Common	Latin	Bayer code
Giant foxtail	*Setaria faberi* Herrm.	SETFA
Johnsongrass	*Sorghum halepense* (L.) Pers.	SORHA
Lettuce	*Lactuca sativa* L.	
Mungbean	*Phaseolus aureus* Roxb.	
Nutsedge	*Cyperus* spp.	
Oat	*Avena sativa* L.	AVESA
Pea	*Pisum sativum* L.	
Peanut	*Arachis hypogaea* L.	
Pineapple	*Ananas comosus* (L.) Merr.	
Potato	*Solanum tuberosum* L.	
Quackgrass	*Agropyron repens* (L.) Beauv.	AGRRE
Redroot pigweed	*Amaranthus retroflexus* L.	AMARE
Rice	*Oryza sativa* L.	ORYSA
Sicklepod	*Cassia obtusifolia* L.	CASOB
Sorghum	*Sorghum bicolor* L.	
Soybean	*Glycine max* (L.) Merr.	
Spinach	*Spinacia oleracea* L.	
Squash	*Cucurbita maxima* Duch.	
Sugar beet	*Beta vulgaris* L.	
Tobacco	*Nicotiana tabacum* L	
Tomato	*Lycopersicon esculentum* Mill.	
Wheat	*Triticum aestivum* L.	
	Triticum durum Desf.	
Wild mustard	*Sinapis arvensis* L.	SINAR
Wild oats	*Avena fatua* L.	AVEFA

TABLE 2

Common and Chemical Names of Carbamothioates and
Carbamodithioate Herbicides and Antidotes

Butylate	S-ethyl bis(2-methylpropyl)carbamothioate
CDEC	2-chloro-2-propenyl diethylcarbamodithioate
Cycloate	S-ethyl cyclohexylethylcarbamothioate
Diallate	S-(2,3-dichloro-2-propenyl) bis(1-methylethyl)-carbamothioate
EPTC	S-ethyl dipropylcarbamothioate
Metham	Methylcarbamodithioic acid
Metham-Na	Na salt of carbamodithioic acid
Molinate	S-ethyl hexahydro-1H-azepine-1-carbothioate
Pebulate	S-propyl butylethylcarbamothioate
Thiobencarb	S-[(4-chlorophenyl)methyl]diethylcarbamothioate
Triallate	S-(2,3,3-trichloro-2-propenyl) bis(1-methyl-bis(1-methylethyl)-carbamothioate
Vernolate	S-propyl dipropylcarbamothioate
Dichlormid	2,2-dichloro-N,N-di-2-propenylacetamide
Naphthalic anhydride	1H,3H-naphtho[1,8-cd]-pyran-1,3-dione

A. History and Use

EPTC was the first carbamothioate developed (1956) and has
been followed by others during the last three decades. The carbamothioates are used to control annual and perennial grasses and
nutsedges. Differential grass selectivity has been found, however,
which allows the control of weedy grasses in wheat and rice. Several
dicotyledonous weeds are also highly susceptible to carbamothioates.
Thus, broad general categorizations of susceptibility and resistance
among plants are often inaccurate.

Several characteristics of the carbamothioates have influenced
their utility. Because of their volatility, the carbamothioates are
applied almost entirely as soil-incorporated preemergence herbicides.
Under these conditions, the concentrations of EPTC (6 kg/ha) that

TABLE 3

Structure, Producer, Year of Introduction, and Advanced Wiswesser Line Notation (AWLN) for Carbamothioates and Antidotes

Name	Structure	Company (year)	AWLN
Butylate	$CH_3-CH_2-S-C-N-(CH_2-CH-CH_3)_2$ with $\overset{O}{\underset{\parallel}{}}$ and CH_3 branch	Stauffer (1957)	1Y1&1JV2&1Y1&1
CDEC	$CH_2=C-CH_2-S-C-N-(C_2H_5)_2$ with Cl and $\overset{S}{\underset{\parallel}{}}$	Monsanto (1954)	2N2&YUS&S1YGU1
Cycloate	$CH_3-CH_2-S-C-N$ with CH_2-CH_3 and cyclohexyl, $\overset{O}{\underset{\parallel}{}}$	Stauffer (1966)	(6$) aJ2&VS2
Diallate	$HC\equiv C-CH_2-S-C-N-\left(CH\overset{CH_3}{\underset{CH_3}{}}\right)_2$ with $Cl\ Cl$ and $\overset{O}{\underset{\parallel}{}}$	Monsanto (1961)	GDTG1SVJY1&1Y1&1

TABLE 3 (continued)

Name	Structure	Company (year)	AWLN
EPTC	$CH_3-CH_2-S-\overset{\overset{O}{\|\|}}{C}-N-(CH_2-CH_2-CH_3)_2$	Stauffer (1956)	3J3&VS2
Metham-sodium	$Na-S-\overset{\overset{S}{\|\|}}{C}-NH-CH_3$	Stauffer (1956)	STS&M1
Molinate	$CH_3-CH_2-S-\overset{\overset{O}{\|\|}}{C}-N$ (cycloheptyl)	Stauffer (1965)	(J7$)2VS2
Pebulate	$CH_3-CH_2-CH_2-S-\overset{\overset{O}{\|\|}}{C}-N{\overset{CH_2-CH_3}{\underset{(CH_2)_3-CH_3}{}}}$	Stauffer (1960)	4J2&VS3
Thiobencarb	$Cl\text{-}C_6H_4\text{-}CH_2-S-\overset{\overset{O}{\|\|}}{C}-N-(C_2H_5)_2$	Chevron Kumiai (1971)	GR d1SVJ2&2

Name	Structure	Company	Code
Triallate	$\begin{array}{c}\text{Cl Cl}\\\text{C=C}-\text{CH}_2-\text{S}-\overset{\overset{O}{\|}}{C}-\text{N}-\left(\text{CH}\overset{(CH_3)}{\underset{(CH_3)}{\diagdown}}\right)_2\\\text{Cl}\end{array}$	Monsanto (1961)	GTGTG1SVJY1&1&Y1&1
Vernolate	$\text{CH}_3-\text{CH}_2-\text{CH}_2-\text{S}-\overset{\overset{O}{\|}}{C}-\text{N}-(\text{CH}_2-\text{CH}_2-\text{CH}_3)_2$	Stauffer (1961)	3SVJ3&3
Dichlormid	$\begin{array}{c}\text{Cl}\quad\text{O}\\\overset{\overset{\|}{}}{\text{CH}}-\overset{\overset{O}{\|}}{C}-\text{N}-(\text{CH}_2-\text{CH}=\text{CH}_2)_2\\\text{Cl}\end{array}$	Stauffer (1961)	GYGVJ1DL1DL
Naphthalic anhydride		Gulf Oil (1969)	

are required to control weeds also injured corn. This problem has been circumvented by the discovery of antidotes, which decrease the injury of EPTC to corn but do not alter the susceptibility of the weeds. Recent evidence has shown accelerated degradation of some carbamothioate herbicides in soils with a history of carbamothioate herbicide application. Thus, "extenders" have been developed to prevent rapid degradation of certain carbamothioate herbicides in "primed" soils.

Solving these problems of specific toxicity, selectivity, volatility, leaching, safening, and extending forms the developmental history of many of the carbamothioate herbicides and influences each member of the group. Variations of each factor lead to different utilization patterns; however, several members of this class have significantly lower volatility than the rest (i.e., thiobencarb, diallate, and triallate) (Table 3).

EPTC (Tables 2 and 3) was released in 1956 [1] for use in alfalfa, pulses, citrus, corn, cotton, flax, potato, pineapple, and other crops for the control of johnsongrass seedlings, nutsedge, quackgrass, and many annual grasses [2] after mechanical soil incorporation to a depth of 5 to 7.6 cm immediately after application. In established perennial crops, EPTC may be incorporated by sprinkler or flood irrigation. The range of application concentrations and responses is extensive. Alflafa seed planted in contact with EPTC [3] or encapsulated with porous material containing EPTC (1 mg/seed) was not injured [4]. Alternatively, 1.7 kg/ha EPTC adequately controlled annual grass in alfalfa and birdsfoot trefoil [5,6]. Corn was injured at 6.72 kg/ha EPTC, but dichlormid (Tables 2 and 3) (0.56 kg/ha) applied concurrently protected corn without reducing weed control efficacy [7,8]. Decreased herbicidal efficacy in soils with a previous history of EPTC application was suggested [9] to be due to more rapid microbial degradation. This was later verified and expanded to include butylate and vernolate but not cycloate [10–13] (Tables 2 and 3). R-33865 (O,O-diethyl-O-phenyl phosphorothioate) increased the persistence of butylate, EPTC, and vernolate in soils with previous EPTC application history [12].

Additional crop uses for carbamothioates have been those of pebulate (sugar beet, tomato, and tobacco), vernolate (soybean and peanut), molinate (rice), cycloate (sugar beet and spinach), butylate (corn), thiobencarb (rice), diallate (sugar beet and corn), and triallate (legumes, barley, and wheat) [2,14]. CDEC (no longer registered) was used in vegetables for annual grass control [15]. Metham-sodium is an active soil fumigant [2,14,15].

B. Physical and Chemical Properties

Pure carbamothioates are clear, oily liquids; commercial grades are colored amber to dark yellow. Metham-sodium is an exception and is formulated as a water solution. Carbamothioate water solubility varies from 4 (triallate) to 800 ppmw (molinate) (Table 4) whereas solubility in organic solvents is very high. Thus, cleansing or glassware and other utensils is easily accomplished by organic solvent rinsing. Volatility is high for many carbamothioates and is the basis for most soil incorporation requirements.

The odor of carbamothioates is aromatic and sharp. These compounds are highly stable in mild acidic or alkaline conditions. Hydrolysis is rapid in 96% H_2SO_4 (85°C) but decreases rapidly as the acid concentration is decreased.

C. Toxicological Properties

Carbamothioates do not induce acute or chronic toxicities at low concentrations (Tables 5 and 6) [2,14,15]. The concentration required for toxicity in aquatic species ranges from 1 (vernolate) to 30 ppmw (molinate). But 1 ppmw of any compound in water is equivalent to 2.7 lb ai/acre-ft water, and this concentration of toxicant is difficult to sustain because of volatility, degradation, dilution, and adsorption. Thus, contamination to toxic levels is improbable. Since thiobencarb and molinate are utilized in rice paddies, their activities on aquatic animals were carefully evaluated and were inversely related to solubility in sea water. The thiobencarb 96 hr LC_{50} was 330 ppmw for mysid (*Mysidopsis bahia*) and 1370 ppmw [16] for eastern oyster (*Crassostrea variegatus*); no consistent responses were observed in carp (*Cyprinus carpio* L.) after exposure to 0.03 ppmw for 30 days [17]. Molinate tolerance level mean (TLM) for several aquatic species was relatively high [18] (Table 7). Mammalian toxicity is low, postulated as a rapid degradation in the liver [19,20] resulting in the formation of carbamothioate sulfoxides which are ephemeral compounds very rapidly conjugated to glutathione (GSH). Some carbamothioate sulfoxides are suspected of conversion to acrolein under very limited photolytic conditions or in highly oxygenated media. Haloacroleins are reported to be mutagenic [21–27] in specially selected strains of *Salmonella* following metabolic activation of the herbicides by rat liver microsomes. Conditions required for this metabolic activation are (a) pretreatment of the microsomes and (b) low GSH concentrations [28]. Acroleins and haloacroleins are conjugated to sulfhydryl groups with extreme rapidity [27]. Since these assays were: (a) not conducted with pure materials, (b) a 500× excess of material was utilized for assay, and (c) GSH is present in mM concentrations in chloroplasts and mammalian systems, the production of these photolysis products in highly oxygenated media is unlikely to reach concentrations required to be deleterious to higher plants or animals.

TABLE 4

Physical and Chemical Properties

Name	Formula	MW	State	sg	bp (C) (mmHg)	VP (mmHg) (C)
Butylate	$C_{11}H_{23}ONS$	217.4	L	0.9402 (25/25)	71 (10)	1.3×10^{-2} (25)
CDEC	$C_8H_{14}ClNS_2$	223.8	L	1.16 (25/15.5)	25 ($1.8\ 10^{-4}$)	2.2×10^{-3} (20)
Cycloate	$C_{11}H_{21}NOS$	215.4	L	1.1016 (30/4)	145-6 (10)	6.2×10^{-3} (25)
Diallate	$C_{10}H_{17}Cl_2NOS$	270.2	L	1.188 (25/15.6)	97 (0.15)	1.5×10^{-4} (25)
EPTC	$C_9H_{19}NOS$	189.3	L	0.960 (25/25)	235	3.4×10^{-2} (25)
Metham-Na	$C_2H_4S_2NNa$	128.17	Sol		110	
Molinate	$C_9H_{17}NOS$	187.13	L	1.0643 (20/20)	202 (10)	5.6×10^{-3} (25)
Pebulate	$C_{10}H_{21}NOS$	203.4	L	0.9555 (20/20)	142 (21)	3.5×10^{-2} (25)
Thiobencarb	$C_{12}H_{16}ClNOS$	257.8	L	1.145– 1.180	126–9 (0.008)	1.476×10^{-6} (20)
Triallate	$C_{10}H_{16}Cl_3NOS$	304.7	L	1.273 (25/15.6)	117 (0.3)	1.2×10^{-4} (25)
Vernolate	$C_{10}H_{21}NOS$	203.4	L	0.954 (20/20)	150 (30)	1.04×10^{-2} (25)

			Solubility (ppmw) (C)						
Ether	EtAc		Water	Kerosene	EtOH	Xylene	MIBK	CHCl$_3$	Acetone
			45 (22)	m	m	m	m	m	m
s	s	s	92 (25)	s	s			s	s
			85 (22)	m		m	m		
m	m	m	14 (25)	m	m	m			m
	s	s	370 (20)	m	m	m	m	s	m
			72.2g 100ml	ss	ss	s	ss		ss
m	m	m	800 (20)	m	m	m	m		m
			60 (25)	m	m	m	m		m
			30 (20)		s	s			s
s	s	s	4 (25)		s				s
			90 (20)	m		m	m		

TABLE 4   (continued)

Name	Formula	MW	State	sg	bp (C) (mmHg)	VP (mmHg) (C)
Dichlormid	$C_8H_{11}Cl_2NO$	208.09	L	1.202 (20/20)	130 (10)	6 × 10^{-6} (25)
Naphthalic anhydride	$C_{12}H_6O$	198.2	C	—	270–4 mp	

Abbreviations: L = Liquid; Sol = solution; C = crystal; sg = specific gravity g/ml determined at specific C references to water at specific C; bp = boiling point at referenced mmHg pressure; m = miscible; s = soluble; ss = slightly soluble.

D.   Synthesis and Analytical Methods

1.   Synthesis

Three methods of synthesis for 266 carbamothioates have been reported [29]. The method reported for cycloate [30], molinate [31], EPTC [32], butylate [33], vernolate [34], and pebulate [35] utilizes the chlorothiolformate method (1):

$$R_1-S-\overset{\overset{\displaystyle O}{\|}}{C}-Cl + HN\overset{R_2}{\underset{R_3}{<}} \xrightarrow{NaOH} R_1-S-\overset{\overset{\displaystyle O}{\|}}{C}-N\overset{R_2}{\underset{R_3}{<}} + NaCl \quad (1)$$

Metham-sodium is synthesized from monomethylamine and carbon disulfide in the presence of concentrated NaOH [36] (2):

$$CH_3-NH_2 + CS_2 + NaOH \longrightarrow CH_3-\overset{\overset{\displaystyle H}{|}}{N}-\overset{\overset{\displaystyle S}{\|}}{C}-S-Na + H_2O \quad (2)$$

				Solubility (ppmw)				
Ether	EtAc	Water	Kerosene	EtOH	Xylene	MIBK	CHCl$_3$	Acetone
		5000 (20)	15 g/L	m	m	m		
		13.9 g/L		dimethylformamide				

Diallate, triallate, and CDEC are synthesized from substituted allyl chloride, carbon oxysulfide, and diisopropyl amine in the presence of NaOH (3). (Monsanto Chem. Co., unpublished data.)

$$\begin{array}{l} R_1 \\ \diagdown \\ C = C - CH_2Cl + COS + (CH_3)_2 - NH - CH - (CH_3)_2 \longrightarrow \\ R_2 \diagup | \\ R_3 \end{array}$$

diallate, triallate

$R_1 = R_2 = R_3 = Cl$ triallate

$R_1 = H, R_2 = R_3 = Cl$ diallate   (3)

2. Analytical Methods

Two basic methods for analysis of the carbamothioates have been described for the analysis of technical or formulated materials. Gas-liquid chromatography (GC) is extremely useful. Quantitation is achieved with an internal or external standard [30–35].

TABLE 5

Toxicological Concentrations for Carbamothioates in the Growth Milieu

Compound	Formulation	Bluegill	Rainbow trout	Mosquito fish	Bobwhite quail	Mallard duckling	Oysters[a]	Brown shrimp[b]	Misc.
				LC$_{50}$ (ppmw)					
Butylate	tech	6.9 96 hr	4.2 96 hr		40,000 7d				
	ec	7.2 96 hr	5.2 96 hr	8.5 96 hr	27,000 7d				
CDEC	ec	4.9	9.6		>5000 10–15 days	>5000 10–15 days			c
EPTC	tech	27 96 hr	19 96 hr	17 48 hr					
	ec	27 96 hr	21 96 hr						
Cycloate	ec	5.9 96 hr	7.9 96 hr	>5000 10–15 days		>5000 10–15 days			d
Molinate	tech	2.9 96 hr	1.3 96 hr				>1 96 hr	>1 96 hr	e
	ec			26 96 hr		13,000 5 days			

Compound	Form								
Pebulate	tech						>1 96 hr	>1 48 hr	f
	ec			10					
Thiobencarb	ec	1.6–3.4 96 hr	1.05 96 hr						
Triallate	tech	1.3 96 hr	1.2 96 hr		>2251 >5000 8 days	>5000 8 days			
	ec	4.9 96 hr	9.6 96 hr						
Vernolate	tech		10.8				>1 96 hr	>1 96 hr	g
	ec	9.6		14.5					
Naphthalic anhydride	ec	4.9 96 hr	6 96 hr						

[a] Inhibition of shell growth.
[b] Disorientation.
[c] Blue crab >20 (24 hr).
[d] Harlequin fish 8.2 (2 day).
[e] Goldfish - 30 (96 hr).
[f] Mallard ducklings - tech - >5000 (8 day). Catfish - 2.3–6.1 (96 hr).
[g] Vernolate - tech and ec - three spine stickle fish 1–10.

TABLE 6

Toxicological Concentrations for Carbamothioates in Feed or on Skin

Compound	Formula-tion	Rat	Longnose killfish	Bobwhite quail	Mallard duckling	Misc.
			Acute Oral			
Butylate	tech	4659				a
	ec	3878				
CDEC	ec	850				c
EPTC	tech	1652	20 48 hr	20,000 7 days		
	ec			26,000 7 days		
Cycloate	ec	395				d
Metham-Na	soln	820				
Molinate	tech	720				
	ec	584				
Pebulate	ec	920	7.78 48 hr	9500 7 days		g,h
Thiobencarb	tech	920		>7800	>10,000 8 days	
	ec	1036				
Triallate	tech			>5000 8 days	>5000 8 days	i
	ec					
Vernolate	tech	1780	12,000 7 days			
	ec	1800	14,500 7 days			

The LD_{50} (mg/kg) heading spans the Acute Oral columns.

	Chronic/Oral			Dermal	
Rat	Dog	Misc.	Rabbit	Misc.
>62	>62			b
90 days	>90 days		>2800	
	>200			
			>10,000	
400^e	>600^e	f		
			2000	
			>10	
>16	>20		4640	
90 days	90 days			
>660	>660			
90 days	90 days			
>30	>30			
2 years	2 years			
	>50^d		8200	
	28 days			
			2000	
				k
>32^d	>38^d			
				l

TABLE 6 (continued)

| | | LD$_{50}$ (mg/kg) | | | | |
| | | Acute Oral | | | | |
Compound	Formula-tion	Rat	Longnose killfish	Bobwhite quail	Mallard duckling	Misc.
Dichlormid	2030					
Naphthalic				4100	>6810	
Anhydride	12,300					

[a]Butylate - tech - guinea pig - 1659.
[b]Butylate - ec - rat >4640.
[c]CDEC - pheasant >15,400.
[d]Cycloate - dog - 510.
[e]Weight loss.
[f]Cycloate - rabbit - 2000-2500.
[g]Pebulate - mouse - 1652.
[h]Pebulate - white mullet - 6.25 (48 hr).
[i]Triallate - mammal - 1100 - tech.
[j]Triallate - ec - mammal - 2700.
[k]Vernolate - guinea pig - 4640 - tech.
[l]Vernolate - guinea pig - 10,000 - ec.

From soil or biological samples, two methods of extraction are available. Organic solvents (i.e., dichloromethane, diethyl ether, pentane) or steam distillation of the sample may be used for extraction. Quantitation is then achieved by: (a) GC or (b) hydrolysis in concentrated H_2SO_4 and the addition of cupric dithiocarbamate, which produces colored products that can be measured spectrophotometrically at 440 nm [30-36]. The colorimetric method is sensitive to approximately 5 μg carbamothioate after correction for the extraction of natural amine-forming substances.

E. Chemical Reactions

The carbamothioate herbicides are relatively stable in dilute acidic or basic solutions [30-36]. For example, <5% hydrolysis was observed when 100 ppm EPTC was stored for 300 days at 20°C in 0.1 N HCl, H_2SO_4, NaOH, or distilled water. Hydrolysis was complete in 10 min in 96% H_2SO_4 at 85°C, but the rate of hydrolysis fell off rapidly in more dilute H_2SO_4 [32].

Chronic/Oral			Dermal	
Rat	Dog	Misc.	Rabbit	Misc.
			>4640	
>500	>500			
90 days	90 days			

TABLE 7

Tolerance Levels (mean) (TLM) for Several
Aquatic Organisms to Molinate

	TLM (hr)			
Organism	24	48	72	96
Mosquito fish	30.7	21.4	17.1	16.4
Grass shrimp	22	20	18	15.9
Cray fish	34.7	33.2	23	21.8
Mactrid clam	750	385	290	197

Source: From Ref. 8.

TABLE 8

Formulations Available

Common	Name Commercial	Form	Concentration
Butylate	Sutan 6E	ec	720 g ai/L
	Sutan + E	ec	800 g ai/L + 33 g ai/L dichlormid
	Sutan + G	10G	10%
	Sutan + atrazine	18–6G	18% butylate + 6% atrazine
CDEC	Vegadex	ec	480 g ai/L
	Vegadex	G	20%
Cycloate	Ro-Neet E	ec	720 g ai/L
	Ro-Neet 10G	G	10%
Diallate	Avadex	ec	480 g ai/L
	Avadex	G	10%
EPTC	Eptam E	ec	720, 800, or 840 g ai/L
	Eptam G	G	2.3, 5, 10, or 20%
	Eradicane E	ec	800 g ai/L + 66 g ai/L dichlormid
	Eradicane 10G	G	10%

Metham-sodium	Polefume	soln	382 g ai/L
	Vapam	soln	382 g ai/L
	Vapam B	soln	340 g ai/L
Molinate	Ordram GE	ec	720 g ai/L
	Ordram G	G	5, 10%
	Arrosolo 3 + 3E	ec	30 g ai molinate + 30 g ai propanil/L
	Arrosolo 36 + 36E	ec	360 g ai molinate + 360 g ai propanil/L
Pebulate	Tillam 6E	ec	720 g ai/L
	Tillam G	G	10 or 25%
	Tillam Devrinol 4-1E	ec	480 g ai pebulate + 120 g ai napropamide/L
Triallate	Far-Go	ec	480 g ai/L
	Far-Go	G	10%
	Avadex BW	ec	480 g ai/L
	Avadex BW	G	10%
Vernolate	Vernam E	ec	720 or 840 g ai/L
	Vernam G	G	5 or 10%
	Vernam atrazine	10–5G	10% vernolate + 5% atrazine
	Surpass E	ec	800 g ai vernolate + 66 g ai dichlomid/L

Abbreviations: ec = emulsifiable concentrate; G = granular; soln = solutions.

F. Formulations

Except for metham-sodium, which is marketed as a solution, the carbamothioates are formulated as emulsifiable concentrates or granular formulations (Table 8). Concentrations may vary for different uses, as may the presence or absence of antidotes and extenders.

II. DEGRADATION PATHWAYS

A. Degradation in Plants

The degradation rate of [^{35}S]EPTC in seeds was greater in resistant species than in susceptible species [37] (Table 9). The ^{35}S was incorporated into cysteic acid, cysteine, methionine, methionine sulfone, and two unidentified compounds. When [ethyl-1-^{14}C] EPTC was applied to alfalfa, the ^{14}C was incorporated into fructose, glucose, and several amino acids (aspartic acid, asparagine, glutamic acid, glutamine, serine, threonine, and alanine) [38]. These combinations of incorporation of radioactivity into amino acids suggest a cleavage near the S atom. This concept was substantiated [39] when the EPTC was shown to be oxidized to EPTC-sulfoxide (EPTC-SO), which was rapidly conjugated to glutathione (GSH) (Fig. 1). This

TABLE 9

Absorption and Recovery of ^{35}S from Fractions of Seedlings after 48 hr Exposure to [^{35}S]EPTC

Species	EPTC absorbed (μg/seed)	Recovery (%)			
		EPTC	Inorganic SO_4	Hot water extract	H_2O- insoluble extract
Kidney bean	0.34	10.7	0.4	80.1	8.8
Mung bean	0.21	4.3	0.2	89.7	5.9
Pea	0.16	2.6	1.2	87.4	8.7
Corn	0.30	8.7	0.4	70.7	20.1
Wheat	0.11	8.0	0.6	74.2	17.2
Oat	0.25	23.0	0.3	62.8	13.9

Source: From Ref. 37.

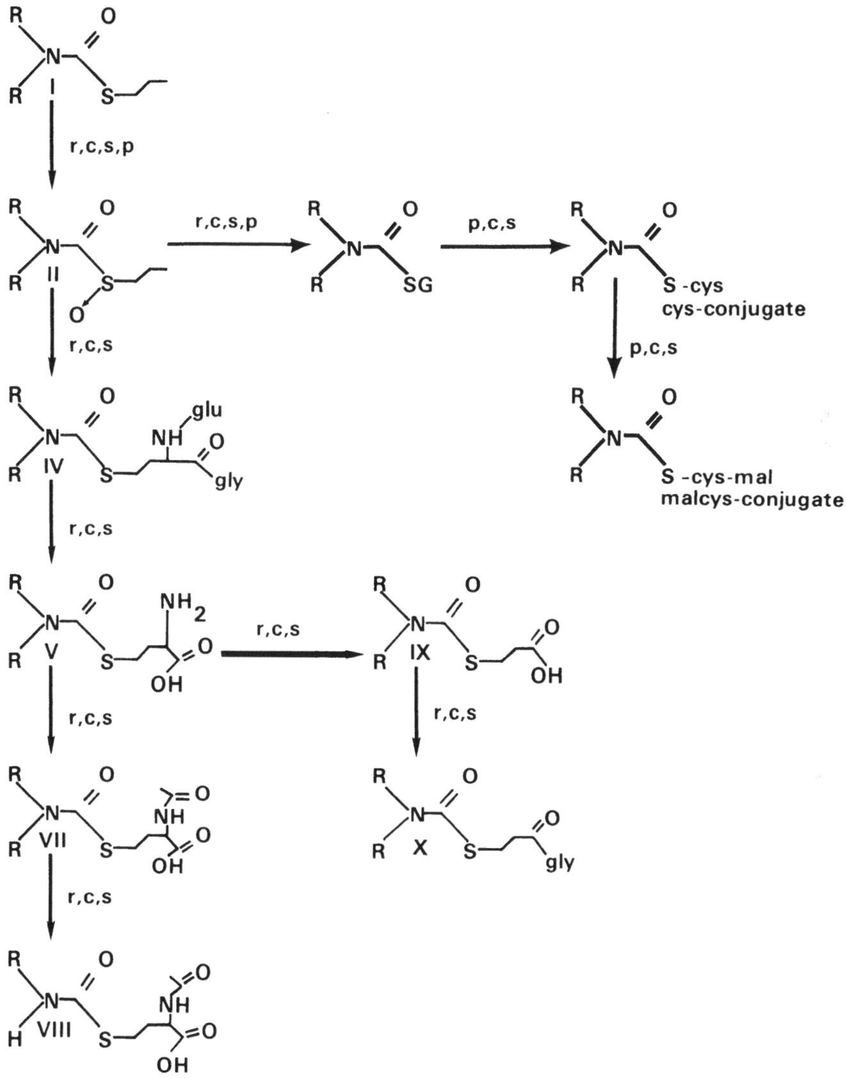

FIG. 1   Degradation of carbamothioates via sulfoxidation.
r = rat, c = corn, s = soybean, p = peanut and cotton.

conjugation was shown to occur nonenzymatically [40]; however, dichlormid induced the synthesis of a novel GSH-S-transferase species which was responsible for carbamothioate detoxification [41]. Some corn varieties are deficient in GSH-S-transferase and are EPTC and atrazine sensitive. But root contents of GSH and GSH-S-transferase activity of EPTC sensitive corn varieties were increased 56 to 95% by treatment with dichlormid [42]; and most of the root applied [^{14}C]dichlormid remained in the root while [^{14}C]EPTC was translocated into the leaves. Later, the glutathione conjugate was shown to be further modified to a malonylcysteine conjugate as the end product [43,44] (Fig. 1) (Table 10).

Carbamothioate research is aided by the existence of antidotes. EPTC (6.72 kg/ha) is highly deleterious to corn, but the addition of 0.56 kg/ha dichlormid prevents injury to the corn. Either (a) the dichlormid stimulates the degradation and(or) detoxification of the EPTC, or (b) the dichlormid alters the mode of activity of EPTC. Both actions appear to occur. Dichlormid enhanced the detoxification of EPTC and pebulate in corn roots by increasing the GSH content and GSH-S-transferase activity [45–47]; several dichlormid analogues were shown to increase corn root GSH content [48], which was directly correlated with antidote activity.

Since selectivity is often associated with detoxification capacity, the suggestion that GSH conjugation and detoxification might be involved in carbamothioate selectivity [49] was substantiated by the GSH conjugation of carbamothioate sulfoxides and increase of GSH content in corn roots [42] which results in a more rapid degradation of the toxicant. Other degradation products, however, were present in soybean pods treated with [^{14}C]vernolate [44]; these unidentified product could not be accounted for by a vernolate-SO conjugation to GSH. Additionally, rapid biosynthetic responses of corn cell suspensions and calli have shown that EPTC inhibited neutral and polar

TABLE 10

Formation of Sulfoxides and Malonylcystine Conjugates

Carbamothioate	Reference	Plant species
EPTC	39,43	Corn, cotton, soybean, peanut
Vernolate	44	Soybean
Molinate	43	Peanut
Butylate	39	Corn

lipid syntheses within 30 min while dichlormid-induced increased
GSH concentration occurred after 85% (2 hr) of the EPTC had been
degraded [50,51]. Thus, the kinetics of these activities could not
depend upon GSH-EPTC-SO conjugation.

Rates of carbamothioate degradation depend upon the herbicide
concentration, species, plant age, and temperature. Metabolism of
[^{14}C]vernolate by soybean seedlings was dependent upon seedling
age [52]. Degradation of pebulate by mung bean seeds was dependent on temperature, but degradation in wheat seeds was not
temperature dependent [53] (Fig. 2). Concentration dependence of
pebulate degradation was evident in wheat but not in mung bean
seedlings [53] (Fig. 3).

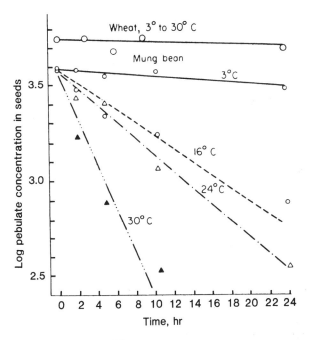

FIG. 2   [^{14}C]Pebulate degradation by mung bean and wheat
seedlings at different temperatures. Initial pebulate concentration:
wheat = 2.44 µg/g, mung bean = 1.70 µg/g. (Reprinted from
P. C. Kearney and D. D. Kaufman (ed.), *Herbicides: Chemistry,
Degradation, and Mode of Action*, 1975, p. 331, by courtesy of
Marcel Dekker, Inc..)

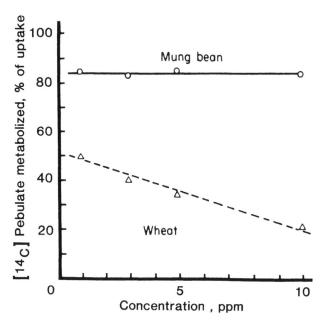

FIG. 3   Influence of pebulate concentration on its metabolism
in mung bean and wheat [53]. (Reprinted from P. C. Kearney and
D. D. Kaufman (eds.), *Herbicides: Chemistry, Degradation, and
Mode of Action*, 1975, p. 331, by courtesy of Marcel Dekker, Inc.)

B. Degradation in Animals

Plant degradation of some carbamothioates has been shown to
proceed, in part, by sulfoxide conjugation to GSH and terminal
product(s) are malonylcysteine conjugates [43] (Fig. 1). In rats,
carbamothioate degradation produces the sulfoxide-GSH conjugate,
which is then converted to *N*-acetylcysteine conjugates, which are
progressively deacylated or modified to a mercaptoacetylglycine conjugate [39,54]. This resulted in 88 to 98% degradation within 48 hr
of oral administration of dosages over 130 mg/kg for EPTC and
butylate [39].   Thiobencarb, cycloate, molinate, and pebulate also
gave mercapturic acid derivatives [39].   About 40% of the administered EPTC and butylate was metabolized by ester cleavage and
$^{14}CO_2$ liberation without conversion to the carbamothioate sulfoxide
intermediates [39].

Halocarbamothioates were oxidized by similar mechanisms, but
the sulfoxides are subject to liver microsomal conversion to halo-
acroleins [55] which are reported to be mutagenic [24] in the Ames
test [28]. As a bioassay, the Ames test is highly variable and

accurate interpretation is dependent upon multiple check cultures as well as concentration-response curves. Correlation of microorganism dose-response curves (100:1, body weight: dose-concentration) to potential applicability in mammalian systems (possible body weight: dose-concentration ratio approximates 10^9:1) is problematical. Additionally, sulfoxides and acroleins are instantly chemically conjugated to *any* available SH group. Thus, consideration of these ephemeral "fugitive" substances as herbicide products, which are to be evaluated as mutagens in mammalian systems, should be documented with great care.

C. Dissipation by Volatility

Persistence of soil-surface applied EPTC was inversely related to wind velocity, soil physical characteristics, and spray adjuvants [56]. Soil moisture, soil organic matter, and soil clay contents are correlated with carbamothioate volatility. Surface-applied EPTC volatilized 20% from dry soil surfaces within 15 minutes; moist soils lost 27%; and wet soils volatilized 44% [57,58]. Losses after 1 day were 23, 49, and 69%, respectively, whereas losses after 6 days amounted to 44, 68, and 90%, respectively [57]. The high volatility of carbamothioates has been utilized for quantitation by distillation from soil [30-36,59]. In field applications, the high dissipation of carbamothioates [2,56] (Table 4) results in decreased efficacy, which is circumvented by soil incorporation (5-10 cm) [5,6], knife-injection [60], or incorporation into irrigation water [61].

Quantities of carbamothioate herbicides lost through volatilization depend upon chemical structure. Persistence was evaluated by trapping the vapors from unincorporated applications to soil surface and by soil analyses. Cycloate, pebulate, and EPTC vapors trapped in 30 min from a 3.36 kg/ha application were equivalent to 4.6, 15.0, and 24.3%, respectively. Soil analyses after 6 hr showed losses of 65, 75, and 86%, respectively [62]. Triallate, however, required 11 days to volatilize 50% of the applied material from Ray silt and 90 days from Drummer silty clay loam [63].

D. Persistence Under Field Conditions

1. Degradation

Soil microorganisms contribute significantly to soil-incorporated carbamothioate degradation, although neither identification of the microorganisms involved nor modes of degradation have been reported. EPTC was inactivated in nonautoclaved soils three times faster than in autoclaved soils [64]; and butylate, cycloate, molinate, and vernolate disappeared from nonautoclaved soils much faster than from autoclaved soils [62]. Presumption of microbial activity was

supported by measuring the release of $^{14}CO_2$ from [^{14}C]EPTC-
treated soil and confirmation of EPTC efficacy by bioassay [65].   The
rate of $^{14}CO_2$ release from the ethyl moiety was slow in comparison
to the rate of inactivation as determined by bioassay.   Similar results were reported when 25% of [^{14}C]EPTC was recovered as $^{14}CO_2$
in 35 days while complete inactivation of the EPTC was revealed by
bioassay [66].   Since these data could have been influenced by
volatilization, cycloate was stored in sealed containers and analyzed
at monthly intervals.   Decomposition approximated 20% and 60% in
autoclaved and nonautoclaved soils, respectively, after 3 months
[67].

Application of EPTC to soils for 3 years resulted in decreased
herbicidal activity and this was attributed to a more rapid microbial
degradation [9].   Corroboration of enhanced microbial degradation
[10] showed $T_{1/2}$ values in soils with and without previous EPTC
applications to be 9 and 18 days, respectively.   Additionally, the
rate of EPTC degradation was highly dependent upon soil moisture
(decreased) below 3% and independent of soil moisture above 3% [10].
This appeared to depend upon the availability of moisture for microbial growth; and, degradation was exponential at 5, 15, and 25°C.
While a single line was the best fit at 5°C, two lines were required
at 15 and 25°C.   In soils without a previous history of EPTC application, however, single lines of correlation were found.

The problem of increased microbial degradation in soils with a
previous history of carbamothiate application was evaluated by the
concomitant application of R-33865 [11].   This extended EPTC persistence from $T_{1/2}$ = 9 days to $T_{1/2}$ = 18 days with a concomitant
decrease in $^{14}CO_2$ evolution.   There was no change in EPTC persistence in soils without a previous history of EPTC application.
EPTC, vernolate, and butylate degradations were increased in soils
with a previous history of EPTC application (Table 11) [12] and one
application of EPTC was sufficient to decrease persistence.   Butylate
degradation rate was not increased as much by one previous EPTC
application as EPTC and vernolate, but the addition of R-33865
greatly extended the $T_{1/2}$ of all three herbicides.   Additionally, at
a herbicide concentration of ∿0.3 ppmw, the degradation rate decreased due to herbicide adsorption on soil particles, which lowered
availability to the microbes.   The rate of increased degradation depended upon which carbamothioate had been utilized as a pretreatment [13].   EPTC and vernolate degradation was enhanced in soils
pretreated with EPTC, vernolate, or butylate; butylate degradation
was enhanced in soils with a previous history of butylate treatment.
Cycloate degradation was not enhanced in soils with a history of
pretreatment with EPTC, vernolate, or butylate.   These degradation
rates were measured as $^{14}CO_2$ evolved from the soils, and in 49-day
incubation periods with isotopic herbicides recovery of butylate (51%),
cycloate (75%), EPTC (83%), and vernolate (87%) were found.

TABLE 11

Influence of EPTC Pretreatment and R-33865 on Butylate,
EPTC, and Vernolate Degradation Rates in Soils

| | $T_{1/2}$ (days) | | |
Compound	No pretreatment	+EPTC pretreatment	+EPTC pretreatment +R-33865
Butylate	35	10	23
EPTC	13	3	10
Vernolate	17	3	14

Source: From Ref. 12.

Diallate and triallate were microbially degraded (50% and 40%, respectively) within 4 weeks in Canadian soils at moisture levels in excess of the wilting point [47] with the degradation rate dependent upon soil type. Triallate residues did not accumulate in soils treated with 1.78 kg/ha/yr for 6 years [68].

Molinate and thiobencarb are utilized in rice culture and degradation rates may be altered under flooded conditions. Under normal soil conditions, molinate had a $T_{1/2} = 3$ weeks, whereas under flooded conditions the $T_{1/2} = 10$ weeks [69]. Under flooded conditions, very little molinate was degraded and volatilization was the primary mode of dissipation. In nonflooded conditions, hydroxylation of the azepine ring, oxidation to sulfoxides, chain cleavage, and acetylation were all found. Thiobencarb degradation was slow under anaerobic flooded conditions [70] and in oxidative flooded conditions $^{14}CO_2$ was evolved. When soil amendments or fertilizers were added to the soil, degradation rates were influenced by the amounts of phosphorus added [71]. Thus, microbial degradation of the carbamothioates in soils forms a complete ecological system dependent upon herbicide, concentration, water content, temperature, soil constituents, microorganisms, and previous conditioning history. Rapidly evolving ecological studies will be the subject of many research reports in the near future.

2. Adsorption and Leaching

Phytotoxicity and persistence of carbamothioates was greatest in nutrient solution and mineral soils and was greatly reduced in

organic soils [72] when evaluated by a barnyardgrass bioassay.
Pebulate was adsorbed most by soils and EPTC was adsorbed least.
Adsorption was directly correlated with organic matter, whereas clay
content was associated, but not directly correlated, with adsorption
[73]. Thus, leaching of pebulate, EPTC, diallate, and cycloate was
directly related to water solubility and inversely correlated with
organic matter [73] which substantiates previous data [74]. In
mineral soils, the depth of leaching depended upon water solubility
[molinate (800 ppmw) > EPTC (370 ppmw) > vernolate (90 ppmw) >
pebulate (60 ppmw) > cycloate (85 ppmw)] (Table 4) [74]. Depth
of leaching decreased as clay and organic matter contents increased;
and in peat soils (35% o.m.) there was no movement out of the treated
zone for any of the five carbamothioates tested when leached with
20.3 cm water [74]. Lateral diffusion was restricted more than
vertical leaching [73].

By using a charcoal barrier method, root, shoot, and seed exposures were found to give greatly different responses between
species [75]. Thus, the potential for leaching into deep root zones
may be a problem. Eradicane is a mixture of EPTC + dichlormid
(Table 8), and these compounds leach through soil at different rates
[76]. When they become separated, injury to corn occurs. When
2-, 4-, 6-, or 8-week-old corn was treated with 12.5 or 25.0 ppmw
EPTC + dichlormid and incorporated with water, 2-week-old corn
had outgrown the EPTC influence at 56 days, whereas the corn
treated at 4 weeks of age was injured [77]. These responses have
been shown to be dependent upon temperature [78], soil moisture
[78], crop cultivar [75-79], herbicide concentration [78], herbicide
placement [78-81], and soil type [72-74]. Addition of "extenders"
to decrease the microbial degradation rate will compound the
problem.

E. Degradation by Photodecomposition

Cleavage of carbamothioate molecules by light has been ignored
until quite recently because separation of the quantity degraded from
the quantity volatilized was virtually impossible. Recently, EPTC,
pebulate, and cycloate dissolved in hexane were reported to be degraded by ultraviolet light [22]. These data were followed by reports of solar photooxidation of molinate and thiobencarb in dilute
(100 μM) hydrogen peroxide [82]. Since irrigation water and other
surface waters reportedly contain H_2O_2 at concentrations of 30 μM
[83], this photodegradation could explain the loss of carbamothioate
herbicides in aquatic systems. Photodegradation of molinate ($T_{1/2} =$
180 hr) and thiobencarb ($T_{1/2} = \sim 35$ hr) [83] was not rapid, but
the products were reported to be sulfoxides, aldehydes, hydroxylated
aromatic rings, N-dealkylated materials, and oxidation of saturated

carbon atoms. Irradiation (300 nm) of diallate in aqueous solution, on pyrex, or on silica gel reportedly produced acroleins or halo-acroleins, which are transient substances [27] and are highly chem-ically reactive with any sulfhydryl group [2,27,40]. Photodegrada-tion rates in oxygenated solutions was 1%, and the acroleins were reported in pmol quantities from solutions starting with 0.5 mM con-centrations [27]. Thus, photodegradation is probably not a major mode of degradation and(or) a mode for the production of compounds with great biological activity.

III. MODE OF ACTION

A. Uptake and Translocation

Studies of carbamothioate absorption and translocation are in-extricably combined with selectivity, growth rates, morphology, degradation, volatility, etc. Charcoal barrier studies on EPTC absorption by shoot, root, and seed absorption [75] demonstrated a differential response among species. Differential vernolate degrada-tion by soybean seedlings was dependent upon the age of the seed-lings [49]. An excellent review on carbamothioate uptake has been published [84].

CDEC was absorbed by roots and translocated apoplastically to leaves [15] but was not absorbed by leaves and was easily washed off of the foliage by precipitation [15]. This lack of leaf absorption was probably due to a lack of cuticular penetration because CDEC inhibited [14C-UL]sucrose accumulation in sorghum leaves that were abraded with carborundum to remove the epicuticular wax [85].

[14C]Diallate was not appreciably absorbed by roots or trans-located to leaves of seedling wild oat or flax [86]; however, diallate was absorbed and translocated symplastically and apoplastically when applied to coleoptiles. Diallate was translocated apoplastically when applied to the roots of emerged plants. [14C]Diallate was found in all parts of the plant when applied to emerged leaves. Similar to CDEC, diallate inhibited the absorption of [14C-UL]sucrose in abraded sorghum leaves [85]. Diallate and triallate absorption was found to be principally through the coleoptilar region in *Avena* seedlings, and phytotoxicity was greatest when these compounds were applied 10–15 mm above the coleoptile node [87]. Thus, depth of seeding and depth of incorporation were important in the selectivity between wild oat and wheat or barley [85], which was determined by the relative growth rates of coleoptilar tissues [86,87].

[35S]EPTC was highly mobile in plants. Distribution from foliar application was to growing tissues while distribution from root appli-cations was to all tissues with accumulation in growing tissues [88, 89]. Sites of [14C]EPTC absorption determined the quantity of

herbicide present in the tissue and the physiological response [90].
Barley was most tolerant of root-absorbed EPTC; wheat, oat, sorghum, and giant foxtail increased in sensitivity in the order named.
Roots were the major site of absorption by barley; but, injury to
the other species from root exposure to EPTC was equal to or slightly
less than that from shoot exposure. Thus, differences in tolerance
were associated with site of uptake.

[^{14}C]Vernolate was absorbed by all tissues of wheat seedlings
[91]; [^{14}C]molinate moved acropetally from leaf application and
throughout the plant from root treatments [92]. Because carbamothioates are normally applied as incorporated materials for absorption
by roots, coleoptiles, and lower stems, few studies have been reported on carbamothioate absorption by aerial tissues. [^{14}C]Verno-
late was rapidly absorbed by soybean pods [44]. Data from the
Herbicide Handbook [2,15] show that absorption by leaves generally
results in acropetal translocation while absorption by plant tissues
exposed to soil results in apoplastic (acropetal) movement. Symplastic translocation from leaf application is slow.

When EPTC and CDEC were applied preemergence-incorporated
to tomato seedlings irrigated with saline water, greater concentrations of total and residual herbicide were found in the seedlings [93]
than was present in plants irrigated with nonsaline water. Combination of EPTC or CDEC plus saline irrigation water resulted in synergistic growth inhibition of tomato or lettuce [94]. These effects
were explained as increased EPTC or CDEC root absorption induced
by salts in the water.

B. Physiological and Biochemical Effects

Biochemical and physiological understanding of carbamothioate
mode of action has lagged far behind the development of their application technology. Carbamothioates have major influence on many
biochemical processes, and designation as to primary or secondary
responses is difficult. Second, separation of morphological responses
and lethality is problematic. Third, these responses change when
different species are utilized. Fourth, antidote modifications of
responses are dependent upon plant species, herbicide concentration
and application technique, experimental technique, and length of
period of exposure.

At sublethal dosages of carbamothioates, grass morphological
responses include a decreased leaf growth that progresses until the
primary leaf fails to penetrate through the coleoptile. At lesser concentrations, primary leaves that penetrate through the coleoptiles
do no unroll. At concentrations that inhibit penetration of the
primary leaf through the coleoptile, one of two conditions may develop. Either the chlorophyll in the enclosed primary leaf is photo-
oxidized and the plant dies or, secondly, after a few days a secondary

leaf will burst through the side of the coleoptile and growth will
ensue, quite often, with the tip of the growing leaf trapped in the
coleoptile; this results in the development of a typical "looped" leaf.
At sublethal dosages, wrinkled and deformed leaves develop. Finally,
a lag period is present in the observation of physiological responses
of plants to carbamothioates.

1. Peroxidase Activity

EPTC stimulated peroxidase activity in corn seedlings and enhanced lignin deposition [95]. This increase in peroxidase (Table
12) was concomitant with inhibition of growth, and both responses
were annulled by the addition of dichlormid but not IAA or GA3.
Enhanced lignin synthesis would freeze primary leaf cells into a
nonexpandable condition. When the EPTC concentration in the tissue was degraded to "inactive" concentrations, growth could resume
by the activity of buds. Alternatively, a portion of this peroxidase
activity could be due to the chloroplast ascorbate-GSH cycle [96].

2. Sulfoxide Formation

Carbamothioates are oxidized to sulfoxides (Fig. 1) [19,22,26,
27] and these sulfoxides are more active than the parent herbicide
molecule [19] (Table 13) (thiobencarb, butylate, cycloate, EPTC,
molinate, pebulate, and vernolate) in curly dock, redroot pigweed,
and wild mustard. The sulfoxides were not injurious to corn (Table

TABLE 12

Peroxidase Activity (μmol H_2O_2
Consumed/min/g FW) at Varying
Times After Shoot Emergence

Days	EPTC (kg/ha)			
	0	2	4	8
	(μmol H_2O_2/min/g FW)			
2	55	55	55	60
3	50	60	75	70
4	52	58	80	90
8	60	110	175	158

Source: From Ref. 95.

TABLE 13

Comparative Injury of Carbamothioates and Their
Sulfoxides on Broadleaf Weeds and Corn

	Control (%)			
	Broadleaf weeds		Corn injury	
Herbicide	Carbamothioate	Sulfoxide	Carbamothioate	Sulfoxide
Thiobencarb	0	0	0	0
Butylate	23	68	0	0
Cycloate	23	77	50	0
EPTC	58	90	70	0
Molinate	17	33	10	0
Pebulate	17	63	50	0
Vernolate	37	95	80	0

Source: From Ref. 19.

13) [19], whereas five of the seven parent herbicides injured corn.
Sulfoxidation increased the injurious influence of the herbicides to
the weeds by 2-3 times. Sulfoxides are rapidly conjugated to
glutathione SH (GSH), and this conjugation was initially proposed
to occur by activity of glutathione-S-transferase (GST) [19]. Corn
root concentrations of GSH and GST were increased after treatment
with dichlormid [97] but similar increases in oat tissue did not occur.
The sulfoxides were proposed as the active herbicidal compounds
[98] and degradation mechanisms from rats and corn roots were reported [39]. Subsequently, sulfoxides have been reported to be
produced by CDEC, diallate, and triallate [25]. This concept of the
carbamothioate activation was subjected to intense evaluation. Reportedly, GST was not requisite to the conjugation because EPTC-
sulfoxide (EPTC-SO) and GSH reacted nonenzymatically [40] and the
response to EPTC and butylate in GST-deficient corn cultivars,
which were highly susceptible to atrazine (6-chloro-N-ethyl-N'-
(1-methylethyl)-1,3,5-triazine-2,4-diamine) (detoxified by GSH conjugation with GST), was not altered by the addition of dichlormid
[99] (Table 14). These atrazine-sensitive varieties were induced by
dichlormid to produce GSH and GST [42], however, after it was
shown that corn produces novel GST enzyme species after treatment

TABLE 14

Height (cm/plant) of Atrazine-Susceptible,
GSH-S-transferase Deficient Corn
(Inbred Line GT-112) to EPTC,
Butylate, and Atrazine

	Dichlormid (kg/ha)	
	0.0	1.12
Control	21.7 c[a]	23.7 c
Butylate	23.4 c	22.1 c
Atrazine	6.9 a	8.9 a
Control	41.0 b	34.5 b
EPTC	39.0 b	36.6 b
Atrazine	14.0 a	17.4 a

[a]Means in a box followed by the same
letter are not significantly different at
the 5% level.
Source: From Ref. 99. Reprinted with
permission from J. Agric. Fd. Chem.,
27:533 (1979). Copyright 1979, American
Chemical Society.

by the antidote [41]. Subsequently, the direct phytotoxicity in
corn was shown to be EPTC-sulfone > EPTC > EPTC-SO [100] and
it was concluded: (a) sulfoxidation and GSH conjugation were
equally important in EPTC detoxification, (b) detoxification was an
important factor in the antidote effect, and (c) phytotoxicity of
EPTC was not due to sulfone activity and only partly due to EPTC-
SO. Later studies [101] showed that dichlormid induced increased
levels of GSH and cytosolic GST in corn while microsomal levels of
GST were increased by both dichlormid and NA.

Light is requisite to the induction of leaf growth inhibition in
corn [102] and wheat [103], and these responses are highly dependent upon high light intensity. Since photooxidation of carbamothioates has been reported in water [27] and dilute H_2O_2 [26], the
increase in peroxidase activity in EPTC-treated corn [95] would
indicate increased concentrations of H_2O_2 in EPTC-treated corn.

Hydrogen peroxide causes very rapid peroxidation of unsaturated fatty acids, which are major constituents of membranes. This peroxidation results in the loss of membrane functionality. But the ascorbate-GSH cycle removes superoxide and H_2O_2 from chloroplasts [96] (Fig. 4); and total ascorbate (13–14 mM) and GSH (mM) concentrations in spinach and pea chloroplasts [104,105] are: (a) 100× the possible sulfoxide concentrations within chloroplasts that have never penetrated through the coleoptile or been involved with accumulation of carbamothioate via transpiration, and (b) total carbamothioates are present in μM concentrations in tissue that has ceased growing [106] (Fig. 5). Additionally, EPTC inhibited incorporation of precursors into lipids of corn calli within 2 hr, whereas dichlormid stimulated GSH levels after 15 hr [50,51].

Thus, the current consensus accepts dichlormid and NA as factors in increased carbamothioate detoxification in corn, however, the concentration of SH-bearing substances in corn leaves and the time scale of EPTC effects on cells does not fit a role for EPTC-SO as a major phytotoxic agent.

3. Photosynthesis and Respiration

$^{14}CO_2$ fixation was not inhibited in kidney bean by 50 μM EPTC after a 50 hr exposure and O_2 uptake was not greatly influenced by 0.53 mM EPTC applied to excised embryos of corn and mung bean; however, oxidative phosphorylation was inhibited by 10 mM but not 0.1 mM EPTC [107]. Diallate (<50 μM) inhibited electron transport between PSII and PSI but did not influence PSI and acted as an uncoupler with inhibition of photophosphorylation [108].

EPTC and diallate influence fatty acid synthesis and gibberellin precursor synthesis in similar fashions. The influence of both compounds has a lag period before physiological responses develop.

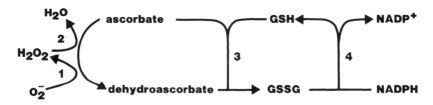

FIG. 4   Ascorbate-GSH cycle for removal of superoxide and H_2O_2 in chloroplasts. Reactions involved: (1) direct non-enzymatic reaction of O_2 and ascorbate, (2) chloroplast ascorbate peroxidase, (3) nonenzymatic at pH 8 and by chloroplast dehydroascorbate reductase, (4) chloroplast glutathione reductase [96].

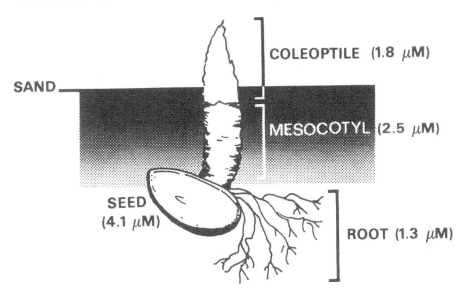

FIG. 5   [^{14}C]Vernolate concentration in wheat seedlings at stasis [91].

Electron transfer modification might explain many of the physiological responses of carbamothioates if it developed after a lag period. And $^{14}CO_2$ fixation and respiratory data for EPTC were obtained using tissue not grown in EPTC. Therefore, evaluation of ferricyanide reduction, DCPIP reduction, and photophosphorylation was undertaken in isolated wheat chloroplasts from : (a) EPTC-treated and (b) untreated plants + EPTC in the reaction. Figures 6 through 8 [91] show increased electron transport but very little influence on photophosphorylation in chloroplasts from EPTC-treated wheat; chloroplasts from untreated wheat were not affected by EPTC added to the reaction chamber. Thus, EPTC per se does not influence photosynthesis (PS). EPTC does, however, induce other physiological responses that indirectly influence PS.

Respiration ($\mu\ell/O_2$/hr/g FW) was decreased in sorghum exposed to 55 g EPTC/ha/cm during growth [109]. Since the pyruvate dehydrogenase complex (PDC) in wheat was 80% inhibited by EPTC (0.01 μM) [110], the inhibition of respiration in tissues from plants grown during exposure to EPTC is explicable.

These responses are corroborated by: (a) a very low influence of EPTC on root growth at concentrations that greatly inhibit shoot growth [111] in plants grown in nutrient solution, and (b) the light

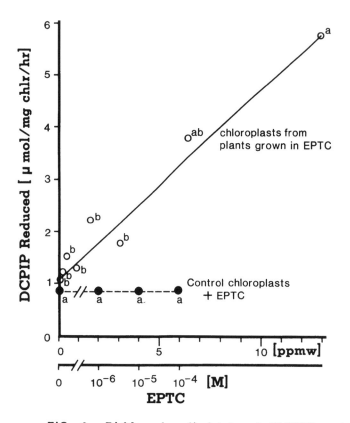

FIG. 6   Dichlorophenolindolphenol (DCPIP) reduction (μmol/mg chlr/hr) in isolated wheat chloroplasts.   o-o plants grown exposed to EPTC, •-• chloroplasts from untreated plants with EPTC added to the reaction mixture. Plants were grown in nutrient solution for 14 days [91].

requirement for the development of EPTC-induced morphological response in corn [102] and wheat [103] leaves.

4. Lipids

EPTC inhibited leaf wax synthesis in cabbage [112,113], sickle-pod petioles [114,115], and corn leaves and roots [116]. Addition-ally, diallate inhibited epicuticular wax biosynthesis in pea [117] and sicklepod leaves [118]. These activities were presumed to be in-hibitions of fatty acid elongation [114,115,117,118]. Inhibition of fatty acid synthesis and desaturation by EPTC in spinach chloro-plasts and reversal by dichlormid and NA led to the hypothesis of

membrane disruption by alteration of desaturation levels in complex
lipids [119,120].   This hypothesis has been extended to butylate,
pebulate, and vernolate in spinach chloroplasts [121].   Alteration of
fatty acid desaturation levels was demonstrated in "aged" beed discs
[122], wheat root phospholipids [123], soybean leaf complex lipids
[124], wheat leaf galactolipids [125], and wheat chloroplast lipid
[126].   Concentration levels required for these activities were in the
μM range and short-term studies on corn cells showed an inhibition
of EPTC on fatty acid synthesis and desaturation within 2 hr that
was reversed by dichlormid [50,51] while the major influence was on
neutral lipids.   Turnover time for the fatty acids in complex membrane lipids would match the lag time for the development of morphological responses; but, enzymatic proof of alteration of enzyme
activity by lipids with a decreased desaturation ratio has not been
reported.

Acetate (Ac^-) incorporation into chloroplast lipids was com-
petitively reversed by EPTC (Fig. 9) [91].   Acetate is produced

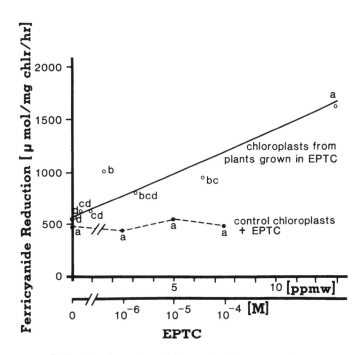

FIG. 7   Ferricyanide reduction (μmol/mg chlr/hr) in isolated
wheat chloroplasts.   ○-○-plants grown exposed to EPTC, ●-● chloroplasts from untreated plants with EPTC added to the reaction mixture.   Plants were grown in nutrient solution for 14 days [91].

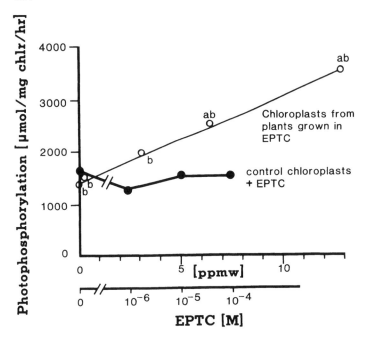

FIG. 8   Photophosphorylation (μmol/mg chl/hr) in isolated wheat chloroplasts.   o-o-plants grown exposed to EPTC, •-• chloroplasts from untreated plants with EPTC added to the reaction mixture.   Plants were gorwn in nutrient solution for 14 days [91].

by PDC and β-oxidation of triglycerides.   β-Oxidation is a major source of Ac⁻ in the seedling stage but essentially ceases to function after 8–14 days and is highly dependent upon triglyceride storage in seeds.   Since PDC is considered to be a mitochondrial enzyme, a means of transfer of Ac⁻ from mitochondria to chloroplasts was requisite to explain carbamothioate inhibition of fatty acid synthesis. Proof of presence of PDC in chloroplasts [127] eliminated this problem, and inhibition of chloroplastic PDC (Fig. 10) [91] explained why EPTC inhibited fatty acid synthesis.

Since linoleic acid desaturation (18:2 → 18:3) takes place in the chloroplast [127], some influence on H^+ transfer must occur.   Inhibition of zeaxanthin [103] and quinone oxidation [128], as well as conversion of geranylgeranyl-chlorophyll → phytol-chlorophyll [129], may be symptoms of modified H^+ transfer.

5.   Isoprenoid Synthesis

Six-day-old cucumber cotyledons treated on day 2 with EPTC had 35–45% more protochlorophyll and 30–40% more glycolipids than

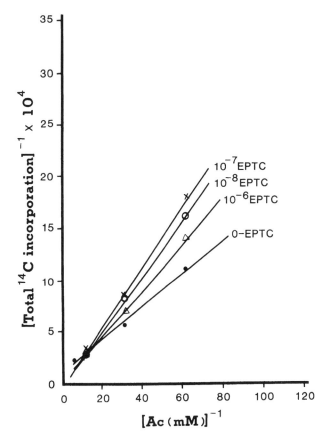

FIG. 9   Influence of EPTC on the incorporation of $[^{14}C]$-acetate into fatty acids by isolated wheat chloroplasts [91].

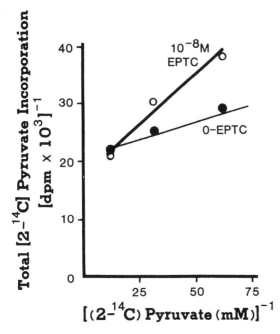

FIG. 10   Influence of EPTC on the incorporation of [2-^{14}C]-pyruvate into fatty acids by isolated wheat chloroplasts [91].

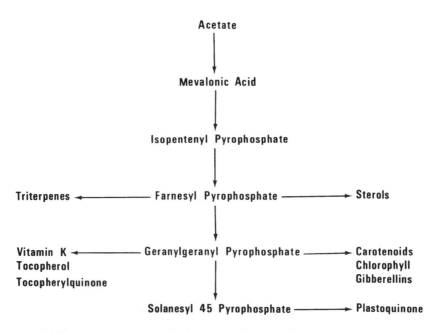

FIG. 11   Terpenoid biosynthesis [131].

the controls, which resulted in increased chlorophyll accumulation
[130]. Increased chlorophyll, carotene, and quinone concentrations
were found in EPTC-treated wheat [103,128]. In terpenoid synthesis
[125] (Fig. 11), geranylgeranylpyrophosphate (GGPP) is a branch
point between chlorophyll, carotene, gibberellin, and quinone syntheses. Inhibition of the synthesis of any terpenoid product results in the accumulation of precursor(s) and other products. Gib-
berellins (GA) are synthesized by cyclization of GGPP → kaurene
[(-)-kaur-16-ene] oxidation of kaurene → kaurenoic acid [(-)]-kaur-
16-en-19-oic acid] [132,133] (Fig. 12). Since GA controls grass
leaf and coleoptile extension, modification of GA synthesis was a
logical site of carbamothioate activity. EPTC decreased GA content
in wheat and inhibited kaurene synthesis (KS) and kaurene oxidation
(KO) [134]. Corn leaf growth and total GA content was decreased
by EPTC, whereas added GA reversed the EPTC responses [135].
In the lettuce hypocotyl test, EPTC (0.01–1.0 mM) decreased the

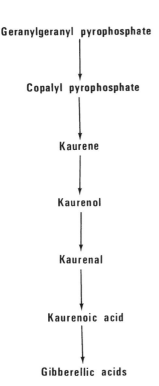

Geranylgeranyl pyrophosphate

↓

Copalyl pyrophosphate

↓

Kaurene

↓

Kaurenol

↓

Kaurenal

↓

Kaurenoic acid

↓

Gibberellic acids

FIG. 12   Conversion of geranylgeranyl phyrophosphate to
gibberellin via kaurene synthesis and kaurene oxidation.

response to added GA [136], and this response was postulated to
be due to an EPTC influence on GA metabolism. But, the GA influence on cell expansion is dependent on functional membranes, which
are disrupted by high (1.0 mM) EPTC concentrations [106]. In
cell-free enzyme preparations from etiolated, unruptured sorghum
coleoptiles, EPTC inhibited KS and some reaction in GA metabolism
so that kaurenol decreased but kaurenal and kaurenoic acid were
not altered (Fig. 13) [137]. Dichlormid appeared to modify the inhibition of KS and the combination of EPTC + dichlormid reversed
the KS and GA metabolism inhibitions. Total [^{14}C] mevalonate incorporation into GA precursors was not altered by EPTC or dichlormid.
The concept of an inhibition of GA synthesis and modification as a
major site for EPTC-induced morphological responses was corroborated by concomitant application of EPTC and GA where the activity
of these compounds was competitive in the growth in length of
sorghum leaves (Fig. 14) [91]. This corroborates previous work
with EPTC [135,136] and molinate [92].

Whether GA synthesis and metabolism are *primary* sites of
carbamothioate activity is problematic. The biochemistry of GA
metabolism is virtually unknown. Kaurenoid synthesis and oxidation
require ATP, NADPH or NADH, and O$_2$. EPTC inhibits utilization
of NADPH utilization so that zeaxanthin is accumulated [103] and the

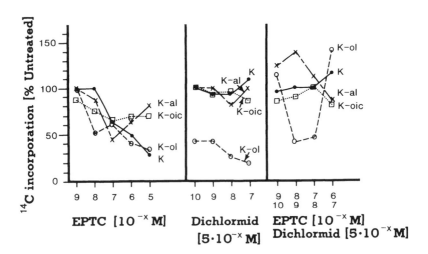

FIG. 13   Influence of EPTC, dichlormid, and EPTC + dichlormid
on incorporation of [^{14}C]mevalonic acid into kaurenoids by a cell-
free enzyme system from unruptured, etiolated sorghum coleoptiles.
K = kaurene, K-ol = daurenol, K-al = kaurenal, and K-oic = kaurenoic
acid [137].

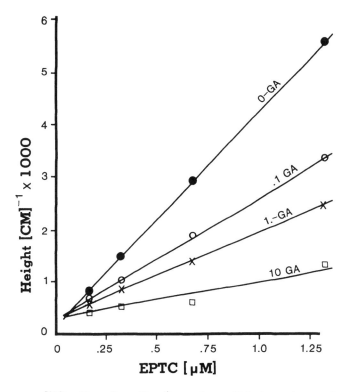

FIG. 14   Growth of sorghum (14 days in sand containing EPTC and gibberellic acid (GA3) [91].

saturation reaction GG-chlorophyl → phytol-chlorophyll [130] is inhibited.   Thus, some as yet unknown reaction requiring H^+ and(or) electron transfer that is highly responsive to carbamothioates may be involved in these processes.

6.   Nucleic Acid Synthesis

In 3-day-old soybean hypocotyl sections (0.5 to 1.5 cm below cotyledons), EPTC inhibited growth and nucleic acid synthesis [138]. EPTC (1 mM) inhibited ribosomal-RNA (r-RNA), DNA like RNA, and tenaciously bound RNA syntheses.   In r-RNA, EPTC inhibited synthesis of 18S-r-RNA more than 25S r-RNA.   [^{14}C]Uridine incorporation into "aged" potato tuber discs was increased by EPTC (Table 15) [91].

TABLE 15

Incorporation of [14C]Leucine and [14C]Uridine
into "Aged" Potato Discs After
24 hr Exposure to EPTC[a]

EPTC (μM)	DPM	
	[14C]Leucine	[14C]Uridine
0	609.3 a[b]	278.8 d
1	425.6 b	362.7 c
10	395.2 b,c	503.5 b
100	368.0 c	674.0 a

[a]Discs were washed with 10% TCA, MeOH, 70%
EtOH, 95% EtOH, and boiling 95% EtOH prior
to counting.
[b]Values in a column followed by the same
letter are not significantly different at the 5%
level.
Source: From Ref. 91.

7. Proteolytic Activity and Protein Synthesis

Proteolytic activity of 3-day-old squash seedlings was 75% inhibited by CDEC (100 μM) [139]. EPTC (600 μM) and CDEC (200 μM)
inhibited α-amylase and RNA synthesis activity in germinating barley
seeds [140]. [14C]Leucine incorporation was inhibited 33% by CDEC
and 24% by EPTC. [14C]Leucine incorporation into "aged" potato
tuber discs was decreased by EPTC after 16-hr exposure (Table 15)
[91].

8. Mode of Action Summary

Multiplicity of physiological and biochemical responses to carbamothioates depends upon plant species, environmental conditions,
specimen age, herbicide, and herbicide concentration. Thus, a commonality of response is difficult to document. Empirical observation
supports: (a) the separation of morphological responses from lethality, and (b) a time lag in both morphological and lethal responses.
A light intensity requirement for EPTC activity in wheat and corn,
but not sorghum or vernolate in wheat, does not support the sulfoxide hypothesis (and accompanying SH conjugation) as a mode of

action. This hypothesis fits most data as a means of degradation.
Morphological responses are explicable as influence(s) on GA metabolism, but these may be secondary responses. Inhibition of the PDC
in mitochondria and chloroplasts may be involved in morphological
and biochemical responses, but failure of lethality at 90% inhibition
of PDC is not consistent. Modification of electron transport may,
ultimately, develop into a coherent hypothesis that fits all of the
data, but insufficient data exists currently. Thus, all current hy-
potheses on the mode of action of carbamothioates suffer from insufficient knowledge of general plant physiology and biochemistry,
as well as specific information on the activity of the various herbicides.

C. Selectivity

The degradation rate is a major factor in carbamothioate selectivity [92] (Table 9). EPTC placed in the row with alfalfa seeds
does not injure the alfalfa [3]. Alfalfa seed encapsulated with porus
material containing EPTC was not injured [4], primarily because of
the very rapid EPTC degradation [141]. Vernolate is rapidly degraded by soybean [44,142]. Pebulate is degraded in mung bean
within 8 hr [143] and in tobacco [144,145]. EPTC is rapidly degraded by corn [39,40,43-46,48,49,52,53], and the rate of degradation is increased in corn when dichlormid is applied concomitantly.
Planting depth however, was a major selectivity factor between wheat
and wild oat [80]. The absorptive organ is a major factor in selectivity because of differences in absorption, etc., between shoot and
root [75,90,146]. Selectivity in response to seed exposure was
species dependent; some seeds were almost totally resistant and other
seeds were highly susceptible [75]. Tissue age influenced resistance
of soybean to vernolate [49] or corn to EPTC or EPTC + dichlormid
[77]. Thus, resistance seems to be determined by the ability of
the plant to escape deleterious concentrations.

D. Factors Affecting Toxicity

Light is requisite for the development of morphological and
lethal responses of wheat and corn to EPTC [102,103]; light intensities required are very different. Corn response requires full
summer sunlight whereas wheat responses are developed at much
lower intensities. Salinity induces an increased absorption of EPTC
and CDEC in tomato and lettuce, which results in a synergistic reduction of growth [93,94]. Temperature and soil moisture influenced
corn response to EPTC and the number of days before emergence of
the coleoptile was critical to the injury index [78,79]. Consequently,
every factor in the ecosystem that influences toxicity and species
biochemistry may be of primary importance.

E. Interactions

Antagonistic response of 2,4-D [(2,4-dichlorophenoxy)acetic acid] and EPTC combinations on corn growth [109] were not based on changes of absorption or movement [109] and 2,4-D reversed an EPTC inhibition of respiration. Later work reported an EPTC inhibition of RNA in soybean hypocotyls that was reversed by 2,4-D [138]. Combinations of EPTC ± dichlormid or butylate ± dichlormid and tebuthiuron on corn produced synergistic inhibitions of height at 24 days after treatment [147].

Crop rotations and the associated herbicide application patterns may induce problems. Where alfalfa (EPTC-treated) was planted on land utilized for corn (atrazine-treated) production the previous year, atrazine injury on the alfalfa was observed [148]. This resulted from an EPTC-induced increase in water utilization, with an associated increased atrazine absorption [111,149-151]. Atrazine degradation by alfalfa was not sufficient to detoxify the atrazine to sublethal levels, and injury resulted. Similar interactions between herbicides based on physiological responses were documented with the inhibition by EPTC of epicuticular wax development in cabbage [112], which resulted in increased susceptibility of the cabbage to secondary applications of alkanolamine salts of 4,6-dinitro-o-sec-butylphenol (DNBP). The influence on herbicide cuticular penetration following preemergence soil-incorporated carbamothioate applications has been demonstrated. Sicklepod control by several secondary postemergence herbicides was greater in plots with than without soil-incorporated applications of vernolate [152].

Antidotes decreased toxicity in certain species without modifying the activity of the herbicide in other species. Thus, a seed treatment of naphthalic anhydride (NA) (70 g/ha) completely protected corn from EPTC injury at concentrations (6.72 kg/ha) that were lethal [153,154] but did not influence the control of weed species. NA did not decrease EPTC absorption [155]. Dichlormid (0.56 kg/ha) applied concomitantly with EPTC (6.72 kg/ha) produced excellent weed control [156-158] with minimum injury to corn [7,102,159]. Extensive studies on the relationships of antidote structure-activity, mode of application, herbicides structure and concentration, and plant species [159-164] showed that the dichlormid response was: (a) physiological and not related to soil factors or direct interaction of dichlormid and EPTC outside the plant; (b) corn was completely protected by dichlormid, sorghum was slightly protected, and 10 other species were not protected at all; (c) dichlormid and NA have different mechanisms as antidotes; (d) dichlormid protected corn from seven carbamothioates, and (e) antidote activity was most efficient when the structure of the antidote was very close to the structure of the herbicide. For each of the carbamothioate herbicides, the most active antidote was a compound most similar to it with respect to the N,N-disubstituted alkyl groups [159,165]. These responses would appear to be physiological influences on degradation [40]. But short-term biochemical responses do not fit a degradation pattern as the "complete" or "major" mode of activity for the carbamothioate antidotes [50,51,137].

Degradation of some carbamothioates is enhanced in soils previously exposed to carbamothioate herbicides [9-13]. Extenders which decrease carbamothioate degradation rate in soils are currently under development.

IV. CONCLUSIONS

Carbamothioate application technology exceeds our understanding of mode of action, selectivity, or degradation. These herbicides influence many physiological and biochemical responses in plants. Consequently, a cohesive clear description of the mode(s) of activity of these compounds has not been developed. Since selectivity and mode of action are inextricably bound to degradation, an understanding of the biochemical bases for selectivity may assist in development of answers to other carbamothioate problems.

Field utilization is a total ecosystem approach. Plant species, plant age, soils, and physical environment influence the results obtained. Possibly these are but reaffirmation of insufficient basic plant biochemistry knowledge to permit recognition of their activity. One clear fact remains, research in this highly active area will continue.

REFERENCES

1. J. Antognini et al., *Proc. Northeast Weed Control Conf.*, 1957 p. 2.
2. *Herbicide Handbook of the Weed Science Society of America*, 5th ed., Weed Sci. Soc. Am. Champaign, IL, 1983.
3. J. H. Dawson, *Weed Sci.*, *28*, 607 (1980).
4. J. H. Dawson, *Weed Sci.*, *29*, 105 (1981).
5. D. L. Linscott, R. R. Seaney, and R. D. Hagin, *Weed Sci.*, *15*, 259 (1967).
6. D. L. Linscott and R. D. Hagin, *Weed Sci.*, *15*, 264 (1967).
7. F. Y. Chang, J. D. Bandeen, and G. R. Stephenson, *Can. J. Plant Sci.*, *52*, 704 (1972).
8. L. J. Rains and O. H. Fletchall, *Proc. North Cent. Weed Contr. Conf.*, *26*, 42 (1971).
9. A. G. Rahman, G. C. Atkinson, and J. A. Douglass, *N. Z. J. Agric.*, *139*, 47 (1979).

10. T. Obrigawitch, R. G. Wilson, A. R. Martin, and F. W. Roeth, *Weed Sci.*, *30*, 175 (1982).

11. T. Obrigawitch, F. W. Roeth, A. R. Martin, and R. G. Wilson, *Weed Sci.*, *30*, 417 (1982).

12. T. Obrigawitch, A. R. Martin, and F. W. Roeth, *Weed Sci.*, *31*, 187 (1983).

13. R. G. Wilson, *Weed Sci.*, *32*, 264 (1984).

14. *The Pesticide Manual. A World Compendium.* The British Crop Protection Council, 7th ed., 1983.

15. *Herbicide Handbook of the Weed Science Society of America*, 4th ed., Weed Sci. Soc. Am., Champaign, IL, 1979.

16. S. C. Schimmel, R. L. Garnas, J. M. Patrick, Jr., and J. C. Moore, *J. Agric. Food Chem.*, *31*, 104 (1983).

17. H. Hirata, S. Yamasaki, T. Hisadome, and T. Makita, *Int. Assoc. Theor. Appl. Limnol.*, *21*, 1314 (1981).

18. S. Chaiyarach, V. Ratonanum, and R. C. Harrell, *Bull. Environ. Contam. Toxicol.*, *14*, 281 (1975).

19. J. E. Casida, R. A. Gray, and H. Tiles, *Science*, *184*, 573 (1974).

20. Y. S. Chen, I. Schuphan, and J. E. Casida, *J. Agric. Food Chem.*, *27*, 709 (1979).

21. I. Schuphan, J. D. Rosen, and J. E. Casida, *Science*, *205*, 1013 (1979).

22. A. C. DeMarco and E. R. Hayes, *Chemosphere*, 5, 321 (1979).

23. J. D. Rosen, I. Schuphan, Y. Segall, and J. E. Casida, *J. Agric. Food Chem.*, *28*, 880 (1980).

24. J. D. Rosen, Y. Segall, and J. E. Casida, *Mutat. Res.*, *78*, 113 (1980).

25. P. J. Marsden and J. E. Casida, *J. Agric. Food Chem.*, *30*, 627 (1982).

26. W. M. Draper and J. E. Casida, *J. Agric. Food Chem.*, *32*, 231 (1984).

27. L. O. Ruzo and J. E. Casida, *J. Agric. Food Chem.*, *33*, 272 (1985).

28. B. N. Ames, J. McCann, and E. Yamasaki, *Mutat. Res.*, *31*, 347 (1975).

29. H. Tilles, *J. Am. Chem. Soc.*, *81*, 714 (1959).

30. G. G. Patchett and G. H. Batchelder, *J. Anal. Meth. Pestic. Plant Growth Regul.*, 5, 491 (1967).

31. G. G. Patchett and G. H. Batchelder, *J. Anal. Meth. Pestic. Plant Growth Regul.*, 5, 469 (1967).

32. G. G. Patchett, G. H. Batchelder, and J. J. Menn, *J. Anal. Meth. Pestic. Plant Growth Regul.*, 4, 117 (1964).

33. J. E. Barney, W. Y. Ja, and G. G. Patchett, *J. Anal. Meth. Pestic. Plant Growth Regul.*, 7, 641 (1973).

34. G. G. Patchett and G. H. Batchelder, *J. Anal. Meth. Pestic. Plant Growth Regul.*, 5, 537 (1967).

35. G. G. Patchett, G. H. Batchelder, and J. Menn, *J. Anal. Meth. Pestic. Plant Growth Regul.*, *4*, 343 (1964).
36. R. A. Gray, *J. Anal. Meth. Pestic. Plant Growth Regul.*, *3*, 177 (1964).
37. S. C. Fang and T. C. Yu, *West. Weed Control Conf. Res. Prog. Rep. 1959*, 91.
38. J. D. Nalewaja, R. Behrens, and A. R. Schmid, *Weeds*, *12*, 269 (1964).
39. J. P. Hubbell and J. E. Casida, *J. Agric. Food Chem.*, *25*, 404 (1977).
40. R. D. Carringer, C. E. Rieck, and L. P. Bush, *Weed Sci.*, *26*, 157 (1978).
41. T. J. Mozer, D. C. Tiemeir, and E. G. Jaworski, *Biochemistry*, *22*, 1068 (1983).
42. M.-M. Lay and A. M. Niland, *Pestic. Biochem. Physiol.*, *23*, 131 (1985).
43. G. L. Lamoreaux and D. G. Rushness, *Pesticide Chemistry Human Welfare and the Environment*: Vol. 3. *Mode of Action, Metabolism and Toxicology* (J. Miyamoto and P. C. Kearney, eds.), Pergamon Press, New York, 1983, p. 295.
44. R. E. Wilkinson, *Pestic. Biochem. Physiol.*, *20*, 347 (1983).
45. M.-M. Lay and J. E. Casida, *Pestic. Biochem. Physiol.*, *6*, 442 (1976).
46. R. D. Carringer, C. E. Rieck, and L. P. Bush, *Weed Sci.*, *26*, 167 (1978).
47. A. E. Smith, *Weed Res.*, *10*, 331 (1970).
48. G. R. Stephenson, A. Ali, and F. M. Ashton, *Pesticide Chemistry, Human Welfare and the Environment*: Vol. 3, *Mode of Action, Metabolism, and Toxicology* (J. Miyamoto and P. C. Kearney, eds.), Pergamon Press, New York, 1083, p. 295.
49. R. H. Shimabukuro, G. L. Lamoreaux, and D. S. Frear, *Chemistry and Action of Herbicide Antidotes* (F. M. Pallos and J. E. Casida, eds.), Academic Press, New York, 1978, p. 133.
50. G. Ezra and J. Gressel, *Pestic. Biochem. Physiol.*, *17*, 48 (1982).
51. G. Ezra, H. M. Flowers, and J. Gressel, *Pesticide Chemistry Human Welfare and the Environment*: Vol. 3, *Mode of Action, Metabolism, and Toxicology* (J. Miyamoto and P. C. Kearney, eds.), Pergamon Press, New York, 1983, p. 225.
52. J. B. Bourke and S. C. Fang, *Weed Sci.*, *16*, 290 (1968).
53. S. C. Fang and M. George, *Plant Physiol. Suppl.*, *37*, XXVI (1962).
54. I. Schuphan and J. E. Casida, *Pesticide Chemistry Human Welfare and the Environment*: Vol. 3, *Mode of Action, Metabolism, and Toxicology* (J. Miyamoto and P. C. Kearney, eds.), Pergamon Press, New York, 1983, p. 287.

55. P. J. Marsden and J. E. Casida, *J. Agric. Food Chem.*, *30*, 627 (1982).
56. L. L. Danielson and W. A. Gentner, *Weed Sci.*, *12*, 92 (1964).
57. R. A. Gray and A. J. Weierich, *Weed Sci.*, *13*, 141 (1965).
58. S. C. Fang, P. Theisen, and V. H. Freed, *Weeds*, *9*, 569 (1961).
59. R. A. Gray, *Weed Sci.*, *13*, 138 (1965).
60. O. B. Wooten, J. T. Holstun, Jr., and R. S. Baker, *Weed Sci.*, *14*, 92 (1966).
61. T. W. Waldrep and J. F. Freeman, *Weed Sci.*, *12*, 315 (1964).
62. R. A. Gray and A. J. Weierich, *Proc. 9th Br. Weed Contr. Conf.*, 94 (1968).
63. G. B. Beestman and J. M. Deming, *Weed Sci.*, *24*, 541 (1976).
64. I. J. MacRae and M. Alexander, *J. Agric. Food Chem.*, *13*, 72 (1966).
65. T. J. Sheets, *Weeds*, *7*, 442 (1959).
66. D. D. kaufman, *J. Agric. Food Chem.*, *15*, 582 (1967).
67. J. Antognini, R. A. Gray, and J. J. Menn, *Proc. Selective Weed Conf. Beetcrops, 2nd Int.*, *1*, 293 (1970).
68. J. D. Fryer and K. Kirkland, *Weed Res.*, *10*, 133 (1970).
69. V. M. Thomas and C. L. Holt, *J. Environ. Sci. Health*, *B15*, 475 (1980).
70. S. Duah-Yentumi and S. Kuwatsuka, *Soil Sci. Plant Nutr.*, *26*, 541 (1980).
71. S. Duah-Yentumi and S. Kuwatsuka, *Soil Sci. Plant Nutr.*, *28*, 19 (1982).
72. E. Koren, C. L. Foy, and F. M. Ashton, *Weed Sci.*, *16*, 172 (1968).
73. E. Koren, C. L. Foy, and F. M. Ashton, *Weed Sci.*, *17*, 148 (1969).
74. R. A. Gray and A. J. Weierich, *Weed Sci.*, *16*, 77 (1968).
75. R. A. Gray and A. J. Weierich, *Weed Sci.*, *17*, 223 (1969).
76. C. A. Buzio and G. W. Bert, *Weed Sci.*, *28*, 241 (1980).
77. G. W. Burt and C. A. Buzio, *Weed Sci.*, *27*, 460 (1979).
78. G. W. Burt and A. O. A. Kinsorotan, *Weed Sci.*, *24*, 319 (1976).
79. G. W. Burt, *Weed Sci.*, *24*, 327 (1976).
80. S. D. Miller and J. D. Nalewaja, *Weed Sci.*, *24*, 134 (1976).
81. A. G. Ogg and S. R. Drake, *Weed Sci.*, *30*, 446 (1982).
82. W. M. Draper and D. C. Crosby, *J. Agric. Food Chem.*, *32*, 231 (1984).
83. W. M. Draper and D. C. Crosby, *Arch. Environ. Contam. Toxicol.*, *12*, 121 (1983).
84. R. A. Gray and G. K. Joo, *Chemistry and Action of Herbicide Antidotes* (F. M. Pallos and J. E. Casida, eds.), Academic Press, New York, 1978, p. 67.

85. R. E. Wilkinson, *Pestic. Biochem. Physiol.*, *23*, 95 (1985).

86. J. D. Nalewaja, *Weed Sci.*, *16*, 309 (1968).

87. C. Parker, *Weed Res.*, *3*, 259 (1963).

88. S. Yamaguchi, *Weeds*, *9*, 374 (1961).

89. A. S. Crafts and S. Yamaguchi, *The Autoradiography of Plant Materials, Manual 35, Calif. Agric. Exp. Sta.* (1964).

90. L. R. Oliver, N. Prendeville, and M. M. Schreiber, *Weed Sci.*, *16*, 534 (1968).

91. R. E. Wilkinson (unpublished data).

92. T. M. Chen, D. E. Seaman, and F. M. Ashton, *Weed Sci.*, *16*, 28 (1968).

93. S. Acosta-Nunez and F. M. Ashton, *Weed Sci.*, *29*, 548 (1981).

94. S. Acosta-Nunez and F. M. Ashton, *Weed Sci.*, *29*, 692 (1981).

95. B. M. R. Harvey, F. Y. Chang, and R. A. Fletcher, *Can. J. Bot.*, *53*, 225 (1975).

96. B. Halliwell, *Biochemical and Physiological Mechanisms of Herbicide Action* (S. O. Duke, ed.), Symp. So. Sect. Am. Soc. Plant Physiol., 1984, p. 31.

97. M.-M. Lay, J. P. Hubbell, and J. E. Casida, *Science*, *189*, 287 (1975).

98. M.-M. Lay and J. E. Casida, *Pestic. Biochem. Physiol.*, *6*, 442 (1976).

99. J. R. C. Leavitt and D. Penner, *J. Agric. Food Chem.*, *27*, 533 (1979).

100. F. Dtuka and T. Komives, *Pesticide Chemistry, Human Welfare and the Environment*: Vol. 3, *Mode of Action, Metabolism and Toxicology* (J. Miyamoto and P. C. Kearney, eds.), Pergamon Press, New York, 1983, p. 213.

101. T. Komives, M. Balazs, V. Komives, and F. Dutka, *Cytochrome P-450: Biochemistry, Biophysics, and Induction* (L. Vereczkey and K. Magyar, eds.), Akademiai Kiado, Budapest, 1985, p. 451.

102. R. E. Wilkinson, *Chemistry and Action of Herbicide Antidotes* (F. M. Pallos and J. E. Casida, eds.), Academic Press, New York, 1978, p. 85.

103. R. E. Wilkinson, *Bot. Gaz.*, *138*, 270 (1977).

104. C. H. Foyer, *Phytochemistry*, *16*, 1347 (1977).

105. M. Y. Law, S. A. Charles, and B. Halliwell, *Biochem. J.*, *210*, 899 (1983).

106. R. E. Wilkinson, *Pestic. Biochem. Physiol.*, *24* (in press) (1985).

107. F. M. Ashton, *Weed Sci.*, *11*, 295 (1963).

108. S. J. Robinson, C. F. Yocum, H. I. Kumara, and F. Hayashi, *Plant Physiol.*, *60*, 840 (1977).

109. C. E. Beste and M. M. Schreiber, *Weed Sci.*, *18*, 484 (1970).

110. T. H. Oswald and R. E. Wilkinson, *Abstracts, Weed Sci. Soc. Am.*, 288 (1985).

111. R. E. Wilkinson and P. Karunen, *Weed Res.*, *17*, 335 (1977).

112. W. A. Gentner, *Weed Sci.*, *14*, 27 (1966).

113. J. A. Flore and M. J. Bukovac, *J. Am. Soc. Hort. Sci.*, *101*, 586 (1976).

114. R. E. Wilkinson and W. S. Hardcastle, *Weed Sci.*, *17*, 335 (1969).

115. R. E. Wilkinson and W. S. Hardcastle, *Weed Sci.*, *18*, 125 (1970).

116. I. Barta, T. Kŏmives, and F. Dutka, *Radiochem. Radioanal. Lett.*, *58*, 357 (1983).

117. G. G. Still, D. G. Davis, and G. L. Zander, *Plant Physiol.*, *46*, 307 (1970).

118. R. E. Wilkinson, *Plant Physiol.*, *53*, 269 (1974).

119. R. E. Wilkinson and A. E. Smith, *Weed Sci.*, *23*, 100 (1975).

120. R. E. Wilkinson and A. E. Smith, *Weed Sci.*, *23*, 90 (1975).

121. R. E. Wilkinson, *Phytochemistry*, *15*, 841 (1976).

122. R. E. Wilkinson and A. E. Smith, *Weed Sci.*, *24*, 335 (1976).

123. P. Karunen and R. E. Wilkinson, *Physiol. Plant*, *35*, 228 (1975).

124. R. E. Wilkinson and A. E. Smith, *Plant Physiol.*, *60*, 86 (1977).

125. P. Karunen and L. Eronen, *Physiol. Plant*, *40*, 101 (1977).

126. L. Eronen and J. Bahl, *Biochemistry and Metabolism of Plant Lipids* (J. F. G. M. Wintermans and P. J. C. Kuiper, eds.), Elsevier Biomedical Press, New York, 1982, p. 217.

127. G. Roughan and R. Slack, *Biogenesis and Function of Plant Lipids* (P. Mazliak, P. Benveniste, C. Costes, and R. Douce, eds.), Elsevier/North Holland Biomedical Press, New York, 1980, p. 11.

128. R. E. Wilkinson, *Pestic. Biochem. Physiol.*, *8*, 208 (1978).

129. R. E. Wilkinson, *Pestic. Biochem. Physiol.*, *23*, 289 (1985).

130. M. B. Weinberg and P. A. Castelfranco, *Weed Sci.*, *23*, 185 (1975).

131. T. W. Goodwin, *Biosynthetic Pathways in Higher Plants* (J. B. Pridham and T. Swain, eds.), Academic Press, New York, 1965, p. 37.

132. M. O. Oster and C. A. West, *Arch. Biochem. Biophysics*, *127*, 112 (1968).

133. D. T. Dennis and C. A. West, *J. Biol. Chem.*, *242*, 3203 (1967).

134. R. E. Wilkinson and D. A. Ashley, *Weed Sci.*, *27*, 270 (1979).

135. W. W. Donald, R. S. Fawcett, and R. G. Harvey, *Weed Sci.*, *27*, 122 (1979).

136. W. W. Donald, *Weed Sci.*, *29*, 490 (1981).

137. R. E. Wilkinson, *Pestic. Biochem. Physiol.*, *19*, 321 (1983).

138. C. E. Beste and M. M. Schreiber, *Weed Sci.*, *20*, 8 (1972).

139. F. M. Ashton, D. Penner, and S. Hoffman, *Weed Sci.*, *16*, 169 (1968).

140. D. E. Moreland, S. S. Malhotra, R. D. Gruenhagen, and H. Shokrah, *Weed Sci.*, *17*, 556 (1969).

141. J. D. Nalewaja, R. Behrens, and A. R. Schmid, *Weeds*, *12*, 269 (1964).

142. J. B. Bourke and S. C. Fang, *J. Agric. Food Chem.*, *13*, 340 (1965).

143. S. C. Fang and M. George, *Plant Physiol.*, *Suppl.*, *37*, XXVI (1962).

144. J. W. Long, L. Thompson, Jr., and C. E. Rieck, *Weed Sci.*, *22*, 42 (1974).

145. J. W. Long, L. Thompson, Jr., and C. E. Rieck, *Weed Sci.*, *22*, 91 (1974).

146. G. N. Prendeville, L. R. Oliver, and M. M. Schreiber, *Weed Sci.*, *16*, 538 (1968).

147. K. K. Hatzios, *Weed Sci.*, *29*, 601 (1981).

148. W. B. Duke, V. S. Rao, and J. F. Hunt, *Proc. North East Weed Cont. Conf.*, *26*, 258 (1972).

149. R. E. Wilkinson and P. Karunen, *Ann. Bot.*, *40*, 1043 (1976).

150. P. Karunen and R. E. Wilkinson, *18th Swedish Weed Conf.*, J2 (1977).

151. P. Karunen and R. E. Wilkinson, *Proc. Eur. Weed Res. Sci. Sympos.*, p. 87 (1977).

152. M. E. Sherman, L. Thompson, Jr., and R. E. Wilkinson, *Weed Sci.*, *31*, 622 (1983).

153. O. L. Hoffmann, *Abstr.*, *Weed Sci. Soc. Amer.*, p. 12 (1969).

154. O. L. Hoffman, *Chemistry and Action of Herbicide Antidotes* (F. M. Pallos and J. E. Casida, eds.), Academic Press, New York, 1978, p. 1.

155. J. T. Murphy, *Chem.-Biol. Interactions*, *5*, 284 (1972).

156. L. J. Rains and O. H. Fletchall, *Proc. North Cent. Weed Cont. Conf.*, *26*, 42 (1971).

157. F. M. Pallos, R. A. Gray, D. R. Arneklev, and M. E. Brooke, *J. Agric. Food Chem.*, *23*, 821 (1975).

158. F. M. Pallos, R. A. Gray, D. R. Arneklev, and M. E. Brooke, *Chemistry and Action of Herbicide Antidotes* (F. M. Pallos and J. E. Casida, eds.), Academic Press, New York, 1978, p. 15.

159. G. R. Stephenson and F. Y. Chang, *Chemistry and Action of Herbicide Antidotes* (F. M. Pallos and J. E. Casida, eds.), Academic Press, New York, 1978, p. 35.

160. F. Y. Chang, J. D. Bandeen, and G. R. Stephenson, *Can. J. Plant Sci.*, *52*, 707 (1972).

161. F. Y. Chang, J. D. Bandeen, and G. R. Stephenson, *Weed Res.*, *13*, 399 (1973).

162. F. Y. Chang, G. R. Stephenson, and J. D. Bandeen, *Weed Sci.*, *21*, 292 (1973).

163. F. Y. Chang, G. R. Stephenson, G. W. Anderson, and J. D. Bandeen, *Weed Sci.*, *22*, 546 (1974).

164. A. M. Blair, C. Parker, and L. Kasasian, *PANS*, *22*, 65 (1976).

165. F. M. Pallos, R. A. Gray, D. R. Arnekelev, and M. E. Brooke, *J. Agric. Food Chem.*, *23*, 821 (1975).

Chapter 6

ALACHLOR

DEXTER B. SHARP*

Monsanto Agricultural Company
St. Louis, Missouri

*Current affiliation: Retired; Consultant.

I. INTRODUCTION

Published information about the degradation and fate of the 2-chloro-acetamide herbicides has increased substantially since Jaworski's 1975 [1] review of alachlor [1], propachlor [2], allidochlor (CDAA) [3], and butachlor [4]. As reported by Hamm [2], propachlor (2-chloro-N-[1-methylethyl]-N-phenylacetamide) [2] was the second 2-chloroacetamide commercially developed by Monsanto Company, following the successful introduction of CDAA (2-chloro-N,N-[diallyl]-acetamide) [3], and prior to the advent of alachlor (2-chloro-N-[2,6-diethylphenyl]-N-[methoxymethyl]-acetamide) [1] and butachlor (2-chloro-N-[2,6-diethylphenyl]-N-[butoxymethyl]acetamide) [4].

In the past decade, dramatic advances have been realized in the design, sophistication, and availability of chromatographic and spectrometric instruments and their associated technologies. These improved tools have greatly expanded the capabilities of scientific investigators to isolate and identify unequivocally the structures of individual compounds in complex mixtures of pesticide metabolites in a variety of sample matrices. Such metabolite identifications have been accomplished with remarkably small amounts of isolated chemical by comparisons of mass spectra with those of authentic synthesized metabolites.

The mode of action of the 2-chloroacetamides remains elusive, although recently uncovered details about the degradation of these herbicides in plants, animals, and other biological systems may encourage increased efforts toward answering this question. This review is designed to consolidate the more recent information on alachlor degradation [1] and to provide leads to definition of the biochemical bases for, and resistance to, the phytotoxic responses induced in plants treated with the 2-chloroacetamides.

Alachlor [1] is the active ingredient in Lasso herbicide manufactured by Monsanto Company [3]. It has been registered for use since 1969 as a preemergence, early postemergence, or shallowly incorporated herbicide for control of most annual grasses and certain broadleaf weeds. It is marketed primarily as several liquid formulations Lasso, Lasso EC, and Lasso MicroTech; the latter is microencapsulated alachlor suspended in water. Additional products include Lasso 15% granule, and package mixes with atrazine (Lasso atrazine granule (15%), and a Lasso/atrazine flowable), and with glyphosate (Bronco). The alachlor emulsifiable liquid formulations are used in tank mixes with other herbicides including: linuron, paraquat, trifluralin, dicamba, cyanazine, simazine, propazine, bifenox, chloramben, chlorpropham, and metribuzin.

Broad-spectrum weed control or suppression includes grasses such as foxtails (*Setaria* spp.), crabgrass (*Digitaria* spp.), barnyardgrass (*Echinochloa crus-galli*) and seedling Johnsongrass (*Sorghum halopense*), and several broadleaf weeds including pigweed

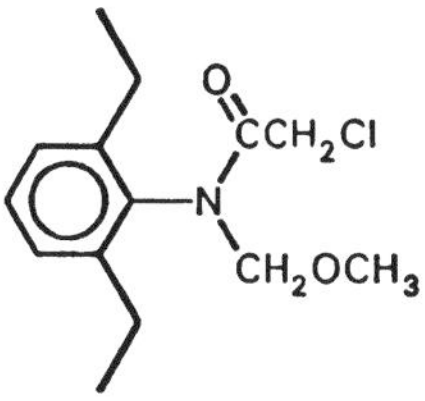

[1]
alachlor

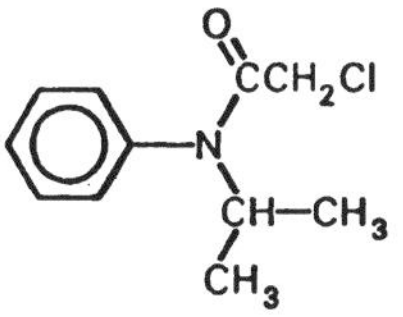

[2]
propachlor

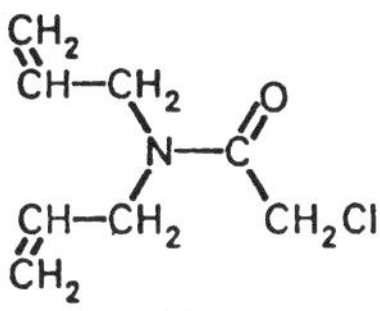

[3]
allidochlor (CDAA)

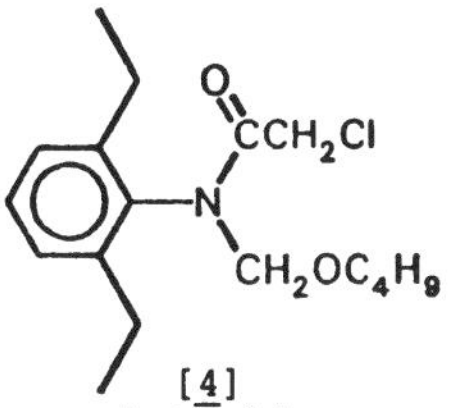

[4]
butachlor

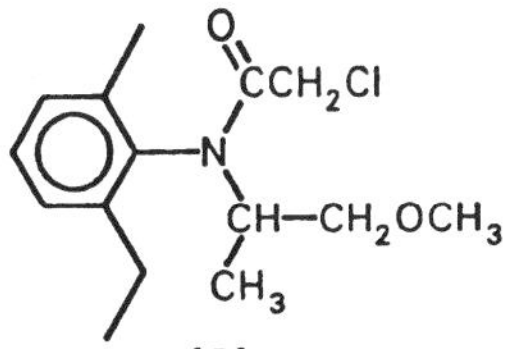

[5]
metolachlor

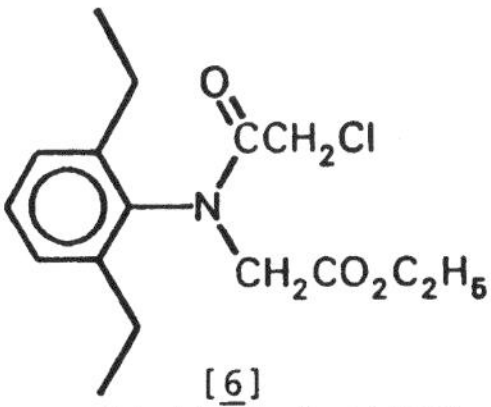

[6]
diethatyl ethyl

FIG. 1 2-Chloroacetamide structures.

(*Amaranthus* spp.) and black nightshade (*Solanium nigrum*) [4]. The various combinations of alachlor with other herbicides in tank or package mixes broaden the weed control spectra [4].

Currently, alachlor is registered for use worldwide on a number of agronomic crop sites including corn, soybeans, peanuts, and dry beans [4,5].

Alachlor, with molecular weight of 269.8, is a crystalline solid with relatively low m.p. of 39.5–41.5°C. The b.p. is 100°C at 0.02 mmHg and 135°C at 0.3 mmHg. The vapor pressure is 2.2×10^{-5} mmHg at 25°C. Alachlor's specific gravity is 1.133 (25/15.6°C), and its solubility in water is 240 ppm. It is soluble in common organic solvents [6].

Two new 2-chloroacetamides, among others in the class are metolachlor (2-chloro-N-[2-ethyl-6-methylphenyl]-N-[2-methoxy-1-methylethyl]acetamide) [5] (Ciba-Geigy), and diethatyl ethyl (N-chloroacetyl-N-[2,6]diethylphenyl]=glycine ethyl ester) [6] (NOR AM)

[6]. The structures of some 2-chloroacetamides are shown in Figure 1.

II. DEGRADATION

A. Degradation in Plants

Since the first registration of alachlor herbicide in 1969 [3,6], Monsanto has conducted extensive research to update the technical data base to support continued registration. Substantial quantities of new information have appeared covering expanded uses, toxicology, plant and animal metabolism, residue and other studies [4-10]. This new information includes significant new data on metabolite identifications and new analytical methods for total residues [9,10]. In addition, many articles have been published in technical journals by academic and other scientists covering a wide range of studies of the fate and mode of action of alachlor and other 2-chloroacetamides.

In 1971, Lamoureux et al. [11] published the first studies, in which plant metabolites of a 2-chloroacetamide, propachlor [2] were identified. The two metabolites of propachlor formed within two days following propachlor treatment of seedlings and excised leaves of corn (*Zea mays* L.), and of excised leaves of sugarcane (*Saccharum officinarum* L.), sorghum (*Sorghum bicolor* (L.) Moench), and barley. One metabolite was the glutathione conjugate of propachlor, 2-(S-glutathionyl)-N-(1-methylethyl)-N-phenylacetamide [7], and the other metabolite was 2-[S-(γ-glutamylcysteinyl)]-N-(1-methyl-ethyl)-N-phenyl]acetamide [8]. The latter was derived from [7] by hydrolytic loss of the glycinyl moiety from the glutathionyl group. Both were transitory metabolites and were not found in the plant material several days after the initial observation. Subsequently, Bakke [19] reported malonyl moieties [9] as present in metabolites of 2-chloroacetamides. The transitory nature of the primary glutathionyl conjugates is confirmed by the information on alachlor plant metabolites described below.

Following the work of Lamoureux et al. [11], primarily qualitative characterization studies were published on plant metabolites of 2-chloroacetamides. These usually involved reporting numbers of metabolites discernible by thin layer chromatographic (TLC) separations.

Leavitt and Penner [12] reported studies of in vitro reaction of [^{14}C]alachlor [1], [^{14}C]metolachlor [5], and [^{14}C]diethatyl ethyl [6] with [^{3}H]glutathione (glycine-2-^{3}H). By TLC separation, they characterized individual dual-labeled spots as the respective conjugates with glutathione, adapting the structure designation proposed by Lamoureux et al. [11] in the case of propachlor. Each spot gave a positive ninhydrin response indicating a free amino group, did not

react with nitroprusside (free thiol test), and did not cochromato-
graph with reactants, results that fit a glutathionyl conjugate.
Positive structural identification was not made, however.

Rubin's group [13] studied several safeners or herbicide antidotes and other compounds which protect sorghum against the phytotoxic effects of the 2-chloroacetamides. They reported that the
metabolism of [^{14}C]alachlor by sorghum was unaffected by the herbicide antidotes based on analyses of plant extracts seven days after
treatments with alachlor either alone or with antidote. No parent
alachlor was found, but, by TLC separations, 4 major and 2 minor
metabolites were observed in roots, and 3 or 4 major metabolites in
shoots. No qualitative or quantitative differences in TLC patterns
were found among the different treatments. Autoradiography revealed that the ^{14}C was distributed throughout the plant at 7 days.

Crop metabolism studies were reported by Monsanto [9] and the
EPA [7,8,10], showing that alachlor is converted by plants into two
major classes of metabolites. One class contains 2,6-diethylaniline
[10] and the other class contains 2-ethyl-6-(1-hydroxyethyl)aniline
[11] as the aniline moieties, hereafter referred to as DEA and HEEA,
respectively. It was reported that acid hydrolysis of alachlor metabolites at elevated pressure and temperature released both DEA and
HEEA, but the HEEA was converted to 2-ethylaniline [12] under
these conditions. The 1-hydroxyethyl group is sufficiently labile to
be replaced by a proton under strong acid and high temperature
conditions. In the metabolism study of soybeans, 50% and 30%, respectively, of the total ^{14}C in foliar and bean residues were converted to 2-ethylaniline [12] by acid hydrolysis. In the same study,
only 10% and 8%, respectively, of the total soybean ^{14}C foliar and
bean residue was found to contain DEA [7,8]. This was a semiquantitative assessment of relative residue levels by the ^{14}C distribution data
in a metabolism study, and is fairly representative of the actual ratio
of metabolite classes, although not of what actual residue might be.

Structures of 11 alachlor plant metabolites are shown in Figure
2. These compounds are in the complex mixture comprised of the
two major classes of alachlor plant metabolites [9,10]. Nine have
2-position sulfur linkages (compounds [15] to [23]). These sulfur-
linked metabolites establish clearly that one major detoxification pathway of alachlor in plants is by initial conjugation with glutathione,
with further metabolic transformations of the sulfur-linked and other
moieties. Therefore, alachlor follows a metabolic pathway in plants
like that suggested by Lamoureux et al. for propachlor [11], and by
Leavitt and Penner for alachlor [12].

Metabolite [13] results from hydrolysis of the 2-chloro group
in alachlor [1] to form the 2-hydroxyacetamide [13], and oxidation
of the 2-position of the acetyl moiety has converted that carbon to
a carboxyl group to produce the oxalamide [14].

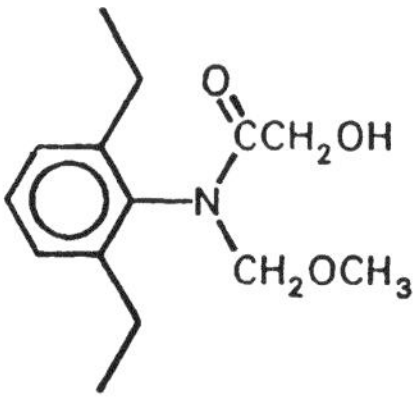

FIG. 2   Structures of alachlor plant metabolites.

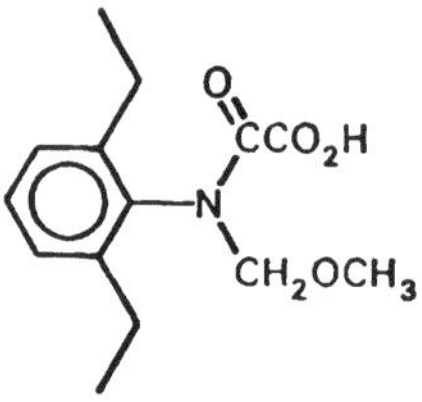

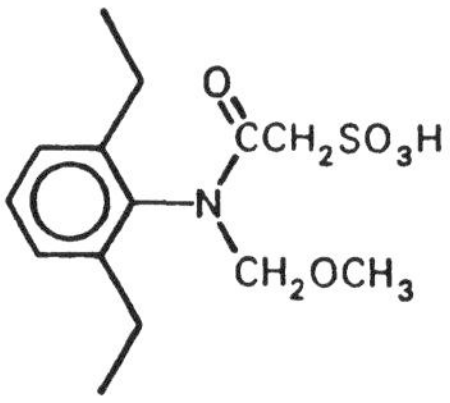

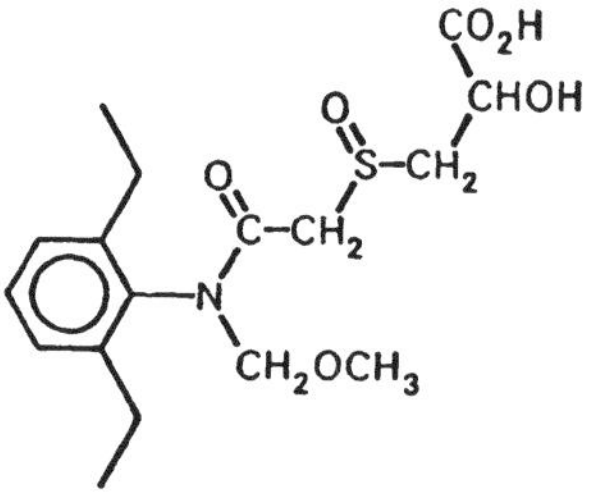

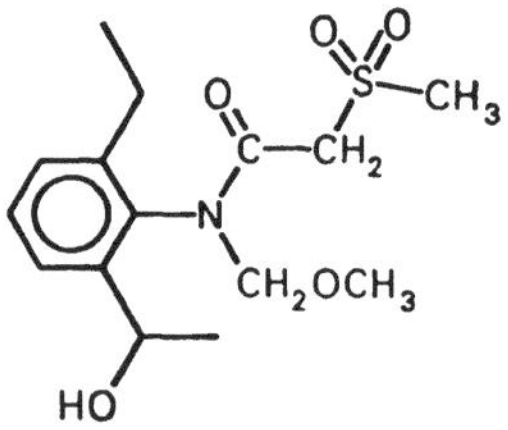

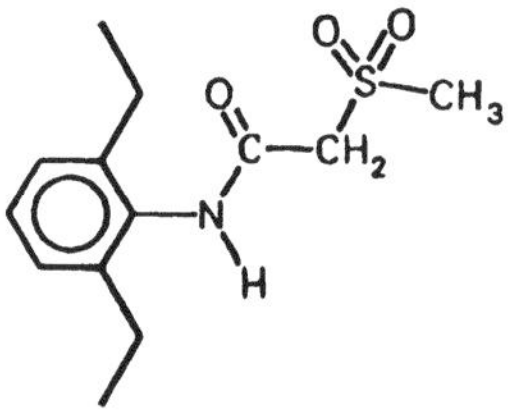

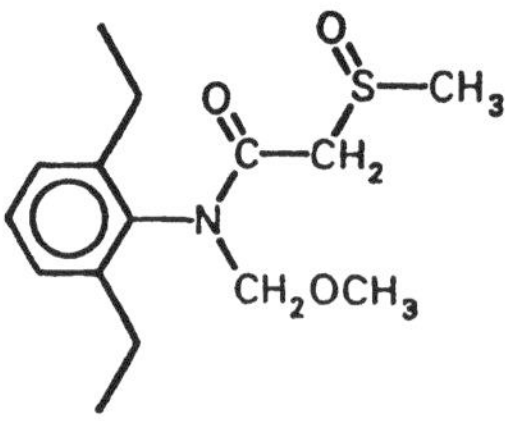

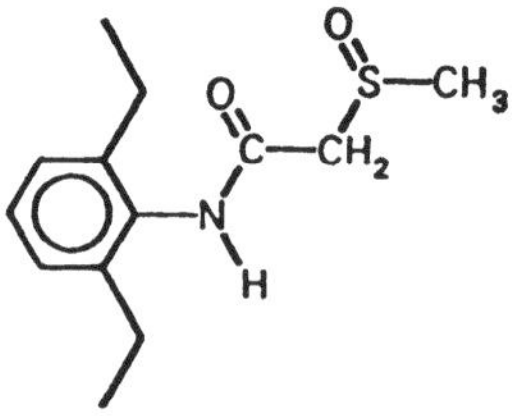

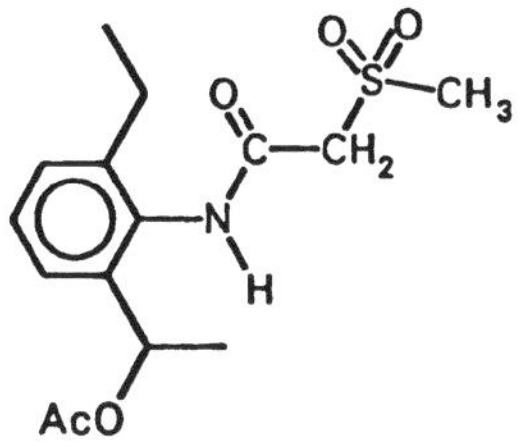

[21]

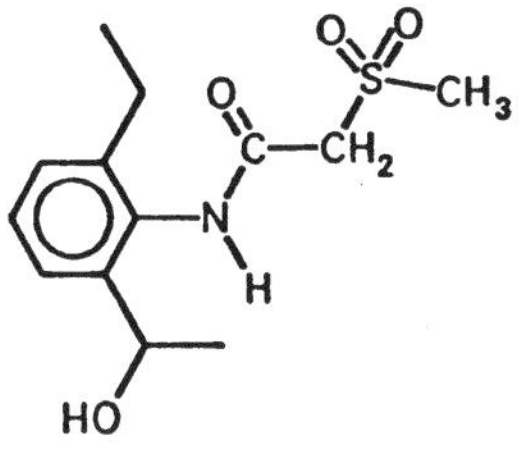

[22], [39] Fig. 3

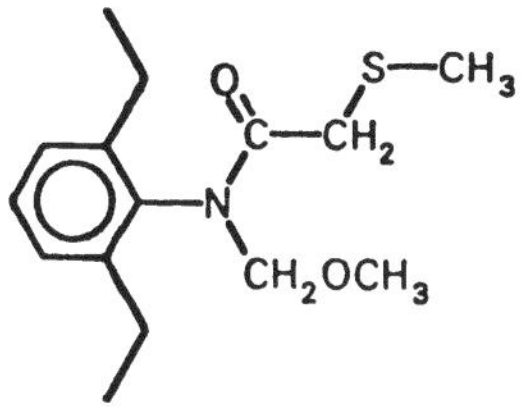

[23]

FIG. 2   (continued)

Compounds [17], [21], and [22] result from an oxidation reaction placing a hydroxyl group at the active benzylic 1-position of
an ethyl group attached to the phenyl ring giving rise to HEEA-
containing metabolites.

Other oxidized metabolites are represented by the sulfonic acid
[15], the metabolites with mono-oxidized sulfinyl links [16], [19],
and [20], and metabolites which have the sulfonyl linkage [17], [18],
[21], and [22]. One sulfide metabolite [23] was identified.

These results for alachlor parallel the glutathione conjugate
information in the recent excellent review by Lamoureux and Rusness
[14] (Xenobiotic Conjugation in Higher Plants). The formation of
HEEA shows that part of the metabolic conversions of alachlor in
plants involves oxidation, in this instance, hydroxylation of an ethyl
group side chain on the phenyl ring.

The first residue data for alachlor were based on hydrolysis
of metabolites to DEA as the only chemophore measured by gas
chromatography to estimate the residues of alachlor metabolites in
crops. The original DEA analytical procedure could not be used to
detect the HEEA, probably attributable to the dipolar nature of
HEEA with amino and hydroxyl groups in the same molecule.

The discovery of the two major alachlor metabolite classes in
plants prompted initiation of research on analytical methods which
would account for both classes. Analytical methods were developed
in which the two major classes of metabolites comprising the alachlor

residues are measured via release and quantitation of DEA from one
metabolite class, and release and quantitation of HEEA from the
other class. The two anilines are calculated as alachlor equivalents,
and the sum represents total residues. The analytical methodolo-
gies were developed for a wide variety of sample matrices from
crops and from animal tissues and products. These methods give
more accurate results for residues than those obtained by DEA
measure alone. Results of these methods parallel the values for
ratios of DEA:HEEA representing the two metabolite classes from
the ^{14}C distributions observed in metabolism studies for soybeans
and corn.

   One of the new methods is described by Cowell and Dubelman
[15]. Necessary modifications in sample preparation or cleanup are
made, depending upon the sample matrix. The sample containing
the alachlor metabolite residues is extracted, cleaned up as necessary, solvent is removed, and the residue is hydrolyzed under
pressure with strong alkali at 150°C for 30 minutes. After cooling,
the DEA and HEEA are extracted and, in one method adaptation,
converted to their respective heptafluorobutyryl derivatives. These
were assayed by negative ion quantitation of two mass fragments
characteristic of the two fluorobutyrated anilines by gas liquid
chromatography/mass spectrometry (GLC/MS) in the selective ion
mode. The fluorinated derivatives for DEA and HEEA are, respectively, N-(2,6-diethylphenyl)heptafluorobutyramide [24] and
N-(2-ethyl-6-[1-{heptafluorobutyryloxyethyl}]-phenyl)-heptafluoro-
butyramide [25]. N-(2,6-diethyl-4-fluorophenyl)heptafluorobutyram-
ide [26] is used as the internal standard, forming from 2,6-diethyl-
4-fluoroaniline [27] which is added in known amount to the sample
being analyzed; [27] is added to the raw sample and is derivatized
at the same time as are the DEA and HEEA released by hydrolysis.

   In a second method reported by Cowell et al. [16], the sample
is hydrolyzed as above, then extracted, solvent exchanged, buffered in water, separated by high performance liquid chromatography (HPLC), and the DEA and HEEA are quantified by electrochemical technique. Reference compounds used to validate both
methods were synthetic samples of alachlor metabolites [15] as the
sulfonic acid sodium salt, and the 2-methylsulfone [17], representing
DEA- and HEEA-containing alachlor metabolites, respectively.

   The new methods provide significantly more accurate estimations
of total residues by measurement of both HEEA and DEA chemo-
phores. In addition, the new methods by electrochemical or GLC/MS
sensing have significantly lower detection limits than were obtain-
able with the DEA assay method.

   From new residue trials for corn, soybeans, and peanuts, new
residue data were obtained for both DEA- and HEEA-containing
metabolites representing the total residue, calculated as the sum of

the alachlor equivalents from the respective DEA and HEEA chemo-
phores found for each major class of metabolites [9]. This permits
better assessments of the alachlor residue contributions to diet for
foods derived from alachlor-treated corn, soybeans, and peanuts
[9,10]. The ratio of DEA- to HEEA-containing metabolite residues
is about 1:3 in soybeans and peanuts based on the new residue
data. Total residue levels in corn are less than one-tenth the levels
found in soybeans, and corn plants do not appear to produce as
much of the HEEA-containing metabolites as observed for legumes.

New residue data for peas, beans, and other legumes show
low, mostly undetectable residues. Combined residues of DEA- and
HEEA-containing metabolites were below 0.1 ppm alachlor equivalents,
except for a single pea location with slightly over 0.1 ppm total
residue [17].

Adaptations of the new assay methods were developed also for
animal tissues and products. The residue samples were obtained
from lactating cows, swine, and laying hens. The animals were fed
a representative mixture of synthetic samples of five major plant
metabolites in simulation of expected exposures to alachlor residues
in treated field crops. The synthetic mixture included the principal
metabolites found in corn and soybeans as well as representatives
of both DEA- and HEEA-containing metabolites. As a preliminary
step, metabolism studies were conducted with lactating goats and
laying hens using a mixture of synthetic ^{14}C-labeled samples representing the same plant metabolites [9,10]. These ruminant and
poultry metabolism studies followed the protocols in the Guidelines
for Registration [18] for determination of the nature of residue in
livestock.

The plant metabolites were fed to lactating cows, laying hens,
and swine at three exaggerated levels; the highest dose was 70
times the maximum exposure expected under realistic field conditions.
Alachlor and DEA were not included in the mixture because neither
has ever been found as a component of crop residues. The new
methods for DEA and HEEA have limits of detection of 0.002 ppm
for each aniline, in contrast to the 0.02–0.05 limits of detection for
DEA in the old DEA method. The new data provide a more reliable
and accurate basis for estimating alachlor residue contributions to
human diet via foods from alachlor-treated crops, and from meat,
milk, eggs, and poultry products derived from domestic animals
consuming alachlor-treated feeds. The results from the new animal
residue studies show detectable residues at the three exaggerated
feeding levels which permit linear extrapolation from high dosage
down to the much lower maximum level of residues expected in crops
and crop components actually used as animal feeds [9,10]. Because
of the greater sensitivities of the new methods, the estimates of
alachlor residue contributions to human diet from animal-derived

foods were lowered almost 100-fold [9,10] as compared to estimates based on the less sensitive (0.02 ppm) DEA assays [8].

B. Degradation and Elimination by Animals

1. Rats

Monsanto [9] instituted extensive programs to define more clearly the metabolism of alachlor in rats in an attempt to establish comparisons with crops and to determine the possible relevance of the rat metabolism to the long-term toxicology results. Initially, efforts were directed toward definition of the pharmacokinetics, dose–response relationships, and metabolism of alachlor in rats as clearly as possible. When administered orally to rats, alachlor is extensively metabolized to a wide variety of metabolites. Subsequently, investigations were focused on determination of similarities and/or differences in metabolism and distribution of alachlor in rodents and primates. The ultimate goal of this work was to develop data to allow a more meaningful assessment of risk to humans from use of alachlor in agriculture [9,10].

The rat metabolism studies reported by Monsanto [9] demonstrated that alachlor was rapidly and extensively metabolized in rats. The extensive metabolism was reflected in the wide variety of urinary metabolites that were identified (Fig. 3); these included mercapturates [28]–[31], methylsulfoxides [32]–[35] and sulfones [36]–[43], monohydroxylated [29], [31]–[33], [39]–[41] and dihydroxylated compounds [36,37], glucuronides, [38, 44, 45], and a phenol sulfate [46]. Mercapturates are commonly found in mammalian metabolism involving xenobiotic chemicals, a subject reviewed recently by Bakke [19]. Based on a close analogy with results reported for metabolism of propachlor in rats by Bakke, Larsen, and others [20–25], the alachlor metabolites probably resulted from (1) glutathione (GSH) conjugation of alachlor in the liver, (2) cytochrome P-450 oxidation and subsequent glucuronidation of alachlor and metabolites in the liver, (3) biliary excretion, (4) metabolism by microflora in the intestines, including peptidases (which convert GSH conjugates to cysteine conjugates), cysteine-β-lyase activity (leading to thiols) and glucuronidase activity, (5) enzyme-catalyzed methylation of the thiols by S-adenosyl-methionine, and (6) enterohepatic circulation of metabolites.

Of interest is that five rat metabolites are identical to metabolites found in plants; from plants these are [17], [18], [19], [20], and [22] (Fig. 2), respectively, identical to rat metabolites [40], [50], [35], [34], and [39] (Fig. 3). None of these metabolites common to plants and rats were found from monkeys. Similarities between animal and crop plant metabolism reside primarily in the

A. Mercapturates

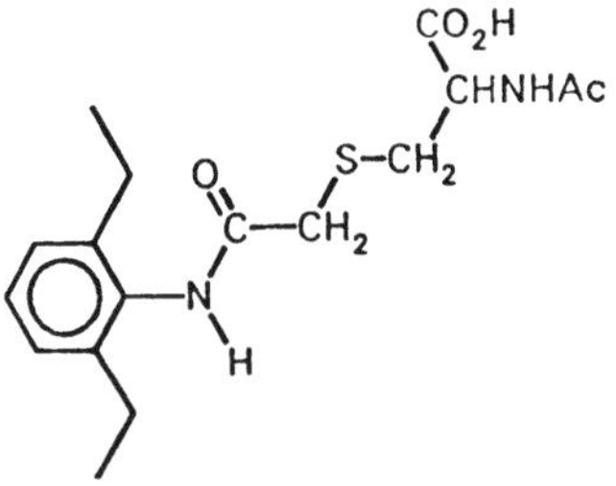

[28]

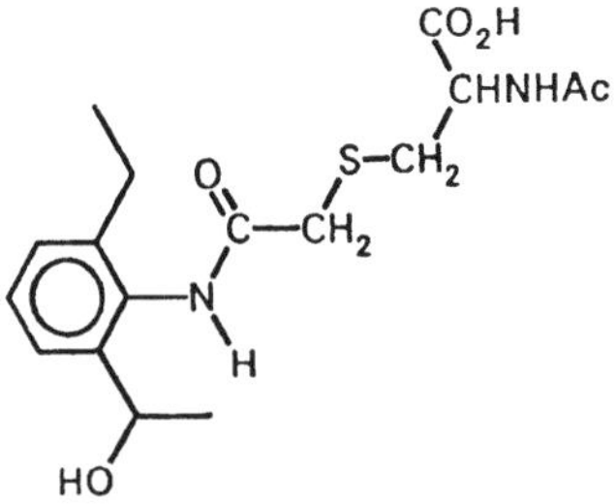

[29]

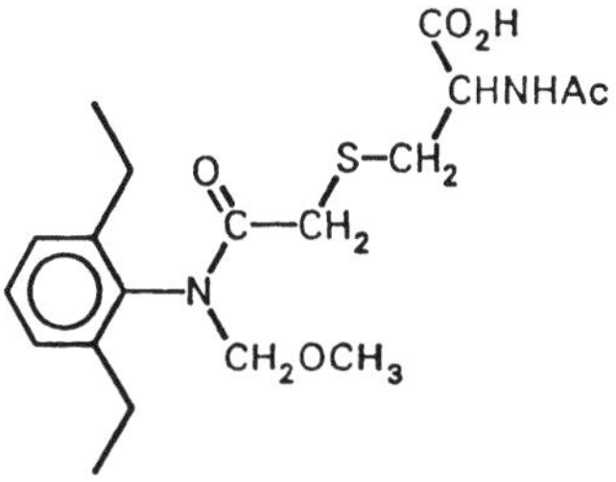

[30]

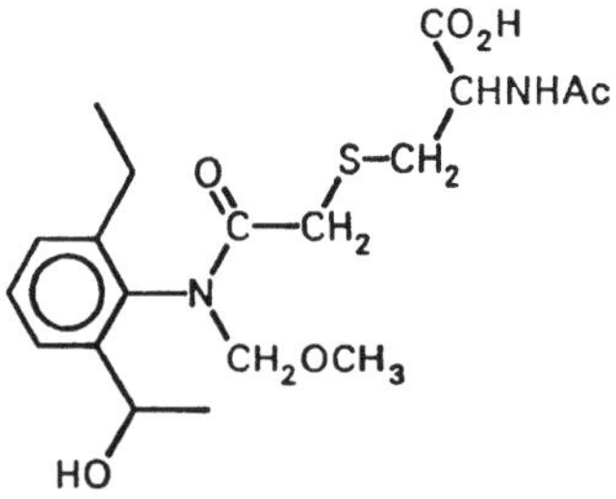

[31]

B. Sulfoxides

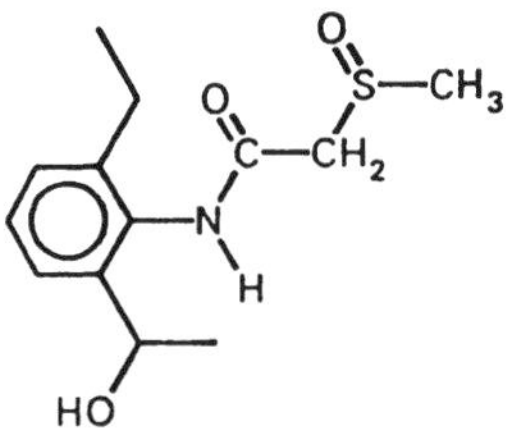

[32]

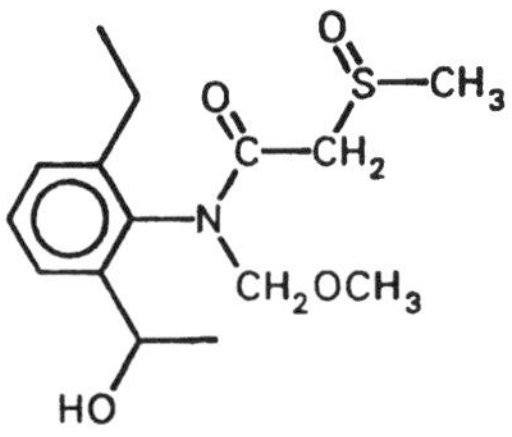

[33]

FIG. 3   Structures of alachlor rat metabolites.

B. Sulfoxides (Cont'd.)

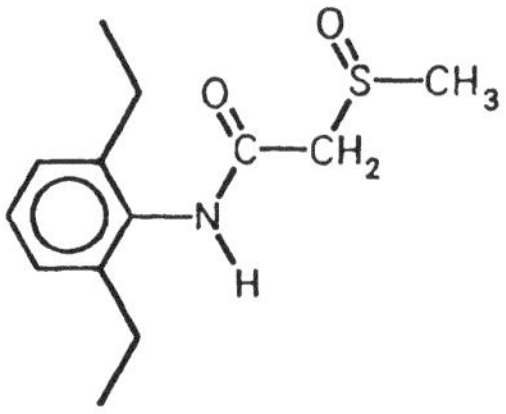

[34], [20] Fig. 2

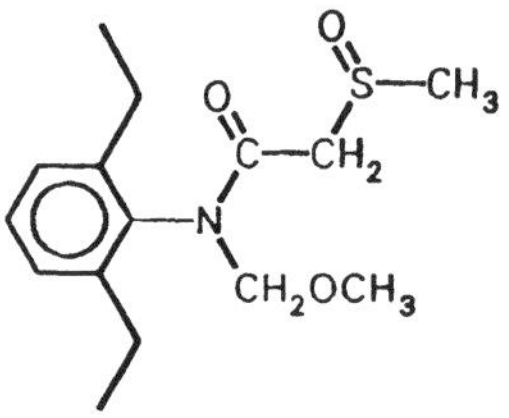

[35], [19] Fig. 2

C. Sulfones

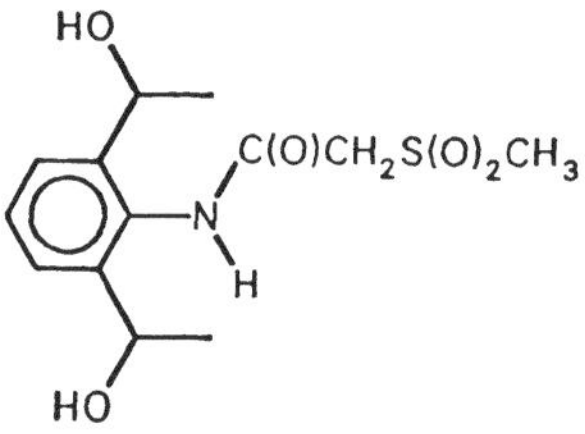

[36]

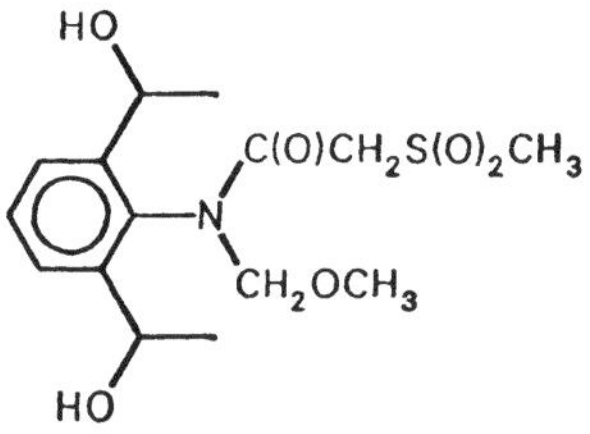

[37]

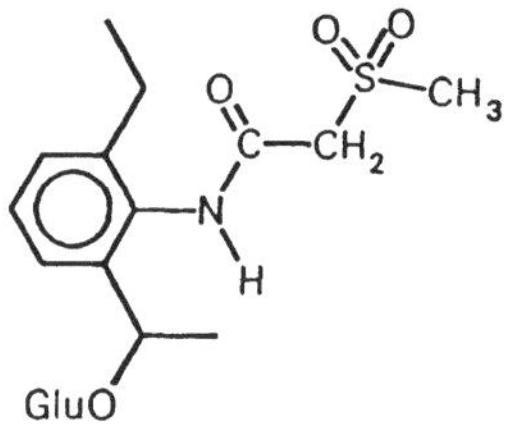

[38]

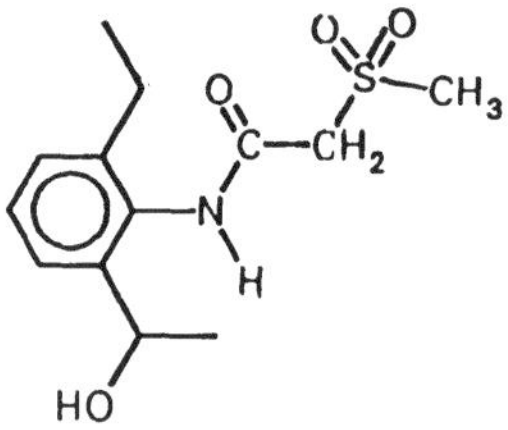

[39], [22] Fig. 2

Glu = glucuronidyl

FIG. 3   (continued)

C. Sulfones (Cont'd.)

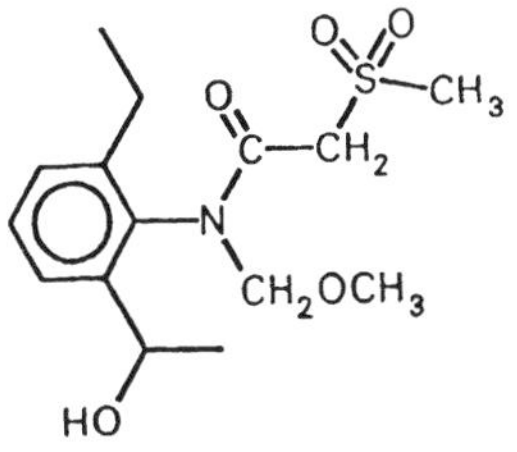

[40], [17] Fig. 2

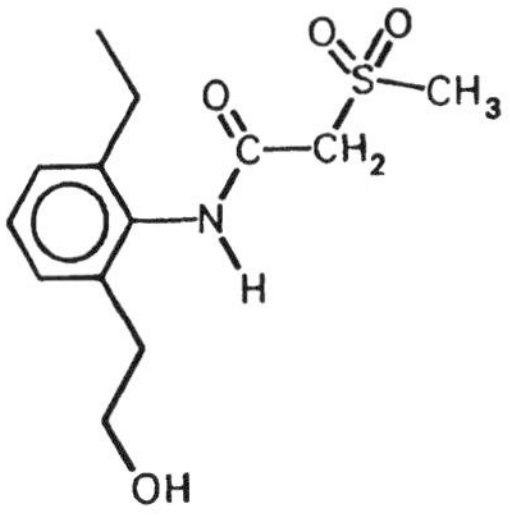

[41]

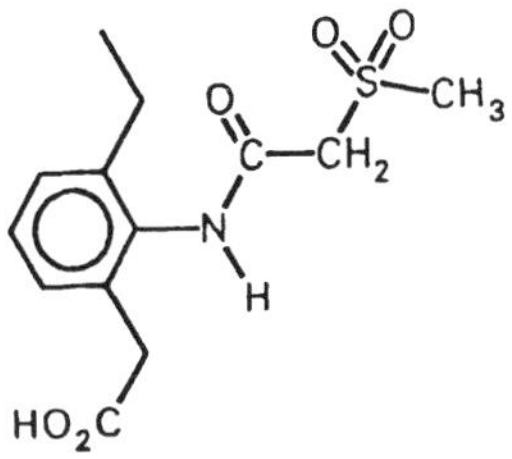

[42]

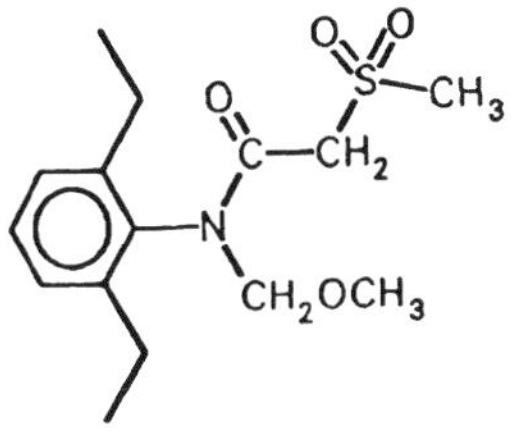

[43]

D. Mono-hydroxylated Compounds

See: [29], [31], [32], [33], [39], [40], [41]

E. Di-hydroxylated Compounds

See: [36], [37]

FIG. 3 (continued)

F. Glucuronides

See: [38]

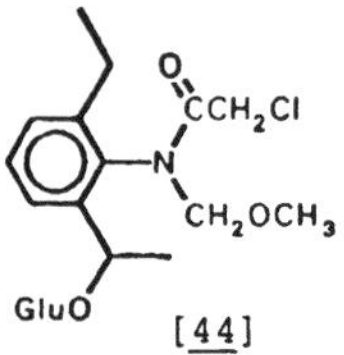

[44]

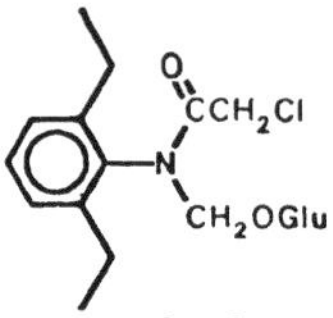

[45]

G. Phenol Sulfate

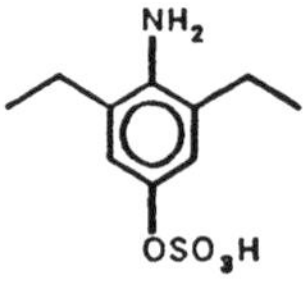

[46]

H. Blood Metabolites

See: [34], [35], [40], [43]

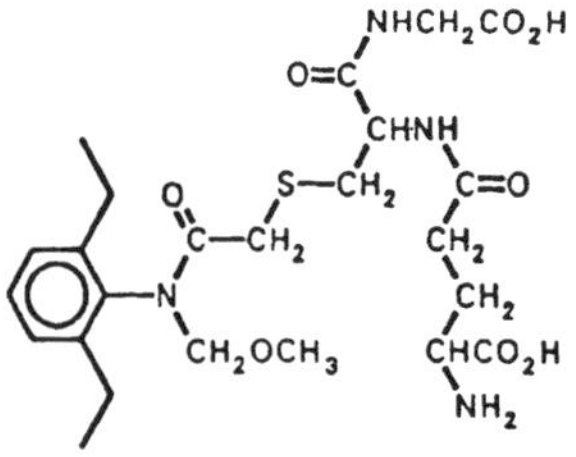

[47]

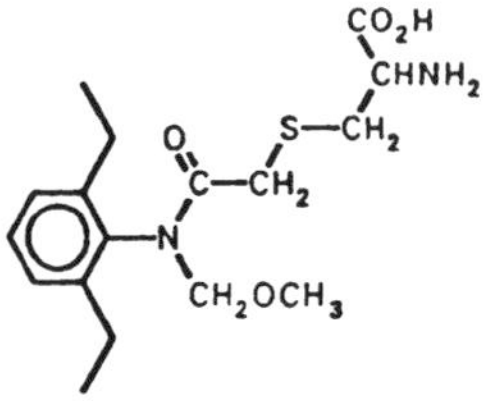

[48]

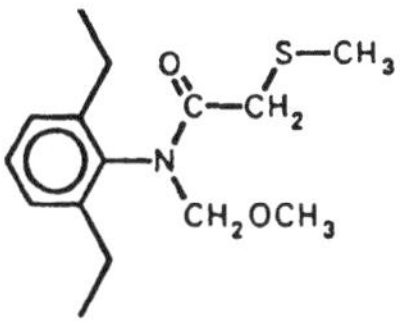

[49]

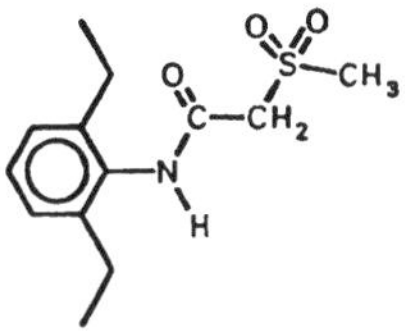

[50], [18] Fig. 2

FIG. 3   (continued)

unquestionably important role played by GSH as a major detoxification mechanism in both plants and animals exposed to alachlor.
For the reader's convenience, identical metabolite structures in
Figures 2 and 3 for plants and rats are numbered twice, and
identified as such in the figures.

In the alachlor pharmacokinetic study in rats, analyses of blood
from female Long-Evans rats dosed with single oral doses of radioactive alachlor revealed that up to 3.5% of the dose was present in
the whole blood at various sampling times, with 2% of the dose still
in the blood after 10 days. The majority of the activity was in the
red blood cells, with about 0.1–1.0% of the administered dose in
the plasma. The alachlor residues in blood consisted of both freely
circulating metabolites and those bound covalently to plasma proteins and hemoglobin. Metabolites in the plasma cleared rapidly.
The free metabolites at early time points after dosing consisted of
alachlor [1], its known amino acid conjugates [47], [48], and [30]
and more highly metabolized sulfur derivatives [48], [34], [35],
[50], [43], and [40] (Fig. 3) [9]. The covalently bound residues
in the rat blood cells declined at very slow rates. They appeared
to be linked to cysteinyl moieties with free thiol groups in hemoglobin, because degradative reactions released acetanilide moieties
(nickel desulfuration). Compound [47] is the tertiary glutathione
conjugate of alachlor, the primary product of the reaction of GSH
with alachlor. As detailed above, a total of 23 metabolites of alachlor
have been identified in rat urine and blood (Fig. 3). The complex
metabolism of alachlor in the rat is illustrated further by the HPLC
separation pattern shown in Figure 4.

The rat eliminated alachlor metabolites in about equal amounts
in urine and feces. Subsequent studies with other animal species
showed marked differences among species in the excretory split
for metabolite elimination.

2. Mice

The metabolism and excretion patterns of alachlor in CD-1 mice
were studied in order to gain information applicable for other animal
species [9]. Ten animals of each sex were administered single oral
doses of 800–900 mg/kg of isotopically labeled alachlor. Urine and
feces were collected daily for seven days, the animals were sacrificed, and blood samples taken. The metabolic pattern profiles were
obtained by high performance liquid chromatography (HPLC) and
compared to the separation patterns for rats dosed similarly.

Unlike the rat, the mouse feces accounted for the major amount
of radioactivity excreted. In rats, the alachlor metabolites are
distributed about equally between urine and feces. In mice, about
two-thirds of the ^{14}C metabolites appear in feces, and one-third
in urine.

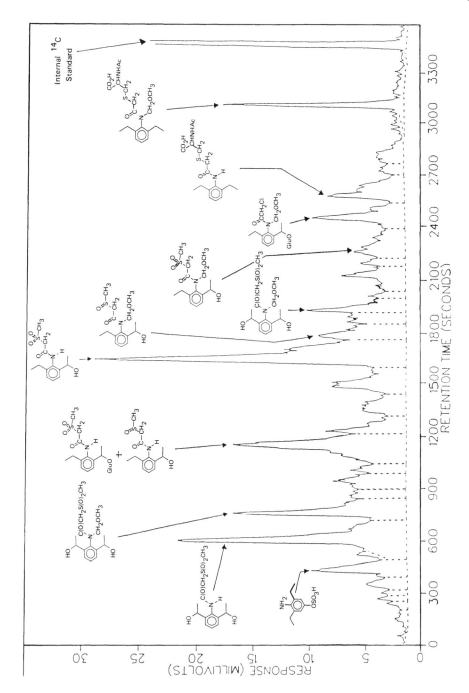

FIG. 4   Alachlor metabolites from the rat.   HPLC chromatogram from female rat urine 24–48 hours after dosing.

With mice, most of the radioactivity was excreted in 4–8 h.
Blood analyses at seven days after dosing showed that less than 0.1%
of the radioactivity was associated with blood; in rats the binding
to hemoglobin accounted for up to 2.0% of the administered dose [9].
The HPLC profile (Fig. 5) appears to show a higher degree of
metabolism in mice than was found in rat studies.

Twelve metabolites in mice are identified in the HPLC trace in
Figure 5, and were found also in the rat; however, relative quantities are markedly different between the species (Figs. 4 and 5).
The mice metabolites include mercapturates [29] and [30], methylsulfoxides [32] and [33], and sulfones [36], [37], [38], [39], and
[40], monohydroxylated metabolites [32] and [33], dihydroxylated
compounds [36] and [37], glucuronides [38], [44], and [45], and
the phenol sulfate [46] (Fig. 3) [9]. Among metabolites from mice,
only [39] and [40] (Fig. 3) are found in plants as, respectively,
[22] and [17] (Fig. 2).

3. Rhesus Monkeys

Rhesus monkey metabolism, pharmacokinetic, excretion and
dermal absorption studies with [^{14}C]alachlor were conducted and
reported by Monsanto [9], Kronenberg et al. [26], and by Carr
et al. [27]. The rhesus monkey excretes [^{14}C]alachlor metabolites
in a ratio of 10 to 1 in urine and feces, respectively. This contrasts with the corresponding ratios of 1:1 for rats and 1:2 for mice.
Most (96%) of the ^{14}C was excreted within 48 hours in the several
studies, which included intravenous and dermal dosing routes. Also,
the degree of alachlor binding by monkey blood was well below that
observed for rat blood [9,10]. Five major metabolites in urine were
identified, and each contained unchanged ethyl groups on the phenyl
ring. However, in raw urine samples containing all metabolites, the
distribution of HEEA to DEA was determined to be about 1:4 by an
acid-catalyzed hydrolysis procedure at elevated temperature and
pressure. Therefore, a minor degree of ring ethyl group hydroxylation occurs in the monkey.

The five major monkey metabolites, in decreasing levels observed, are shown in Figure 6, and include the mercapturates [30]
and [28], the cysteinyl conjugate [48], and the glucuronide [45];
each was found also in the rat studies. The fifth metabolite [51],
a thioacetic acid conjugate, is unique to the rhesus monkey in these
multispecies studies. Structures shown for rats in Figure 3 are
repeated in Figure 6 for the reader's convenience. Oxidation of the
cysteinyl conjugate [48] probably gives the thioacetic acid conjugate
[51]. An oxidation by a mixed function oxidase such as cytochrome
P-450 may effect conversion of alachlor to the N-hydroxymethyl
analog followed by conjugation with glucuronic acid to give [45].

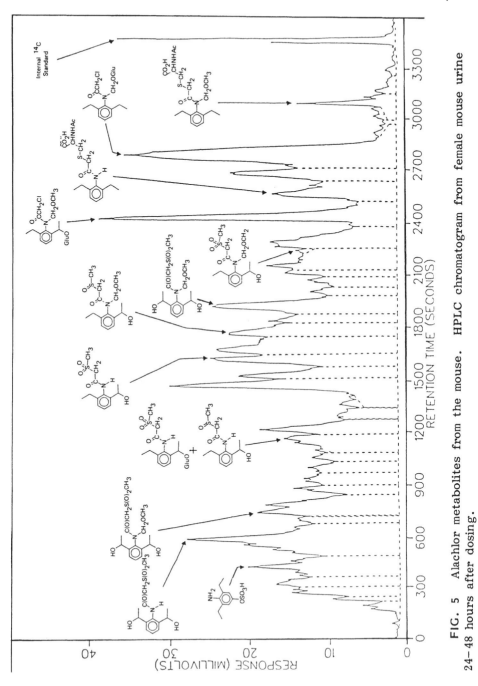

FIG. 5   Alachlor metabolites from the mouse.   HPLC chromatogram from female mouse urine 24–48 hours after dosing.

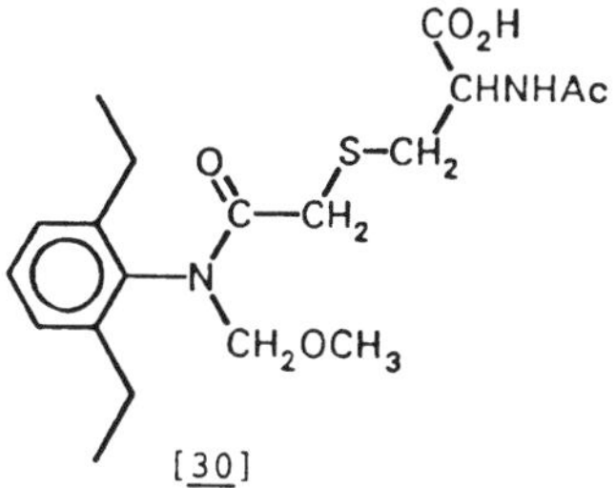

[30]

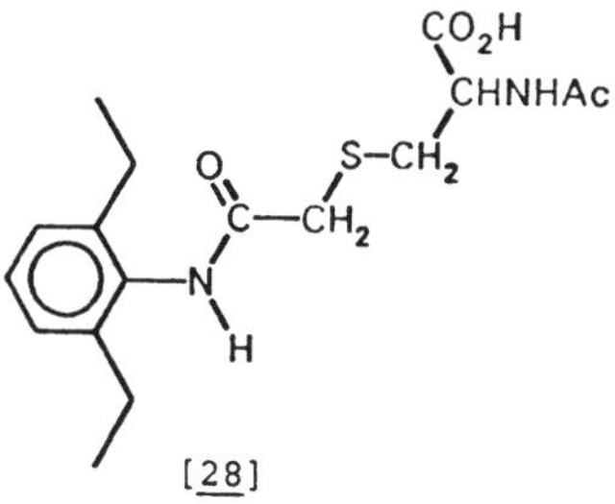

[28]

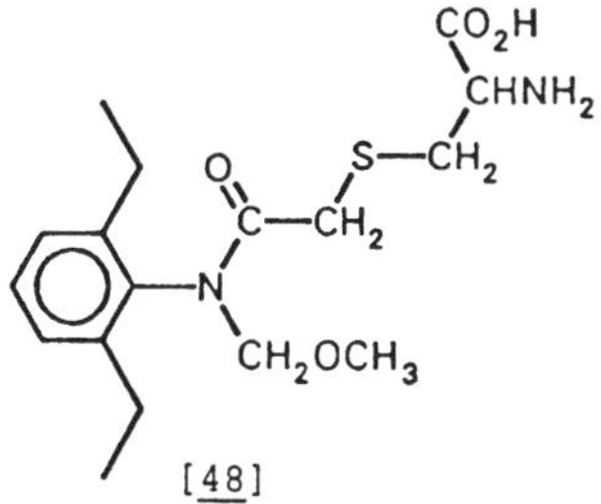

[48]

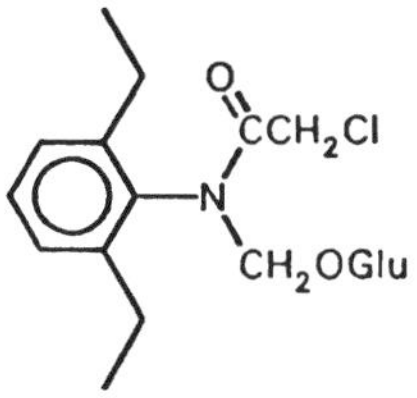

[45]

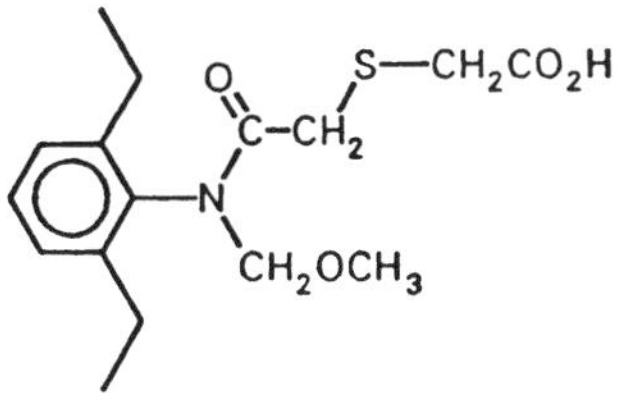

[51]

Glu = glucuronidyl

FIG. 6 Structures of alachlor monkey metabolites.

The high degree of urinary excretion of [^{14}C]alachlor metabolites by the rhesus monkey appears to result from more rapid removal by blood of the alachlor metabolites from the liver than in the rats. Correspondingly, there is much higher biliary excretion of alachlor metabolites from the liver in rodents (rats, mice) than in the monkey. The identification of only five metabolites in significant quantities from the monkey contrasts sharply with the 23 metabolites identified for the rat and 12 for the mouse. A cyclic process called enterohepatic circulation which allows metabolites to be absorbed repeatedly from the intestinal tract, exposed to the liver, and released via the bile back into the intestines is significant in rodents but minimal in monkeys.

4. In Vitro Studies

In vivo studies of animal metabolism of alachlor showed strong binding to blood components in rats. Significantly greater amounts of [14C]alachlor or 14C metabolites are bound to rat hemoglobin as compared to that bound by hemoglobin in human, monkey, or mouse blood [9,10]. Both in vivo and in vitro experiments confirm this preferential binding by rat hemoglobin in comparison to binding by mouse and monkey hemoglobin. In vitro techniques were required for comparative studies using human blood. Alachlor preferentially forms a soluble glutathione conjugate in the red blood cells of human, monkey, and mouse blood, but binds to the protein subunit of hemoglobin in the rat, possibly by a cysteine linkage. This hemoglobin binding is virtually unique in rats [9].

The in vitro metabolism of [14C]alachlor by rat liver and kidney cell-free preparations was also reported [9] in efforts to elucidate the initial and intermediary metabolic pathways. The membrane-bound cytochrome P-450 oxygenases were separated by centrifugation from soluble glutathione transferase (GST) enzymes in order to study the two systems separately. Studies also were conducted with similar preparations from naive rats, mice, and rhesus monkeys and from rats previously exposed to alachlor to compare rates and products of the enzymatic degradation of alachlor among different animal species. HPLC was used to quantitate the rates of alachlor degradation based on radioactivity and to profile and isolate metabolites for mass spectral identification.

GST enzymes from rats, mice, and monkeys rapidly catalyzed complete conversion of alachlor to its glutathione conjugate. Monkey liver preparations readily hydrolyzed the glutathione conjugate [47] to the S-cysteinyl conjugate [48] (see Fig. 3), but rat liver preparations did so very slowly. Rat and monkey kidney enzymes also produced [48] readily. When acetyl coenzyme A (Ac coA) was added to the kidney incubations from rats, [48] was acetylated easily to the mercapturate [30], but this acetylation was much slower with monkey enzymes. Rat liver enzymes catalyzed the acetylation of the cysteinyl conjugate [48] to [30], but the monkey liver preparation did not. The results show that rats, mice, and monkeys can metabolize alachlor via the mercapturic acid pathway, but the rates and sites of the intermediate steps can vary significantly among species.

Incubation of alachlor with liver cytochrome P-450 enzymes with added reduced nicotinamide adenine dinucleotide phosphate (NADPH) led to the formation of the same four metabolites for each species-origin preparation tested. Alachlor underwent O-demethylation, benzylic hydroxylation on one or both ethyl groups, or a combination of these transformations. The major product was the monohydroxyethyl metabolite in all cases. When uridine diphosphate

glucuronic acid (UDPGA) was added with NADPH to these incubations, glucuronides of the hydroxylated metabolites were formed.
The carbinolamide glucuronide [45] was produced, and had also
been observed previously as a metabolite in rat urine and bile and
in monkey urine. Marked differences in rates of alachlor oxidative
transformations were observed and related to both species and
genders in a species. Male monkey cytochrome P-450 activities were
one-half to one-third that from male rats, and female rat liver
material had only a fifteenth the activity of male rat liver preparation
[9].

The metabolism of the alachlor 2-methylthio conjugate [23] by
rat and monkey liver homogenates produced similar oxidation products
which included the sulfoxide [20] and a hydroxylsulfone (see Fig.
2) [9].

A single high dose of alachlor (700 mg/kg) given to rats 24 h
before sacrifice showed a 70% increase in GST activity compared to
controls, but cytochrome P-450 levels remained unchanged. After
two weeks, administration of 300 mg/kg of alachlor per day caused
elevation of both GST and cytochrome P-450 enzyme activities to
150-200% of control levels [9].

Kimmel et al. [28] reported that, among other compounds,
alachlor was metabolized in intraperitoneally treated rats and by
hepatic mixed-function oxidase systems to the free 2,6-diethyl-
aniline [10], and, in turn, to the corresponding 2,6-diethylnitro-
sobenzene [52] (DENB) (Figure 7). This metabolite was reported
to be identified by GC/MS with selective ion monitoring and by
cochromatography. For the set of compounds studied, it was suggested that the nitrosobenzenes and related compounds could possibly contribute to the toxicological profiles of the compounds
examined.

Because 2,6-diethylnitrosobenzene [52] was never observed as
a metabolite of alachlor in the extensive animal metabolism and in vitro
enzymatic studies reported in this chapter [9], the results reported
herein [28] were examined further. The synthesis and properties
of DENB [52] have been reported by Wratten et al. [29]. DENB
[52] was found to decompose rapidly in solution and during GC
analysis. The high degree of instability, combined with the formation of an equilibrating mixture of monomeric and dimeric forms in
solution, resulted in very complex analytical data that appear to
have been misinterpreted in the study cited above [28].

Companion studies were reported by Feng and Wratten [30],
in which 2,6-diethylaniline (DEA) [10] was incubated with NADPH-
fortified rat liver microsomal enzymes. The major product of this
in vitro bio-oxidation was 4-amino-3,5-diethylphenol [53]. Metabolites
resulting from N-oxidation or alkyl group hydroxylation were not
observed, contrary to the report by Kimmel et al. [28]. Efforts to

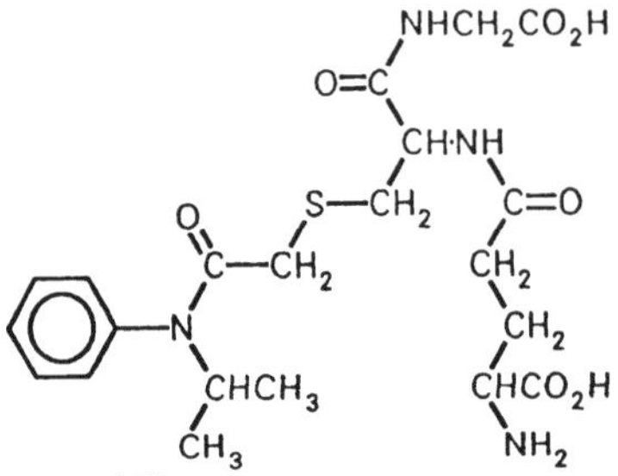

FIG. 7   Miscellaneous metabolites and other compounds.

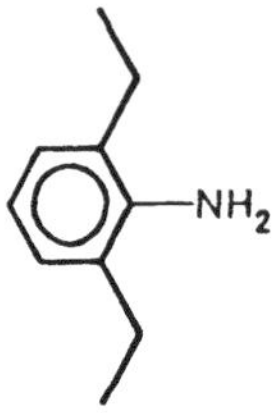

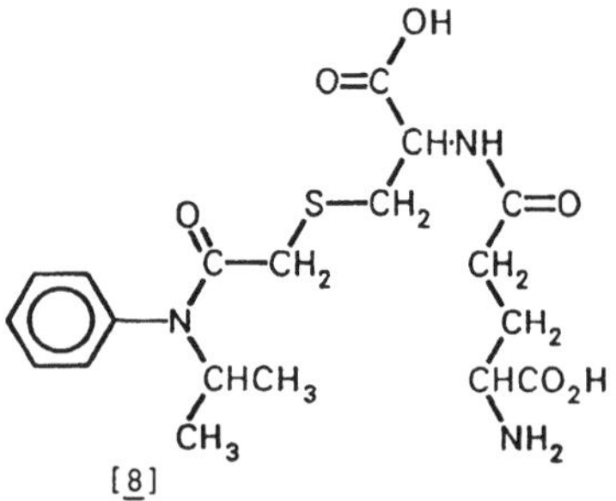

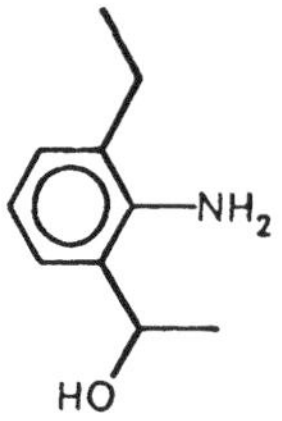

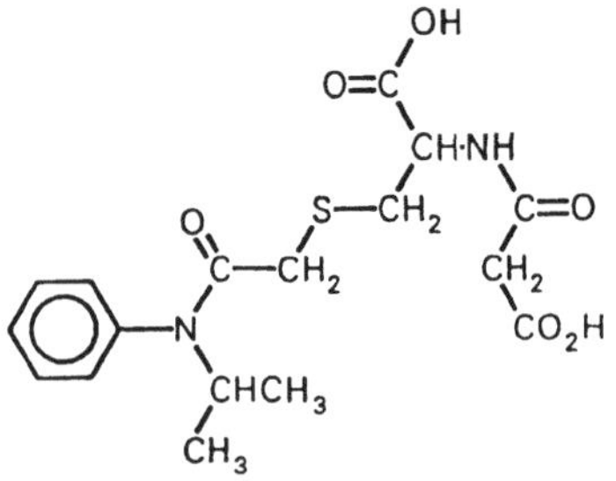

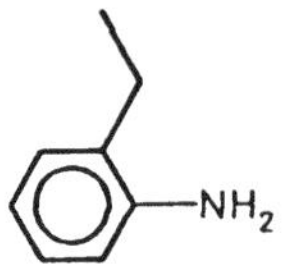

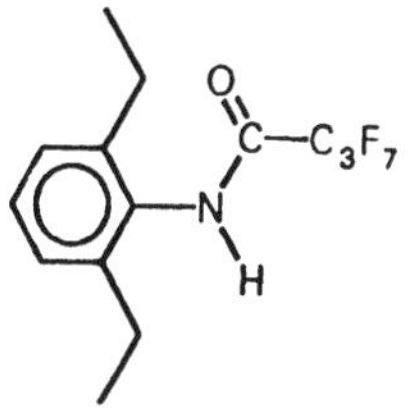

[24]

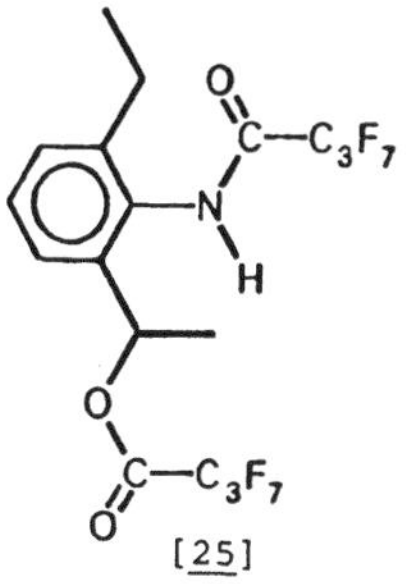

[25]

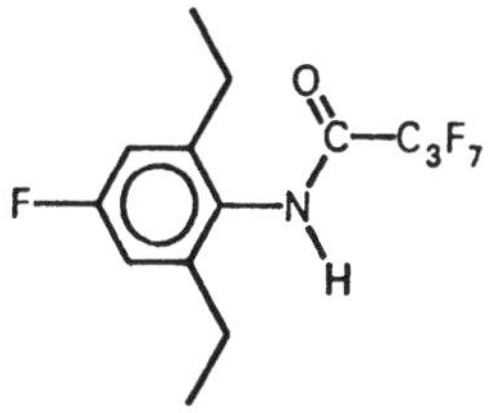

[26]

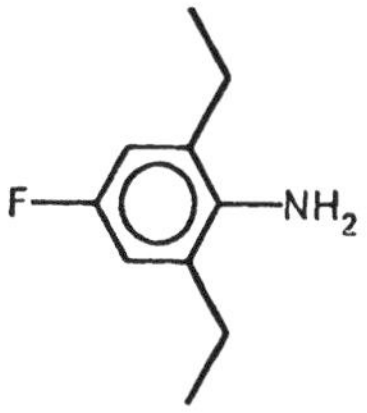

[27]

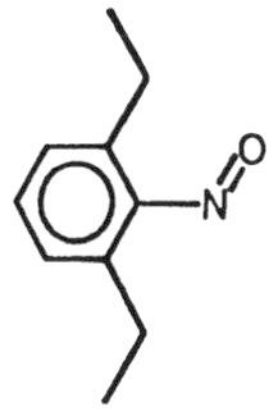

[52]

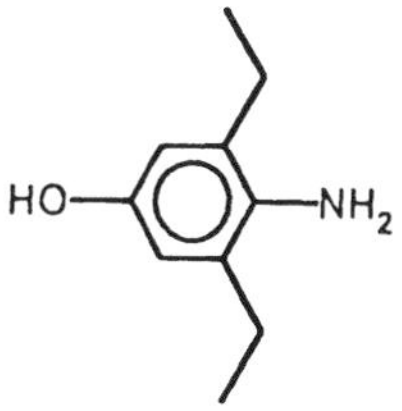

[53]

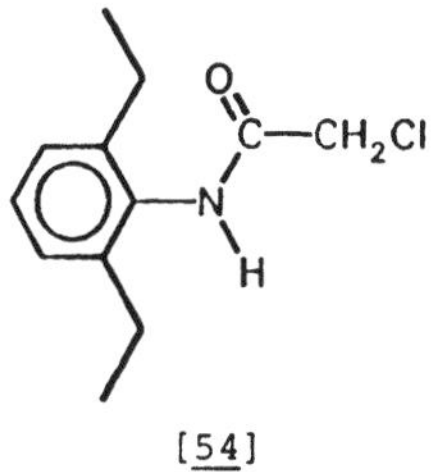

[54]

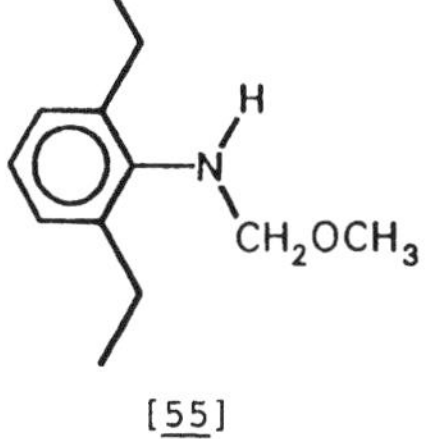

[55]

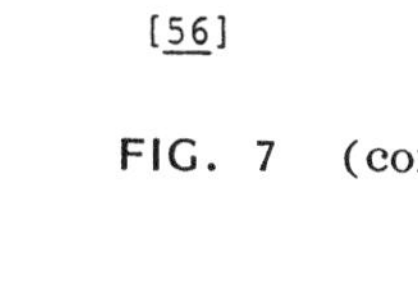

[56]

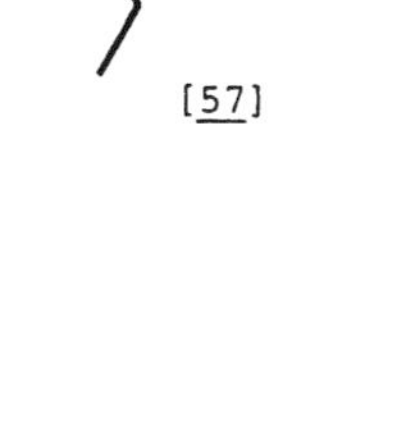

[57]

FIG. 7   (continued)

323

reproduce that work [28] failed to reveal any of the claimed nitroso-
metabolite by HPLC/MS sensitive to the 30 ng/injection level [29,30]
(see Fig. 7).

C. Degradation by Microorganisms

Among early papers published on alachlor, Beestman and
Deming [31] reported results of studies of the dissipation of
2-chloroacetamides from soil, however, no identifications of degradation products were given. They concluded that the primary route
of dissipation was biodegradation by microorganisms.

Tiedje and Hagedorn [32] published the first paper on degradation of alachlor in which some degradation products were identified;
they utilized fungus isolates in incubation studies. Degradation of
alachlor by a soil fungus Chaetomium globosum gave products and
metabolites identified as chloride anion, 2-chloro-N-(2,6-diethyl-
phenylacetamide [54], 2,6-diethyl-N-(methoxymethyl)-aniline [55],
and 1-chloroacetyl-2,3-dihydro-7-ethylindole [56] (Fig. 7). C. globosum further degraded [54], 2,6-diethylaniline [10], and chloro-
acetic acid. Among other fungi tested, a Paecilomyces sp. degraded
alachlor 60% in 7 days; species unable to degrade alachlor included
Chaetomium bostrychodes, Penicillium spp., Fusarium roseum, Phoma
spp., Alternaria spp., and Trichoderma spp.

Chopra and Sethi [33] identified [54] as the degradation
product of alachlor on silica gel, probably a result of hydrolytic
loss of the methoxymethyl group catalyzed by the silica gel. The
same authors [34] identified [54] in soil studies.

Fang [35] studied the photodegradation of alachlor in three
soils, and reported products to be [54], [55], [56], [10], and
N-(2,6-diethylphenyl)acetamide [57] (Fig. 7).

Novick et al. [36,37] have published studies of the degradation of alachlor by microorganisms, but these have been directed
largely at rates of complete mineralization, namely, $^{14}CO_2$ evolution,
using [^{14}C]phenyl-labeled alachlor. Suspensions of soils previously
treated in the field with alachlor were incubated with [^{14}C]alachlor
to test for the ability of the soils to metabolize the ring-labeled
herbicide. No identifications of degradation products were reported
by these authors. However, by thin layer chromatography (TLC)
of extracts from these soil suspensions, four metabolites were reported [36]. From 6-week sewage and lake water incubations of
[^{14}C]alachlor, six and four metabolites, respectively, were detected
by TLC [37]. Alexander states that the biodegradation of alachlor
is largely by cometabolism, because no microorganisms able to
mineralize alachlor could be isolated from these soils.

Zimdahl and Clark [38] studied the dissipation of alachlor,
metolachlor, and propachlor in the laboratory and field using bioassay techniques with sorghum [Sorghum vulgare Pers. 'NB280S')

or annual ryegrass (*Lolium multiflorum* Lam.). A clay loam and a
sandy loam were the two soils used. The half-lives were inversely
proportional to moisture and temperature, as expected, with more
rapid dissipation occurring in the clay loam. The rates of degradation were in the increasing order metolachlor < alachlor ≪ propachlor,
with half-life range of alachlor of about 2–4 weeks. There was good
agreement between laboratory and field results.

The Metcalf model ecosystem [39] was used by Yu et al. [40]
in studies of [^{14}C]phenyl-labeled alachlor and propachlor. They
reported detecting eight alachlor degradation products by TLC and
radioautography at the end of the 33 day test period, with less
than 2% of parent alachlor remaining. Similar results were observed
for propachlor. There was no evidence that these herbicides or
their degradation products were magnified in the food chain of this
simple ecosystem model.

Francis et al. [41] studied the degradation, persistence, and
bioaccumulation of alachlor and other herbicides alone and paired
with paraquat in a terrestrial/aquatic microecosystem. Alachlor and
the other herbicides with paraquat showed only slight interactions
with each other with respect to the microenvironmental fate. Type
of application of a herbicide had greater impact on fate than interactions of paired herbicides.

The two major soil metabolites of alachlor are the sulfonic acid
[15] and the oxalamide [14], both reported as metabolites found in
plant metabolism studies reported by Monsanto [9].

III. MODE OF ACTION, ACTIVITY, AND
       SELECTIVITY

The most recent and best review of research on the mode of
herbicidal action of 2-chloroacetamides is by Fedtke [42]. The review is well described by its title (Herbicidal Germination Inhibitors
with Unknown Mode of Action), and major attention is centered upon
the 2-chloroacetamides. Fedtke compiled a partial list of physiological processes considered to be most important during the germination and growth of seedlings as these processes may be related to
possible actions by a variety of herbicides.

Test systems that have been used with many herbicides for
detection of inhibition of germination and growth include the following: GA$_3$-induced α-amylase synthesis; meristematic activities in
cell division and growth; seedling root growth; IAA-induced coleoptile elongation; cytokinin-controlled cell proliferation; and mitosis
index in root apices. Among these, the first two have been examined for 2-chloroacetamides. Metabolic pathways and processes
associated with germination and growth of seedlings examined for

biochemical susceptibility to inhibition by herbicides include the
following: synthesis and/or regulation of specific proteins; ion
transport and membrane permeability; lipid synthesis and membrane
organization; biosynthesis of coenzymes, cytochromes, etc.; biosynthesis of amino acids; hormonal regulation; biosynthesis of
hormones; effects on respiratory energy; nucleic acid synthesis;
biosynthesis of pyrimidines and purines; and microtubular synthesis
and control systems. Among these, studies have been published
involving the first five metabolic systems with 2-chloroacetamides
[42].

Brief mention is made of other germination-inhibiting herbicides
(e.g., chlorsulfuron), but brief only because far less has been
published about these herbicide examples than about the substituted
acetamides. Fedtke [42] states that the mode of action of the
2-chloroacetamides is unknown as of the date of his review, a point
also made by Jaworski [1] and Ashton and Crafts [43] in their
reviews.

Fedtke [42] emphasizes that concentration dependencies and
time-lag variables observed between herbicidal addition and inhibition
onset must be considered to differentiate between direct and indirect effects. One aspect of this is that after application of
2-chloroacetamides to plants or to plant tissues, growth inhibition
occurs after time lags up to several hours or longer. He also states
that present knowledge and understanding of germination and seedling growth mechanisms are limited and many other systems remain
to be recognized.

In experiments designed to examine and differentiate among
various modes of physiological action of nine herbicides, Fedtke [44]
used *Chlamydomonas reinhardii,* a unicellular algae, as test species
grown under synchronous conditions. Alachlor acts as an indirect
inhibitor of cell division or cell separation in these algae studies.
The inhibition by alachlor is reversible when alachlor is removed,
and appears to involve a metabolic process or reaction required for
normal cell division. It is suggested by Fedtke [44] that ". . .ala-
chlor might specifically interfere with the regulation of cell division. . .," but the primary step of the mechanism of herbicidal
action is yet to be elucidated.

In a series of papers, Wilkinson [45–48] reported that alachlor,
metolachlor, and CDAA reduced mevalonate incorporation into various terpenoid precursors from which gibberellic acid is derived.
Wilkinson speculated that the growth reduction caused by 2-chloro-
acetamides is the result of inhibition of gibberellic acid biosynthesis.

Ebert and Ramsteiner [49] reported that metolachlor inhibited
the synthesis of long chain alcohols and fatty acids in epicuticular
wax in plant samples harvested 4–7 days after treatment. Although
terpenes and waxes are formed via different biosynthetic pathways,

both routes utilize coenzyme A intermediates and substrates, which may indicate a common mechanism of inhibition through actions upon coenzyme A. The effects reported may have little relation to herbicidal primary events in view of the time interval between treatment and harvest.

Molin and colleagues [50] reported inhibition by alachlor of anthocyanin and lignin accumulation in excised sections of 6-day-old etiolated sorghum leaves, postulating that some inhibitory effect on coenzyme A occurs with alachlor as a possible basis for the herbicidal activity.

Warmund et al. [51] proposed that some effect by alachlor which deprives seedlings of energy for germination might be related to a depressed function of coenzyme A by a reduction in lipid catabolism in sorghum seedlings. Alternatively, alachlor may be inhibiting the production of proteins essential for coenzyme A synthesis.

Protein synthesis was measured by [^{14}C]leucine incorporation by Egli et al. [52] in efforts to define a relatively simple screening system to test candidate chemicals for potential herbicidal activity. The test system was comprised of suspension cultured, heterotrophic cells of black nightshade (*Solanum nigrum*). Alachlor at 0.1 mM and higher concentrations was a strong inhibitor of protein synthesis, as were a number of other known herbicides used as positive controls. The results match those of earlier workers cited in the aforementioned review papers.

Mellis and co-workers [53] studied leakage of ^{32}P-labeled orthophosphate from roots of onion (*Allium cepa* L.), a susceptible species, by exposure to 10^{-5} M metolachlor, and observed an increase of 14-fold over leakage from control. In contrast, there was no significant loss of ^{32}P from roots of two tolerant species, soybeans (*Glycine max* (L.) Merr. "Bragg") and corn (*Zea mays* L. "Pioneer 3369A"). By these results they suggested the possibility that the herbicide may be causing a loss of integrity of root cell membrane. Naphthalic anhydride applied at 10^{-7} with metolachlor at 10^{-5} M essentially prevented ^{32}P leakage from onion roots. Because protein synthesis is inhibited by metolachlor only at the high level of 10^{-4} M, effects on synthesis of lipids, the other major constituents of cell wall, were measured. Neither metolachlor nor alachlor at 10^{-4} M inhibited the uptake of acetate-2-[^{14}C] or malonic-2-[^{14}C]acid into excised cotton root tips or the incorporation of these precursors into lipids. Similarly, neither herbicide produced evidence of inhibition of phospholipid synthesis nor of incorporation of choline-1,2-[^{14}C]-chloride into phosphatidylcholine. The authors conclude by stating that there is no evidence under the experimental conditions used that the loss of membrane integrity is caused by inhibition of total lipid, phospholipid, or phosphatidyl choline synthesis by either alachlor or metolachlor.

Glutathione, that is, the reduced glutathione in thiol-(-SH) form (GSH), has been reported by Fedtke to react readily with 2-chloroacetamides via nucleophilic displacement of the 2-chloro-group [42, cited references]. Frear et al. [54] reported that alachlor and propachlor reacted directly with GSH in vitro, forming the respective glutathione conjugates. Leavitt and Penner [55] reported the same for metolachlor. Other herbicides which do not react readily directly with GSH do form glutathione conjugates enzymatically via glutathione S transferase (GST).

Corn GST was characterized by Mozer and co-workers [56], and the structural analysis of a gene encoding corn GST I has been published by Shah et al. [57].

The effects on GSH levels by alachlor (and propachlor) were reviewed briefly by Stephenson et al. [58]. Rubin and colleagues [59] reported twofold and higher GSH increases over controls in 4 and 7 days, respectively, in sorghum roots treated with 10 ppm flurazole (Screen, Monsanto; benzyl 2-chloro-4-(trifluoromethyl)-5-thiazole-carboxylate) a herbicide antidote (safener) for alachlor injury. They also reported positive but lower GSH increases with 10 ppm R-25788 [2,2-dichloro-N,N-[diallyl]-acetamide), Stauffer]. These authors observed similar but lower GSH increases in sorghum roots with several N-substituted maleimides and related compounds, but these required 100 ppm treatment rates.

Sorghum seed treated with cyometrinil (Concep, Ciba-Geigy; α-[(cyanomethoxy)imino]benzene acetonitrile) safens sorghum against metolachlor, reported initially by Ellis et al. [60], and by Nyffeler et al. [61]. Similarly, flurazole-treated sorghum seed is safened against alachlor, reported by Schafer and Brinker with others [62-64]. Both herbicide antidotes are used commercially, and either safener as a seed treatment protects sorghum against both of the 2-chloroacetamides.

The detoxification of 2-chloroacetamides by conjugation with glutathione suggests a possible mode of action of this class of herbicides as a conjugation with thiol-containing enzyme(s), with consequent inhibition of some critical function of the enzyme(s). Fedtke [42], however, cites the lack of inhibition of several SH enzymes tested in the presence of 0.3 mM alachlor reported by Marsh et al. [65] as evidence that the 2-chloroacetamides are not classifiable as general -SH reagents. Nevertheless, Leavitt and Penner [55] reported that [14C]alachlor conjugates with coenzyme A on the basis of detection of 14C in a TLC separation of a conjugate.

IV. CONCLUSIONS

The unequivocal identifications of the 11 major plant alachlor metabolites from corn and soybeans [9,10], metabolites comprised

of a preponderance of sulfur-linked compounds, are among the most
significant developments since Fedtke's 1982 review [42] of the
2-chloroacetamides. The prominent role played by glutathione
(GSH) in plants in detoxifying alachlor and other 2-chloroacetamides
is rivaled by a similar magnitude of activity by GSH in animal
metabolism of these and other susceptible xenobiotic compounds.
Examples for alachlor include compounds [15] through [23] containing the sulfur linkage listed in Figure 2, and the preponderance of
sulfur-linked metabolites from animals shown in Figures 3 and 6.
Not to be overlooked are the five identical metabolites found in/from
alachlor-treated plants and rats, establishing a common link in detoxification pathways between plants and animals.

Of additional importance are the oxidative transformations observed among the plant metabolites of alachlor. The bio-oxidations
by plants of the sulfides derived from glutathione conjugates to
sulfoxides and sulfones are paralleled in animal catabolism of such
glutathione conjugates, but have not received much attention with
respect to herbicide–plant interactions. The conversions of the
aryl-ethyl groups in alachlor to the 1-hydroxyethyl groups by plant
systems is not very well known, possibly because of the difficulties
in isolating and identifying these polyfunctional compounds containing polar groups. The high degree of conversion by legumes in
these oxidations was surprising. The generation of highly complex
metabolite mixtures during the metabolism of alachlor by plants now
is more understandable, considering the permutations of intermediary
transient metabolites that can arise from a variety of combinations
of these catabolic transformations. With the possible exception of
compounds [13] and [23] in Figure 2, the remainder represent
oxidized metabolites. The major plant metabolites of alachlor illustrate this point well. Construction of a metabolic pathway is not
attempted in this review, but several can be postulated by considering the probable interrelationships among the 11 plant metabolites
listed.

The literature reports facile and rapid conversion of alachlor
to the glutathione adducts in plant species resistant to the herbicidal
effects induced by alachlor such as corn and sorghum [11,13,14].

An understanding of the primary mode of herbicidal action of
alachlor and other 2-chloroacetamides remains elusive. The ubiquitous presence of GSH and glutathione transferase (GST) and, presumably, glutathione synthetase, or their analogs in plant and animal
cells still suggests that this biochemical reaction of alachlor with
GSH has an important relationship not only to the 2-chloroacetamide
detoxification, but also to herbicidal activity and the specific plant
selectivities shown by alachlor and the other 2-chloroacetamides.
This remains a major clue from which to design and execute future
studies to elucidate the mechanism of phytotoxic action by the
2-chloroacetamides. Detoxification in plants may be an important

facet because of the herbicide- and herbicide antidote-induced increases in the levels of GSH reported in certain plants [58,59].

Crop safety to 2-chloroacetamides could be either or both a combination of the intrinsic level of GSH in the germinating seedling of the crop plant, and/or of the functional capacity of glutathione-synthetase in that particular crop seedling to respond to the GSH demand caused by exposure to a 2-chloroacetamide. Conversely, weed species susceptible to 2-chloroacetamides may be deficient in the level of, and/or ability to replenish GSH in sufficient quantity to detoxify the 2-chloroacetamide before the herbicide fatally inhibits some essential survival process.

Detoxification of 2-chloroacetamides by conjugation with gluta-thione has given rise to the suggestion that a possible mode of action of this class of herbicides is a conjugation with one or more thiol-containing enzyme(s) critical to germination or survival of seedlings. Fedtke [42], however, cites the lack of inhibition of several SH enzymes tested in the presence of 0.3 mM alachlor reported by Marsh et al. [65]. Nevertheless, additional research in this or a similar area involving effects by alachlor or other 2-chloroacetamides upon enzymatic processes seems attractive. For example, it might be useful to undertake studies, perhaps by necessity on a micro-scale, of the effects of alachlor on germinating seeds of susceptible plant species. Many susceptible species have small seeds, and measurements of their GSH content and reaction with alachlor would be of interest relative to the GSH content and reactivity of the larger seeds of resistant species. Conceivably this could be applicable also to GST measurements, and, if possible, to glutathione synthetase capacities among resistant and susceptible species. Expansion to examination of the relative rates of detoxification of alachlor and other 2-chloroacetamides by the germinating seeds of the several species might provide leads to selection of one or more species which offer promise in a search for effects upon some critical enzyme systems essential for the survival of the susceptible seed/seedling exposed to alachlor.

ACKNOWLEDGMENTS

Contributions from Drs. J. M. Malik and S. Dubelman and their co-workers are gratefully acknowledged. Special thanks are due Dr. R. K. Howe and colleagues for supplying the structural formulae and graphics in the figures.

REFERENCES

1. E. G. Jaworski, in *Herbicides: Chemistry, Degradation and Mode of Action* (P. C. Kearney and D. D. Kaufman, eds.), Marcel Dekker, Inc., New York and Basel, 1975.

2. P. C. Hamm, *Weed Sci.*, *22*, 541 (1974).

3. *Herbicide Handbook* (S. E. Dowdy and M. E. Rohan, eds.), Monsanto Company, St. Louis, MO, 1974.

4. Lasso Herbicide Label, Monsanto Agricultural Company, EPA Reg. No. 524-314.

5. Alachlor; Tolerances for Residues, *40 Code of Federal Regulations, Section 180.249*, 1985.

6. *Herbicide Handbook* (C. E. Beste, ed.), Weed Science Society of America, 5th Edition, Champaign, IL, 1985.

7. Environmental Protection Agency, Alachlor Registration Standard; Guidance for the Reregistration of Pesticide Products Containing Alachlor as the Active Ingredient, November 20, 1984.

8. Environmental Protection Agency, Notice of Initiation of Special Review of Registrations of Pesticide Products Containing Alachlor; Alachlor Position Document 1, December 31, 1984.

9. Monsanto Co., *Rebuttal to the Special Review Position Document 1 for Pesticide Products Containing Alachlor*, 20 Volumes, Monsanto Agricultural Co., April 9, 1985.

10. EPA Notice, Alachlor Special Review, Position Document 2/3, *Fed. Reg.*, *51*, 36166 (1986); Technical Support Document, September 1986.

11. G. L. Lamoureux, L. E. Stafford, and F. S. Tanaka, *J. Agric. Food Chem.*, *19*, 346 (1971).

12. J. R. C. Leavitt and D. Penner, *J. Agric. Food Chem.*, *27*, 533 (1979).

13. B. Rubin, O. Kirino, and J. E. Casida, *J. Agric. Food Chem.*, *33*, 489 (1985).

14. G. L. Lamoureux and D. G. Rusness, in *Xenobiotic Conjugation Chemistry* (G. D. Paulson, J. Caldwell, D. H. Hutson, and J. J. Menn, eds.), American Chemical Society Symposium Series 299, 1986, p. 62.

15. J. E. Cowell and S. Dubelman, *IUPAC Sixth Intl. Cong. Pest. Chem.*, *Poster/Workshop Subtopic 8D*, Exposure Assessment and Monitoring, Poster 8D-01, Ottawa, Canada, August 1986.

16. J. E. Cowell, R. G. Danhaus, J. L. Kunstman, A. G. Hackett, M. E. Oppenhuizen, and J. R. Steinmetz, *Arch. Environ. Contam. Toxicol.*, *16*, 327 (1987).

17. S. Dubelman, private communication.

18. Environmental Protection Agency, *Pesticide Assessment Guidelines, Subdivision O: Residue Chemistry*, Guideline No. 171-4 (a)(3), Nature of Residue in Livestock, U.S. Dept. of Comm., Nat. Tech. Inform. Serv., Springfield, VA, October 1982.

19. J. E. Bakke, in *Xenobiotic Conjugation Chemistry* (G. D. Paulson, J. Caldwell, D. H. Hutson, and J. J. Menn, eds.), American Chemical Society Symposium Series 299, 1986, p. 301.

20. J. E. Bakke, G. L. Larsen, et al., in *Sulfur in Pesticide Action and Metabolism* (J. D. Rosen, P. S. Magee, and J. E. Casida, eds.) American Chemical Society Symposium Series 158, 1981, p. 165.

21. G. L. Larsen and J. E. Bakke, *Xenobiotica, 13*, 115 (1983); ibid., *11*, 473 (1981).

22. J. E. Bakke, G. L. Larsen, et al., *Drug Metab. Dispos., 9*, 525 (1981).

23. J. E. Bakke, J.-A. Gustaffson and B. E. Gustaffson, et al., *Science, 210*, 433 (1980).

24. G. L. Larsen and J. E. Bakke, *J. Environ. Sci. Health, B14*, 495 (1979).

25. J. E. Bakke and C. E. Price, *J. Environ. Sci. Health, B14*, 279 (1979).

26. J. M. Kronenberg, T. W. Fuhremann, and D. E. Johnson, *IUPAC Sixth Intl. Cong. Pest. Chem., Poster/Workshop Subtopic 8D*, Exposure Assessment and Monitoring, Poster 8D-17, Ottawa, Canada, August 1986.

27. K. H. Carr, R. C. Chott, R. K. Howe, and J. M. Malik, *IUPAC Sixth Intl. Congr. Pest. Chem., Poster/Workshop Subtopic 7C*, Metabolism in Mammals, Poster 7C-06, Ottawa, Canada, August 1986.

28. E. C. Kimmel, J. E. Casida, and L. O. Ruzo, *J. Agric. Food Chem., 34*, 157 (1986).

29. S. J. Wratten, H. Fujiwara, and R. T. Solsten, *J. Agric. Food Chem., 35*, 484 (1987).

30. P. C. C. Feng and S. J. Wratten, *J. Agric. Food Chem., 35*, 491 (1987).

31. G. B. Beestman and J. M. Deming, *Agron. J., 66*, 308 (1974).

32. J. M. Tiedje and M. L. Hagedorn, *J. Agric. Food Chem., 23*, 77 (1975).

33. S. L. Chopra and R. K. Sethi, *Indian J. Agric. Chem., 7*, 119 (1974).

34. R. K. Sethi and S. L. Chopra, *J. Indian Soc. Soil Sci., 23*, 184 (1975).

35. C. H. Fang, *J. Chin. Agric. Chem. Soc., 15*, 53 (1977).

36. N. J. Novick, R. Mukherjee, and M. Alexander, *J. Agric. Food Chem., 34*, 721 (1986).

37. N. J. Novick, R. Mukherjee, and M. Alexander, *Appl. Environ. Microbiol., 49*, 737 (1985).

38. R. L. Zimdahl and S. K. Clark, *Weed Sci., 30*, 545 (1982).

39. R. L. Metcalf, G. K. Sangha, and I. P. Kapoor, *Environ. Sci. Technol., 5*, 709 (1971).

40. C-C. Yu, G. M. Booth, D. J. Hansen, and J. R. Larsen, *J. Agric. Food Chem., 23*, 877 (1975).

41. B. M. Francis, R. L. Lampman, and R. L. Metcalf, *Arch. Environ. Contam. Toxicol., 14*, 693 (1985).

42. C. Fedtke, in *Biochemistry and Physiology of Herbicide Action*, Springer-Verlag, Berlin, 1982, p. 148.

43. M. A. Ashton and A. S. Crafts, in *Mode of Action of Herbicides*, 2nd Edition, John Wiley & Sons, New York, 1981.

44. C. Fedtke, in *Biochemical Responses Induced by Herbicides* (D. E. Moreland, J. B. St. John and F. D. Hess, eds.), American Chemical Society Symposium Series 181, Washington, DC, 1982, p. 231.

45. R. E. Wilkinson, *Pest. Biochem. Physiol.*, *16*, 63 (1981).

46. R. E. Wilkinson, *Pest. Biochem. Physiol.*, *16*, 199 (1981).

47. R. E. Wilkinson, *Pest. Biochem. Physiol.*, *17*, 177 (1982).

48. R. E. Wilkinson, *Pest. Biochem. Physiol.*, *23*, 19 (1984).

49. E. Ebert and K. Ramsteiner, *Weed Res.*, *24*, 383 (1984).

50. W. T. Molin, E. J. Anderson, and C. A. Porter, *Pest. Biochem. Physiol.*, *25*, 105 (1986).

51. M. R. Warmund, H. D. Kerr, and E. J. Peters, *Weed Sci.*, *33*, 25 (1985).

52. M. A. Egli, D. Low, K. R. White, and J. A. Howard, *Pest. Biochem. Physiol.*, *24*, 112 (1985).

53. J. M. Mellis, P. Pillai, D. E. Davis, and B. Truelove, *Weed Sci.*, *30*, 399 (1982).

54. D. S. Frear, R. H. Hodgson, R. H. Shimabukuro, and G. G. Still, *Adv. Agron.*, *24*, 327 (1972).

55. J. R. C. Leavitt and D. Penner, *J. Agric. Food Chem.*, *27*, 533 (1979).

56. T. J. Mozer, D. C. Tiemeier, and E. G. Jaworski, *Biochemistry*, *22*, 1068 (1983).

57. D. M. Shah, C. M. Hironaka, R. C. Wiegand, E. I. Harding, G. G. Krivi, and D. M. Tiemeier, *Plant Molec. Biol.*, *6*, 203 (1986).

58. G. R. Stephenson, A. Ali, and F. M. Ashton, in *Pesticide Chemistry: Human Welfare and the Environment*, Vol. 3, Pergamon Press, Oxford, 1983.

59. B. Rubin, O. Kirino, and J. E. Casida, *J. Agric. Food Chem.*, *33*, 489 (1985).

60. J. F. Ellis, J. W. Peek, J. Boehle, Jr., and G. Muller, *Weed Sci.*, *28*, 1 (1980).

61. A. Nyffeler, H. R. Gerger, and J. R. Hensley, *Weed Sci.*, *28*, 6 (1980).

62. D. E. Schafer, R. J. Brinker, and R. O. Radke, *Proc.-North Cent. Weed Control Conf.*, *35*, 67 (1980).

63. D. E. Schafer, R. J. Brinker, and R. O. Radke, *Abstr. Weed Sci. Soc. Am.*, No. 53 (1981).

64. R. J. Brinker, D. E. Schafer, R. O. Radke, G. Boeken, and H. W. Frazier, *Proc. Br. Crop Prot. Conf. Weeds*, *2*, 469 (1982).

65. H. V. Marsh, J. Bates, and P. Trudeau, *Abstr. Weed Sci. Soc. Am. Meet.*, 1975, p. 67.

Chapter 7

METOLACHLOR

HOMER M. LEBARON, JANIS E. MCFARLAND,
and B. J. SIMONEAUX

CIBA-GEIGY Corporation, Greensboro, North Carolina

E. EBERT

CIBA-GEIGY Limited, Basel, Switzerland

I. INTRODUCTION

The chloroacetamide herbicides have likely been the most important and widely used class of chemicals discovered and developed for
control of weeds in crop production.

The major discoveries in this group were not made at one time
or even in the same laboratory, but cover a period of at least 35
years. The first discoveries took place in the Monsanto Company,
beginning in 1952 [1,2]. The major discoveries leading to promising
or commercial herbicides included the following [3,4]:

Year of discovery, first publication, or introduction	
1953	N,N-diallyl-2-chloroacetamide (IUPAC);* 2-chloro-N,N-di-2-propenylacetamide (CA & WSSA);† (CDAA[a], allidochlor[b], CP-6343[c], Randox[d]) Monsanto[e]
1964	2-chloro-N-isopropylacetanilide (IUPAC); 2-chloro-N-(1-methylethyl)-N-phenyl-acetamide (CA & WSSA), (propachlor,[a,b] CP-31393,[c] Ramrod[d] or Bexton[d]) Monsanto Monsanto,[e] Dow[e]
1966	2-chloro-2',6'-diethyl-N-(methoxymethyl)-acetanilide (IUPAC); 2-chloro-N-(2,6-diethylphenyl)-N-(methoxymethyl)acetamide (CA & WSSA); (alachlor,[a,b] CP-50144[c], Lasso[d]) Monsanto[e]
1970	N-(butoxymethyl)-2-chloro-2',6'-diethyl-acetanilide (IUPAC); N-(butoxymethyl)-2-chloro-N-(2,6-diethylphenyl)acetamide (CA & WSSA); (butachlor,[a,b] CP-53619,[c] Machete[d]) Monsanto[e]
1970	2-chloro-N-(ethoxymethyl)-6'-ethyl-o-aceto-toluidide (IUPAC); 2-chloro-N-(ethoxy-methyl)-N-(2-ethyl-6-methylphenyl)-acetamide (CA & WSSA); (acetochlor,[a,b] MON-097[c], Harness[d]) Monsanto[e]

Year of discovery, first publication, or introduction	
1974	2-chloro-6'ethyl-N-(2-methoxy-1-methyl-ethyl) acet-o-toluidide (IUPAC); 2-chloro-N-(2-ethyl-6-methylphenyl)-N-(2-methoxy-1-methylethyl)acetamide (CA & WSSA); (metolachlor,[a,b] CGA-24705,[c] Dual[d]) Ciba-Geigy[e]
1976	2-chloro-N-(2-methoxyethyl)acet-2',6'-xylidide (IUPAC); 2-chloro-N-(2,6-di-methylphenyl)-N-(2-methoxyethyl)acetamide (CA); (dimethachlor,[b] CGA-17020,[c] Teridox[d]) Ciba-Geigy[e]
1979	2-chloro-N-(pyrazol-1-ylmethyl)acet-2',6'-xylidide (IUPAC); 2-chloro-N-(2,6-di-methylphenyl)-N-(1H-pyrazol-1-ylmethyl)-acetamide (CA); (metazachlor,[b], BAS-47900H,[c] Butisan S[d]) BASF[e]
1983	2-chloro-2',6'-diethyl-N-(2-propoxyethyl) acetanilide (IUPAC); 2-chloro-N-(2,6-diethylphenyl)-N-(2-propoxyethyl)aceta-mide (CA); (pretilachlor,[b] CGA-26423,[c] Solnet[d] or Rifit[d]) Ciba-Geigy[e]

*IUPAC, Chemical name by International Union of Pure and Applied Chemistry.
†CA, Chemical name by Chemical Abstracts.
WSSA, Weed Science Society of America.
[a], WSSA Common Name.
[b], BSI and ISO common names.
[c]Company code number
[d]Trade mark or product name(s)
[e]Manufacturer(s)
See Figure 1 for chemical structures.

The earlier commercial chloroacetamide herbicides have been al-most totally replaced for major uses by alachlor and metolachlor due to their improved activity and longer weed control, especially in sandy loam soils [1]. In addition, the more recently developed chloro-acetamides do not cause the skin and eye irritation seen with some of the earlier acetamides.

Metolachlor is one of the most extensively and intensively used herbicides in the chemical class of chloroacetamides and much data are

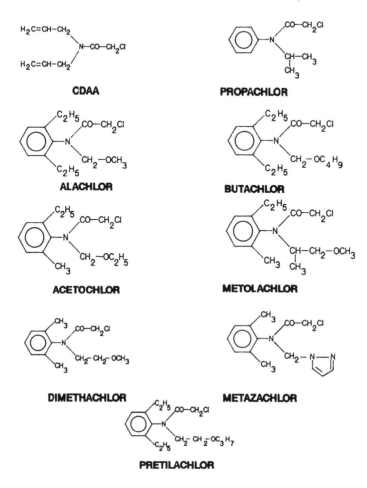

FIG. 1   Structures and common names of chloroacetamide herbicides.

available on metolachlor that have not been included in some earlier chloroacetamide reviews [5,6]. This chapter reviews data from journal articles, reviews, and unpublished reports on the chemistry, biology, degradation, and mode of action of metolachlor.

II. CHEMICAL AND PHYSICAL PROPERTIES

Metolachlor provides excellent control of most annual grass and many broadleaf weeds. Its very high level of biological activity and its

desirable chemical and soil stability [7–13], provide good control of
late germinating grasses, especially under conservation tillage systems
[12,14–16]. In addition it is one of the most effective herbicides for
control of yellow nutsedge (*Cyperus esculentus* L.) [17–19] in crops
where it can be used. Metolachlor is used very successfully in several major crops, including corn (*Zea mays* L.), soybean (*Glycine
max* L. Merr.), sorghum (*Sorghum bicolor* L. Moench), cotton (*Gossypium hirsutum* L.), and peanut (*Arachis hypogaea* L.), and many
minor crops. Some of the biological characteristics of metolachlor can
be explained on the basis of its chemical and physical properties,
presented in Table 1 [3,4].

A. Aqueous Hydrolysis

The hydrolysis of metolachlor was studied at a variety of pH
values in aqueous media at three different temperatures: 30, 50, and
70°C. Rate constants (k) and half-lives ($t_{1/2}$) were calculated using
Arrhenius parameters, assuming first order kinetics. The half-lives
of metolachlor at 20°C were calculated to be greater than 200 days
at pH 1, 5, 7, and 9, and 97 days at pH 13 [20].

Under basic conditions, metolachlor was hydrolyzed to *N*-(2-ethyl-
6-methylphenyl)-2-hydroxy-*N*-(2-methoxy-1-methylethyl) acetamide.
Under acidic conditions, metolachlor hydrolyzed to 2-chloro-*N*-
(2-ethyl-6-methylphenyl)-*N*-(2-hydroxy-1-methylethyl)acetamide, which
was rapidly converted to 4-(2'-methyl-6'-ethylphenyl)-3-methyl-
morpholinone-5. See Figure 2 for the hydrolysis degradation routes
of metolachlor.

B. Volatilization

Metolachlor is considered to be relatively nonvolatile (1.3×10^{-5} mmHg
at 20°C). Burkhard and Guth [21] calculated that metolachlor at a
concentration of 80 µg/g on a wet soil weight basis volatilizes from
moist soil at 20°C at a rate of between 1.5 to 4.5 ng/cm^2 per hr
when the air flow rate is 30 L/hr. In a more detailed laboratory
study, Burkhard [22] reported that the amount of metolachlor
volatilized from three soil types containing 80 ppm of the herbicide,
12% moisture, at 35°C, and with an air flow of 30 L/hr ranged from
0.03 to 0.09 kg/ha per day, with the greatest amount being lost from
a sand soil. A temperature change of 10°C changed the volatilization
rate about 3.8 times, while doubling the air flow approximately doubled
the loss of metolachlor. Burkhard estimated that, under practical field
conditions, only 0.6 to 1.4% of applied metolachlor would volatilize
from the soils he tested within the first 24 hours.

Volatility, however, can cause a minor loss of metolachlor when
applied preemergence to bare soil or in some conservation tillage situations. Parochetti [23] observed that while only 0.1% of metolachlor

TABLE 1

Chemical and Physical Properties of Metolachlor

Properties	Data
Molecular formula	$C_{15}H_{22}ClNO_2$
Molecular weight	283.8 g/mol
Specific gravity	1.085 at 20°C
Form at 25°C	Liquid
Color	Colorless to tan
Boiling point	100°C at 0.001 mmHg
Decomposition temperature	Stable up to 50°C for 2 years. Stable up to 300°C for short period
Vapor pressure	1.3×10^{-5} mmHg at 20°C
Odor	Odorless
Corrosiveness	Noncorrosive
Solubility: Water	
ppm at 20°C	530 ppm
M/l at 20°C	1.868×10^{-3}
Acetone, benzene, chloroform, cyclohexanone, dimethyl formamide, ethanol, ether, ethylene dichloride, hexane, methanol, methyl cellosolve, toluene, and xylene	All are soluble or miscible
Ethylene glycol	Insoluble
Propylene glycol	Insoluble
Photolysis Soil	Stable, but about 50% degradation in 8 days in natural or artificial light
Solution	Stable, but 6% degradation in 30 days in natural sunlight
Hydrolysis	Hydrolyzed at high pH ($t_{1/2}$ = 97 days at pH 13, 20°C)

TABLE 1

(Continued)

Properties	Data
Partition coefficient (Octanol/water)log P_{ow}	1,349 3.13
Henry's law constant (calc. by vapor press/ solubility)	5.35×10^{-3}
Soil half-life Northern states Southern states	 30—50 days 15—25 days
Patent	DOS 2,320,340 USP 3,937,730
Manufacturer	Ciba-Geigy
First described	*Proc. Br. Weed Cont. Conf.,* *12*(2), 787 (1974).

was lost from the soil surface after eight days, about 50% was lost from a glass surface, and from 11.5 to 36.6% volatilized from the straw surface, of various plant residues. Strek and Weber [9] observed in greenhouse studies that wind speed influenced the disappearance of metolachlor, and that volatilization could be an important factor in its disappearance from certain surfaces.

C. Photolysis

Photodegradation is considered of little importance in the loss of metolachlor from soil under field conditions, especially when precipitation occurs soon after soil application. Parochetti [23] reported no photodecomposition of metolachlor. However, when Aziz [24] applied [^{14}C]metolachlor equivalent to 4.6 pounds a.i./A to a thin film of soil on glass slides and exposed them to natural sunlight for 8 days, 50% of the herbicide was photodegraded. Most of this (39%) was identified as CGA-41638 2-chloro-*N*-(2-ethyl-6-methylphenyl)-*N*-(2-hydroxy-1-methylethyl) acetamide. More than 10% had volatilized, most of which was still parent herbicide. Similar results were obtained with artificial sunlight.

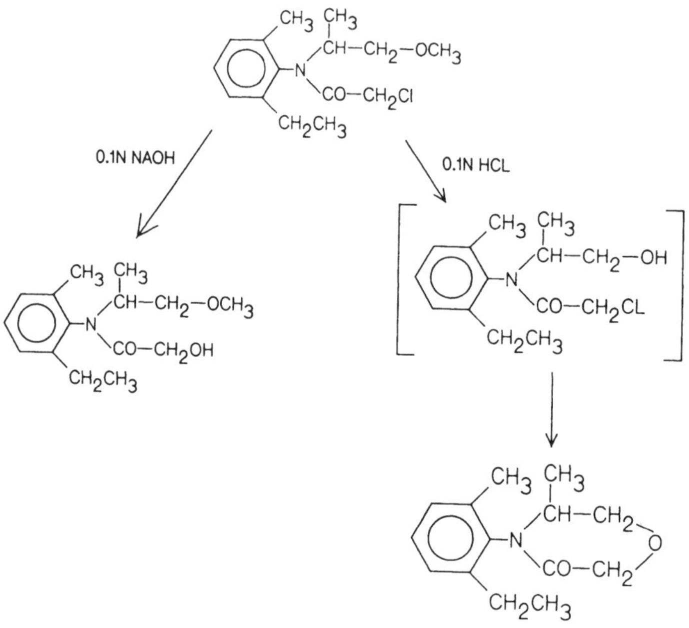

FIG. 2   Hydrolysis pathway of metolachlor.

When metolachlor was exposed to natural sunlight in aqueous solution, very slight (6%) decomposition took place after one month [25]. Under exaggerated conditions of artificial sunlight exposure in aqueous solution, about 60% metolachlor was photodegraded in 15 days.

III. STRUCTURE-ACTIVITY RELATIONSHIPS

The herbicidal activity of the chloroacetamides is dependent on both the presence of an α-halogen as well as the nature of the substituents on the nitrogen side chain. In addition, herbicidal activity has been correlated with different isomers of metolachlor and with the rotameric forms of both bromoacetanilides and iodoacetanilides.

Hamm [1] and Hamm and Speziale [2], in their overviews of the discovery and development of the initial chloroacetamide herbicides, reported that the nonhalogenated acetamides tested were herbicidally inactive. They also reported that when both of the nitrogen substituents are replaced with hydrogens, the resultant chloroacetamide is inactive. If one of the nitrogen substituents is a hydrogen and the other is an unsubstituted aryl or a substituted aryl, the

compound is an effective herbicide. Dialkyl nitrogen substitution is
superior to monoalkyl nitrogen substitution for herbicidal activity.
Most of the marketed chloroacetamide herbicides are actually chloroacetanilides. However, the aromatic ring is not always necessary for
herbicidal activity as evidenced by CDAA.

McFarland and Hess [26] found that nonhalogenated acetamides,
including the fungicide metalaxyl, N-(2,6-dimethylphenyl)-N-methoxy-
acetyl alanine methyl ester, were one to two orders of magnitude
less active in a bioassay of oat (*Avena sativa* L.) root growth compared to their chlorinated derivatives. Other halogenated acetanilides
have also been tested for herbicidal activity. An analog of metolachlor, in which bromine is substituted for the α-chlorine, inhibited oat
root growth at 10 μM, while an analog with fluorine substituted for
the α-chlorine was not inhibitory at 100 μm [27]. While the α-halogen
is important for herbicidal activity, it is not necessarily an important
structural feature of the acetamide fungicides. Instead, the methoxy-
acetyl moiety appears to be one of the important structural features
for the fungicidal activity of the nonhalogenated acetanilide,
metalaxyl [28].

Chupp and Olin [29] and Chupp et al. [30] investigated certain
haloacetanilides (2'-6'-dialkyl-2-halo-N-methylacetanilides) that are
similar in structure to the chloroacetamide herbicides and exist in
two rotameric forms. The rotameric forms were stable for varying
lengths of time, depending on the structures of the hindered acetanilides. Of these two rotameric forms, the major rotamer comprised
about 90% of the equilibrium mixture and a minor rotamer represented
∿10% of the equilibrium mixture. Hamm [31] established that the
minor rotamer of a brominated acetanilide (α-bromo-2',6'-di-t-butyl
N-methylacetanilide) inhibited the growth of ryegrass (*Lolium perenne*
L.), whereas the major rotamer was at least 100 times less active.
Molin et al. [32] found that the major rotamer of α-iodo-2',6'-di-t-
butyl N-methylacetanilide was approximately 100 times less active than
the minor rotamer in an anthocyanin synthesis assay in excised
sorghum mesocotyls.

Metolachlor is a mixture of four stereoisomers, the isomerism of
which is based on a combination of a chiral center in the aliphatic
side chain and a chiral axis between the phenyl and the nitrogen
atom (Fig. 3). The four stereoisomers differ in their ability to inhibit the growth of weeds [33]. The two mixtures of diastereomers
(only separated at the chiral carbon) of metolachlor have been
designated CGA-77101 and CGA-77102 by the CIBA-GEIGY Corporation (Fig. 3). An application of 1000 g/ha of CGA-77101 inhibited
the growth of seven grasses and three broadleaf weeds by an average
of approximately 40%; while CGA-77102 inhibited growth by approximately 95% in greenhouse studies. A difference in the biological activity between the two mixtures of diastereomers can also be detected
in an oat root growth bioassay in intact seedlings [34]. At 10 μM,

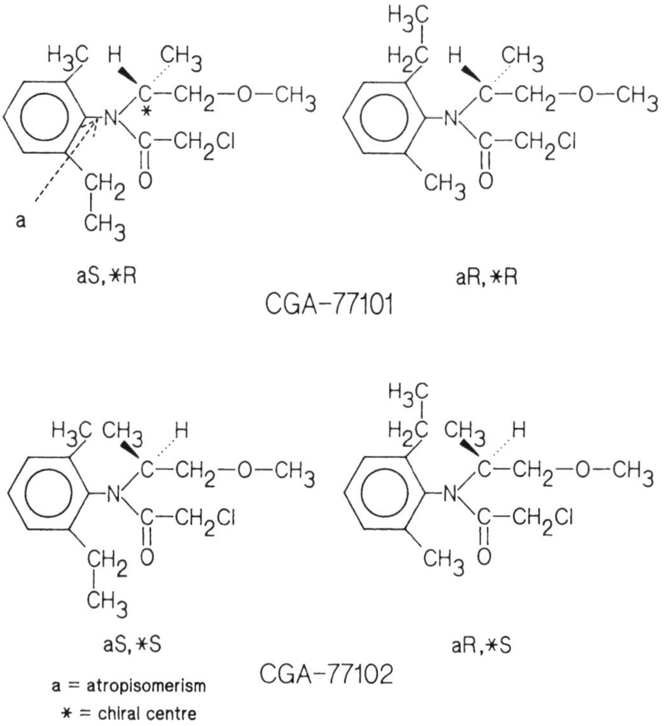

FIG. 3   Mixtures of diastereomers of metolachlor, designated CGA-77101 and CGA-77102 by Ciba-Geigy.

CGA-77101 inhibited root growth by only 6% during a 24-hr treatment, while CGA-77102 inhibited growth by 50% (Table 2).

In addition to structure-activity differences between herbicide analogs, rotamers, and isomers, there are also chemical activity differences between the marketed chloroacetamide herbicides. Although the herbicidal activity of alachlor, metolachlor, propachlor, and other chloroacetamides can be similar in certain weed species in the field, certain chemical reactivities of these herbicides are quite different. For instance, in an in vitro alkylation reaction with cysteine, propachlor had a greater alkylation potential than alachlor, and alachlor had a greater alkylation potential than metolachlor [34]. The information gathered on the structure-activity relationship of the chloroacetamides will greatly aid the progress currently being made in elucidating the mode of action of these compounds.

TABLE 2

Effects of Metolachlor and Its "Active" (CGA-77102) and "Inactive"
(CGA-77101) Mixtures of Diastereomers on Oat Root Growth
in Sand Culture

Treatment	Oat root growth % of control			
	Hours after treatment			
	6	12	18	24
Untreated Control	100a[a]	100a	100a	100a
10 μM CGA-77101	93a	93b	93b	94b
10 μM CGA-77102	91a	70c	56c	50c
10 μM Metolachlor	100a	78c	61c	54c

[a]Values within a sample time and treatment followed by different
letters indicate significant differences (5%) level as tested by
Duncan's multiple range test. Each treatment was replicated at
least 3 times. Each of these mixtures of diastereomers is approximately 85% pure and 15% contaminated with the other mixture.
Source: From Ref. 27.

IV. BEHAVIOR IN SOILS

A. Soil Adsorption, Mobility, Leaching, and Runoff

A thorough understanding of the behavior of herbicides in soil
is essential in order to predict their biological activity and their environmental fate. In recent years, numerous studies have been conducted on the influence of soil properties on the effectiveness,
adsorption, movement, and fate of metolachlor.

Investigating factors that influence metolachlor adsorption in three
Wisconsin soils, Jordan [35] observed that adsorption increased at
lower temperatures, while pH had little influence. Ballerstedt and
Banks [36], and Peter and Weber [37] also found little effect of pH
on the phytotoxicity of metolachlor. Strek and Weber [16] found
that metolachlor was adsorbed equally to high and low organic matter
sandy loam soils, but greater adsorption occurred on montmorillonite
clay. In a later study [9], they noted that soil adsorption and desorption of metolachlor and other acetanilide herbicides were mostly
related to water solubility.

Gerber et al. [38] demonstrated an equal leaching behavior for both alachlor and metolachlor, with less movement of both herbicides in high organic matter soils. They concluded that herbicide adsorption plays a more important role in leaching than water solubility.

Jordon [35] reported that adsorption of alachlor and metolachlor did not differ in 10 different Wisconsin soils, and that mobility of 10 acetanilide herbicides was inversely related to their adsorptivity. Although the distribution coefficients (K_d) of metolachlor in most soils were slightly higher than alachlor, there were no detectable differences in leachability between these herbicides. Soil adsorption was strongly correlated with organic matter and cation exchange capacity. Clay content had little influence. In an extensive study of nine different soils, Weber and Peter [39] found that adsorption isotherms for metolachlor were well defined by linear regression analysis, as seen in Table 3 [37]. Metolachlor adsorption was correlated positively with soil organic matter content, clay content, and surface area as measured by ethylene glycol monoethyl ether (EGME, which measured hydrophilic sites) or benzyl ethyl ether (BEE, which measured lipophilic sites), and inversely correlated with herbicidal activity.

Burkhard [40] reported that the Freundlich adsorption constants (K) for metolachlor ranged from 1.54 to 10.0 µg/g of soil, indicating that it is not very strongly adsorbed to soil particles. It was found that desorption occurred at a slower rate than adsorption and that adsorption may not be completely reversible after three days.

Burkhard [41] also studied the leaching behavior of aged (after 30 days of incubation) [^{14}C]metolachlor in a sand soil and a silty loam in response to 20 cm of total precipitation in 1.25 cm/day increments. At the end of 16 days, 11.8% and 4.8% of the applied dose were found in the eluates from 40 cm columns of the sand and silt loam soils, respectively, most of which was identified as an inactive metabolite. Almost all the remaining applied [^{14}C]metolachlor was recovered from the soil columns.

Comparative leaching studies were conducted on metolachlor, alachlor, and monuron in columns of three soils and Lakeland sand by Guth [42]. He found that metolachlor leached less than monuron (roughly about half as much), and about the same as alachlor, in all three soils. Metolachlor movement in Lakeland sand was about the same as monuron and somewhat less than alachlor. Movement of metolachlor in soils seemed to be mainly influenced by organic matter and/or clay content.

Obrigawitch et al. [43] reported less leaching of metolachlor in a Pullman clay loam when compared to an Amarillo fine sandy loam and Patricia fine sandy loam. They related the metolachlor movement characteristics to the organic matter and clay content of the soil. Increasing the irrigation amount from 9 to 18 cm significantly increased the metolachlor movement in all soils.

TABLE 3

Amounts of Metolachlor Adsorbed at An Equilibrium Concentration of 14 μM and Distribution Coefficients (K_d) on Nine Soils

Soil series	pH	CEC (mEq/100 g)	Organic-matter content	Amount adsorbed (ηmol/g)	Distribution coefficients[a] (K_d)
Augusta	5.7	3.2	0.5	6.9	0.5
Norfolk	5.4	2.3	0.5	7.4	0.5
Goldsboro	5.3	3.3	1.2	19.2	1.4
Appling	6.8	6.9	1.4	15.6	1.1
Lynchburg	5.5	6.6	2.5	32.6	2.5
Cecil	5.4	3.1	1.7	13.4	1.0
Rains	6.0	7.1	1.7	31.6	2.4
Portsmouth	5.4	10.6	4.4	44.5	3.3
Cape Fear	5.1	10.3	8.7	143.8	10.9
Mean	—	—	—	35.0	2.6

[a] Amount adsorbed (ηmol/g)/equilibrium concentration (μM) (mean values for concentrations of 0 to 60 μM).

Source: Modified from Ref. 37.

Diffusion studies by Scott and Phillips [44] reported that as soil moisture was increased, the rate of herbicide diffusion was also increased. They attributed this to the shorter and more continuous diffusion pathway in the saturated soil, as compared to the unsaturated soil.

Metolachlor and other acetanilide herbicides have contributed greatly to the rapid and continuing development of conservation tillage systems. Because of the great benefits, such as reduction in loss of top soil and water, and less labor and energy (i.e., fuel), the various reduced tillage systems are an essential part of the more efficient and high production agricultural technology of the future. It has been predicted that by 2010, at least 95% of our croplands would be under some form of conservation tillage [45]. However, Schertz [45] recently revised this estimated to 63 to 82%, based on a more realistic and accurate definition of conservation tillage systems. The only way for us to achieve this very desirable and ambitious goal is to control weeds by means other than by mechanical tillage. In the foreseeable future, this will require a continuing high dependence on metolachlor and other chemical herbicides. Conservation tillage systems should, in turn, help to keep these herbicides where they are applied.

Movement of metolachlor off the plant residues and into the soil, thus facilitating uptake by weeds, is very important in reduced tillage systems. Rainfall is important in washing the herbicide from the plant residue into the soil, as well as in moving the herbicide down into the zone of weed seed germination. The amount and timing of this rainfall are important since the longer the herbicide sits on the residue, the greater the potential losses will be due to volatilization, photodegradation, and adsorption by the residue. Fawcett [46] has stated that surface crop residue is unlike soil organic matter in that it is not able to bind herbicides as strongly.

The effects of the various quantities of different plant residues and degrees of tillage on the performance and fate of metolachlor have been studied extensively. Straw or decaying plant residues on the soil surface at the time of application reduces soil reception of herbicides.

Washoff studies using herbicide-treated straw showed that metolachlor was readily washed off by simulated rainfall. Also from field studies, Strek [47] concluded that only a minor decrease in grass control was seen with metolachlor under conservation tillage, apparently due to interception of the herbicide by the crop residues.

Crutchfield et al. [48] found that metolachlor concentrations in mulched soil remained lower throughout the entire crop season, due to interception of metolachlor by plant residues, compared to applications to unmulched soil. However, weed control was actually increased with increasing mulch level.

In a Georgia study, Banks and Robinson [49] found that 15 to 20% of the originally applied metolachlor was washed into the soil by overhead irrigation, regardless of straw level. Little water-extractable herbicide remained on the straw. Initial herbicidal activity was reduced by the presence of wheat (*Triticum aestivum* L.) straw at the time of application (see Table 4), but when grain sorghum was planted 10 days after treatment, response was roughly inversely related to the amount of straw mulch, and metolachlor residues in the soil were 11 to 26% of that on Day 0.

All of the herbicides that are washed from the crop residue may not enter the soil directly, but will also be subject to surface erosion or runoff. Those herbicides that do leave the field, either in runoff water or attached to eroded soil particles, do cause environmental concerns. There has been much research in recent years to quantify the amount of herbicides applied to soil or plant residues that are lost by runoff, and how to prevent or minimize the nonpoint source levels of these chemicals in water.

TABLE 4

Response of Grain Sorghum to Metolachlor in Soil as Affected By Straw Mulch After 1.3 cm of Sprinkle Irrigation

Straw level (kg/ha)	Metolachlor Concentration (ppm)[a]		Injury (%)	
	Day 0	Day 10	Day 0[b]	Day 10[c]
0	5.35	0.58	98	70
1120	2.91	0.47	88	60
2240	1.72	0.27	81	63
4480	0.95	0.25	80	20
6720	0.85	0.14	48	30
LSD (0.05)	0.91	0.13	16	25

[a]The herbicide is assumed to all be in the upper 7.5 cm.
[b]Grain sorghum was rated 10 days after planting (planted on day of treatment).
[c]Grain sorghum rated 10 days after planting (planted 10 days after treatment.)
Source: Modified from Ref. 49.

There are a few cases, such as the study by Harvey et al. [50], when greatest herbicide runoff occurred from minimum tillage plots under conditions of a very heavy simulated rainfall (about 7 inches in 90 minutes) resulting in most of the water running over rather than under the packed-down plant residues. However, it can generally be concluded that the various conservation tillage systems will significantly reduce the amount of metolachlor and other herbicides lost in runoff due to reduced loss of water and sediment from treated fields. The increasing trend toward conservation tillage is, therefore, very desirable, not only for the reasons generally cited, but to reduce the loss of herbicides in runoff water.

B. Metabolism, Degradation, and Dissipation in Soils

One of the most essential characteristics contributing to the extensive and valuable use of metolachlor is its nearly ideal length of biological activity. The rate of dissipation of metolachlor is sufficiently rapid under almost all normal field conditions to avoid any phytotoxicity or carryover to rotational crops the next season. It may, however, cause some minor crop injury in some areas when sensitive crops are planted very soon after herbicide application, such as with failure of the treated crop due to weather conditions or when a sensitive crop (e.g., cereal grains) is planted soon after harvest of the treated crop. Burnside and Schultz [7] reported a slight reduction in winter wheat growth three months after application of metolachlor.

There have been very few cases of metolachlor carryover under rather rare or special soil conditions. Braverman et al. [51] reported that although metolachlor was very useful in reducing the red rice (*Oryza sativa* L.) population in a rice-soybean rotation when used in soybeans, it occasionally caused carryover injury to the following rice (*Oryza sativa* L.) crop when metolachlor application was followed by dry weather. Any minor yield reduction could be overcome by increasing the rice seeding rate, decreasing seeding depth, applying a safener to the rice seed, or by using a lower rate of herbicide. In researching this unusual behavior of metolachlor in an Arkansas Taloka silt loam soil, Bouchard et al. [52] observed that its rate of degradation was slower than had been reported in other soils, with a half-life of 10.1 weeks at 10 to 20 cm depth and 23°C. This is about half the rate of dissipation reported for most other field soils [53]. Metolachlor showed even greater persistence at colder temperatures (e.g., 15°C) and at lower depths [52], as would be expected. The 10 to 20 cm depth soil showed higher adsorptivity (K_d = 1.48) than the 40 to 50 cm soil (K_d = 0.92), which was related to differences in organic matter content. No metolachlor degradation occurred during four months of incubation in sterile soils. In a further study, Braverman et al. [54] observed that metolachlor degradation was significantly faster at 40°C than at 30°C, but it was not affected by moisture levels or CO_2 concentrations.

Zimdahl and Clark [11] reported that the rate of degradation of metolachlor was significantly greater at 80% field capacity than at 20%, and was significantly greater at 30°C than at 10°C or 20°C. From field studies, metolachlor was still detected after 120 days in both a clay loam and sandy loam soil. These scientists also noted that, contrary to the case with most herbicides, the degradation rate of metolachlor was faster in clay loam than in sandy loam soils. They concluded that this was due to either (a) the greater amount of water at a given field capacity in the clay loam, or (b) a higher level of microbial activity in the clay loam. The half-life data based on bioassays under various temperatures, moisture levels, and the two soils, ranged from 13 to 100 days for metolachlor, as indicated in Table 5. Field data were consistent with their laboratory data, with metolachlor showing half-life values of 17 and 23 days in a clay loam soil and sandy loam soil, respectively.

TABLE 5

Half-lives for Metolachlor in Clay Loam (CL) and Sandy Loam (SL) as Determined by Bioassay

Storage conditions		Half-life (days)[a]	
Temperature (°C)	Soil moisture (% field capacity)	SL	CL
		Temperature Experiment	
10	50	100.3a	27.4a
20	50	50.2ab	27.4a
30	50	43.0b	13.1b
		Moisture Experiment	
20	20	100.3c	37.6c
20	50	50.2cd	27.4cd
20	80	33.4d	15.8d

[a]Values in both columns followed by the same letter were not significantly different at the 5% level by student's t-test for comparison of two linear regressions.

Source: Modified from Ref. 11.

Walker et al. [10], provided data from field studies under various conditions in the United Kingdom. Calculated half-lives for metolachlor from early May compared to early June applications were 60 days and 55 days, respectively. Walker and Zimdahl [55] studied the effect of soil temperature and moisture content on the rates of degradation of metolachlor and other herbicides, both in the laboratory and field, using soils from three locations in the United States (Colorado, New York, and Mississippi), to determine if a computer program to simulate or predict herbicide persistence could be developed. As expected and as can be seen in Table 6, both temperature and moisture have a significant effect on half-lives. Differences between soil types were significant, but of relatively minor importance. There was a tendency for the model to overestimate the observed soil residues, possibly due to losses from volatilization, photodegradation, or wind erosion.

Walker and Brown [56] conducted an extensive comparative soil persistence study to determine which of five acetanilide herbicides would provide longest or best weed control under a wide range of climatic and soil conditions. Half-life values for metolachlor in a sandy loam soil ranged from about 24 to 108 days as temperatures decreased from 25°C to 5°C, and from about 21–81 days as moisture levels decreased from 1% to 6%. Metolachlor showed greater soil

TABLE 6

First-Order Half-Lives ($t_{1/2}$ Days) for Metolachlor Degradation in Three Soils as Affected by Temperature and Moisture

Temperature (C)	Moisture (%)	Colorado	New York	Mississippi
35	17.3	19	22	15
25	16.8	34	35	24
15	17.9	59	71	61
5	16.9	158	135	126
25	12.4	49	51	37
25	8.3	60	74	52
25	5.1	99	117	81
25	2.5	1303	531	656

Source: Modified from Ref. 55.

TABLE 7

First-Order Half-Lives for Degradation of Five Acetanilide
Herbicides in a Sandy Loam Soil

Temperature (°C)	25	25	25	25	15	5
Moisture (% w/w)	6	9	12	15	12	12
	Half-life for degradation (days)					
Propachlor	7.7	4.6	4.2	3.7	9.2	21.7
Alachlor	23.0	8.3	7.4	5.7	16.5	38.6
Dimethachlor	21.4	9.8	7.4	5.8	14.4	35.7
Metazachlor	24.2	18.1	13.2	11.6	29.2	77.0
Metolachlor	80.6	41.8	23.9	20.9	47.4	107.8

Source: Modified from Ref. 56.

stability compared to all other chloroacetamide herbicides under all
conditions (Table 7).

Gerber et al. [38] found that the soil half-life for metolachlor
under greenhouse conditions was 26 days, and that metolachlor still
retained 85% of its activity 10 weeks after a 4 kg/ha preemergence
application. Skipper et al. [8], estimated the half-life for metolachlor in South Carolina field soils to be 2 to 3 weeks. No activity
was detected after 4 weeks in a Dunbar sandy loam or Cecil sandy
loam soils. They concluded that leaching was the major means of
dissipation. Both Dixon et al. [57], and Cornelius et al. [58], reported that metolachlor had a long field residual activity, and gave
good control of yellow nutsedge, both initially and throughout the
season.

Under normal conditions of field use, the primary factor responsible for soil degradation and loss of metolachlor and other
acetanilide herbicides is microbial degradation [59–61]. McGahen and
Tiedje [62] found that metolachlor was degraded by the soil fungus,
Chaetomium globosum, which had also readily degraded alachlor [63].
Only 55% of the metolachlor remained after six days, but it did not
appear that the fungus was able to utilize these herbicides as carbon
or energy sources. These scientists also observed [64] that metolachlor
was totally degraded by anaerobic microorganisms within 8 weeks,
concluding that the herbicide would not be stable or resistant to degradation in poorly drained fields or if it reached groundwater

and aquatic sediments. Even the small quantities of metolachlor that are lost through runoff or leaching and end up in aquatic environments are dissipated or biodegraded.

The metabolism of metolachlor in the soil has been investigated under a variety of experimental conditions. Ellgehausen [65,66] studied the degradation of [14C]phenyl-labeled metolachlor for up to 12 weeks by incubating the herbicide in aerobic and anaerobic, and autoclaved soil collected at Stein, Switzerland. The soil was a clay loam, with a cation exchange capacity of 23.5 mEq/100 g, a pH of 7.5, and 3.4% organic matter. The treatment rate was approximately 5 ppm based on a dry soil weight. The herbicide was rapidly degraded in nonsterile soil, with less than 8% of the applied [14C]metolachlor remaining unchanged under aerobic, as well as anaerobic conditions. In autoclaved soil, 65% of the radioactivity was still present as metolachlor after 12 weeks of incubation. After a short lag phase, a slow but constant evolution of $^{14}CO_2$ started in the nonsterile aerobic soil, reaching 4.8% of the applied dose in 12 weeks. Liberation of $^{14}CO_2$ was markedly reduced when the aeration was stopped and the system was ventilated with nitrogen. Autoclaving the soil resulted in almost complete suppression of the degradation to $^{14}CO_2$.

At the end of 12 weeks, nearly 45% of the radioactivity was extractable in the aerobic and aged anaerobic (aerobic/nonsterile for the first 4 weeks and anaerobic/nonsterile for the following 8 weeks) experiments, while the nonextractable residues accounted for 36% and 41%, respectively. In an additional experiment, soils containing the insoluble residues from the aerobic and aged anaerobic studies were combined and mixed with fresh soil in the ratio 1:1. When this soil was further incubated, 2.4% of the radioactivity was released as $^{14}CO_2$ within 4 weeks, indicating that the nonextractable residues are susceptible to microbial degradation. In the autoclaved soil, only 5% of the radioactivity remained bound, thus indicating that the nonextractable radioactivity is primarily formed as a result of microbial activity.

Analysis of the extractable radioactivity showed that the oxalic acid derivative of metolachlor (N-(2'-methoxy-1'-methylethyl)-2-ethyl-6-methyl-oxalic acid analide) accounted for 18% of the total radioactivity and was the main degradation product in the nonsterile soil. Several minor degradation products in the order of 1% of the total radioactivity were found in the soil extracts. One of these was characterized as the dechlorinated derivative [N-2-methoxy-1'-methylethyl)-2-ethyl-6-methyl-acetanilide]. In the autoclaved soil, this compound accounted for 30% of the applied dose.

The nonextractable radioactivity was distributed between soil organic matter fractions according to a procedure outlined by Stevenson [67]. The aerobic and aged anaerobic soils had similar profiles at 12 weeks. Less than 10% of the nonextractable radioactivity was found in the acid washings. Between 20 and 27% of the ^{14}C was found in the humin fraction and approximately 21% was found in the alcohol-insoluble humic acid fractions. Approximately 40% of the radioactivity was present in the pH 4.8-soluble fulvic acid fraction.

Guth [68] studied the degradation of metolachlor in a sandy loam soil collected at Les Barges, Switzerland. The soil contained 1.1% organic matter and had a cation exchange capacity of 7.3 mEq/ 100 g and a pH of 7.5. The soil was treated at a concentration of 10 ppm, and incubated at 25°C for 187 days at 75% of field capacity. The herbicide degraded with a half-life of about 100 days. Degradation of metolachlor proceeded via oxidation of the chloroacetyl group. The major metabolite was the oxalic acid derivative [N-(2'-methoxy-1-methylethyl)-2-ethyl-6-methyl-oxalic acid anilide], which amounted to 16.4% after 187 days of incubation. Two other major metabolic fractions, each consisting of approximately 5%, were not further identified. The concentration of nonextractable radioactivity steadily increased during the incubation period, reaching 29.8% of the dose after 187 days.

The soil fungus *Chaetomium globosum* has the ability to metabolize metolachlor and 4 of the 8 extractable metabolites have been identified [62]. The general transformations involved were dehalogenation with hydroxylation, dehydrogenation of the 6'-ethyl, dealkylation, demethoxylation, and indoline or oxoquinoline formation.

An acetinomycete strain isolated from soil also metabolized metolachlor [69]. Eight metabolites were identified and benzylic hydroxylation of the aralkyl side chains and demethylation at the N-alkyl substituent appeared to be the predominant transformation mechanisms. The transformation mechanisms reported for the microbial degradation of metolachlor and other chloroacetamide herbicides also include dechlorination, dehydrogenation, dealkylation, hydroxylation, and indoline ring formation [70–73].

V. BEHAVIOR IN PLANTS

A. Uptake and Distribution in Plants

Early studies related to plant selectivity and mode of action proved that differential uptake was not a significant factor in crop tolerance to acetanilide herbicides. Also, while both susceptible and tolerant plants were able to metabolize the herbicides, Smith et al. [74] concluded that the degree of susceptibility of various weeds to chloroacetamides could be directly related to the length of time required to initiate metabolism of these chemicals. Those species that are able to metabolize the chemical as soon as it is absorbed or within a short time thereafter (within 6 hours), have only small amounts of the chemical present internally at any time.

On the other hand, those species with delayed or slow metaoblic capabilities accumulate relatively higher and, therefore, lethal concentrations. The basis for selectivity of the chloroacetamide herbicides could, therefore, be attributed to the ability of resistant plants to metabolize them at a rate sufficient to keep cellular levels of the herbicides below that required for growth inhibition.

Relatively limited research on plant uptake of metolachlor has been conducted or published. Much research had been conducted earlier on alachlor uptake, and some comparative studies were done which confirmed that the two herbicides act similarly. Chloro-acetamides appear to be most efficiently taken up by the shoots or cotelydons of grasses, while root absorption is very important for uptake by many dicotyledonous plants. Translocation of both herbi-cides in plants is mostly by acropetal movement.

Obrigawitch et al. [75] observed that translocation of [^{14}C]-metolachlor applied to the root or shoot of yellow nutsedge was pri-marily acropetal, with some limited basipetal movement. Application of [^{14}C]metolachlor to the root or shoot of purple nutsedge (*Cyperus rotundus* L.) resulted in mostly acropetal translocation; however, there was relatively little movement of ^{14}C from the treat-ment sites. Increasing the depth of herbicide incorporation did not result in any significant increase in control. In some areas of the world (e.g., South Africa), however, deep incorporation for improved weed control has been recommended.

Dixon and Stoller [76] reported that metolachlor did not inhibit germination of corn seed or yellow nutsedge tubers, and did not kill tubers exposed to 4 ppmw for nine weeks. Corn shoot growth was reduced when 10 ppmw of metolachlor was placed above the seed, but had no effect when placed around or below the seed. Nutsedge shoot growth, on the other hand, showed the greatest inhibition when metolachlor was placed around the tuber, less inhibition when placed above the tuber, and no effect below the tuber. Root-applied [^{14}C]-metolachlor was translocated acropetally to shoots of both species following a 7 to 13-day absorption period, with yellow nutsedge translocating a greater percentage of the absorbed material. In 2-day-old seedlings with roots exposed to [^{14}C]metolachlor for up to 48 hours, both species absorbed and translocated the radioactivity to shoots, but corn absorbed much more than yellow nutsedge. When [^{14}C]metolachlor was applied to shoots of both species, radio-activity was translocated basipetally into roots. Yellow nutsedge absorbed more [^{14}C]metolachlor through shoot tissues than corn. Both corn and yellow nutsedge seedlings readily metabolized [^{14}C]-metolachlor, but corn was able to metabolize it at a faster rate and more completely than yellow nutsedge. This may account for the difference in selectivity, as the I_{50}'s for corn and nutsedge were $>10^{-4}$ M and $<10^{-6}$ M, respectively.

B. Metabolism in Plants

Information on the metabolism of specific chloroacetamide herbicides in plants has been reported previously by Lamoureux et al. [77], Jaworski [6], Shimabukuro et al. [78], Hatzios and Penner [79], Lamoureux and Rusness [80], Breaux [81], and Blattmann et al. [82].

Gross [83] studied the uptake, distribution, and metabolism of metolachlor in corn grown in nutrient solution containing 2 ppm of [^{14}C]metolachlor. Uptake was rapid, and from 65–75% of the radioactivity in the nutrient solution was taken up by the plants during one week of exposure. Translocation resulted in a distribution of 30% of the radioactivity in aerial tissue and 70% in roots. About 90% of the radioactivity in the aerial tissue at all sampling intervals was extractable with methanol/water (8:2). Five weeks after the one-week exposure period, the organic-soluble radioactivity in the methanol fraction of the aerial samples represented only 2% of the total plant radioactivity, whereas the aqueous-soluble fraction amounted to 80% of the residues.

Sumner and Cassidy [84] studied the distribution of metolachlor in corn grown in a field plot treated preemergence with 2 lbs. a.i./A of [^{14}C]metolachlor. At maturity, the stalks contained 0.17 ppm and grain contained 0.02 ppm, equivalent to [^{14}C]metolachlor. Extensive metabolism was evident by extraction, partition, and chromatography data. In plants harvested at 4 weeks and 12 weeks, the radioactive residues consisted of 80% and 90% of water-soluble metabolites, respectively. At maturity, there was a decrease in water-soluble metabolites and a subsequent increase in nonextractable radioactivity. Mature stalks contained 5.9% organic solubles, 47.1% aqueous solubles, and 47.10% nonextractables, using a biphasic extraction solvent. Most of the water-soluble polar radioactivity was acidic in nature and consisted of at least 10 metabolites.

Several key metolachlor metabolites in corn were identified by Gross [83, 85], and Blattmann et al. [82] using corn plants treated in nutrient solution, by stem injection, or under greenhouse or typical field conditions. The structures of metabolites were determined by appropriate chromatographic and/or spectroscopic techniques. The pathway for metabolism of metolachlor in corn is presented in Figure 4. The degradation proceeds via:

Conjugation of the chloroacetyl side chain with glutathione
Breakdown of the glutathione tripeptide to the cysteine conjugate
Oxidative deamination and reduction of the transient α-ketoacid
    to the thiolactic acid conjugate
Oxidation to the corresponding sulfoxide derivatives
Cleavage of the side chain ether group and subsequent conjugation with glucose

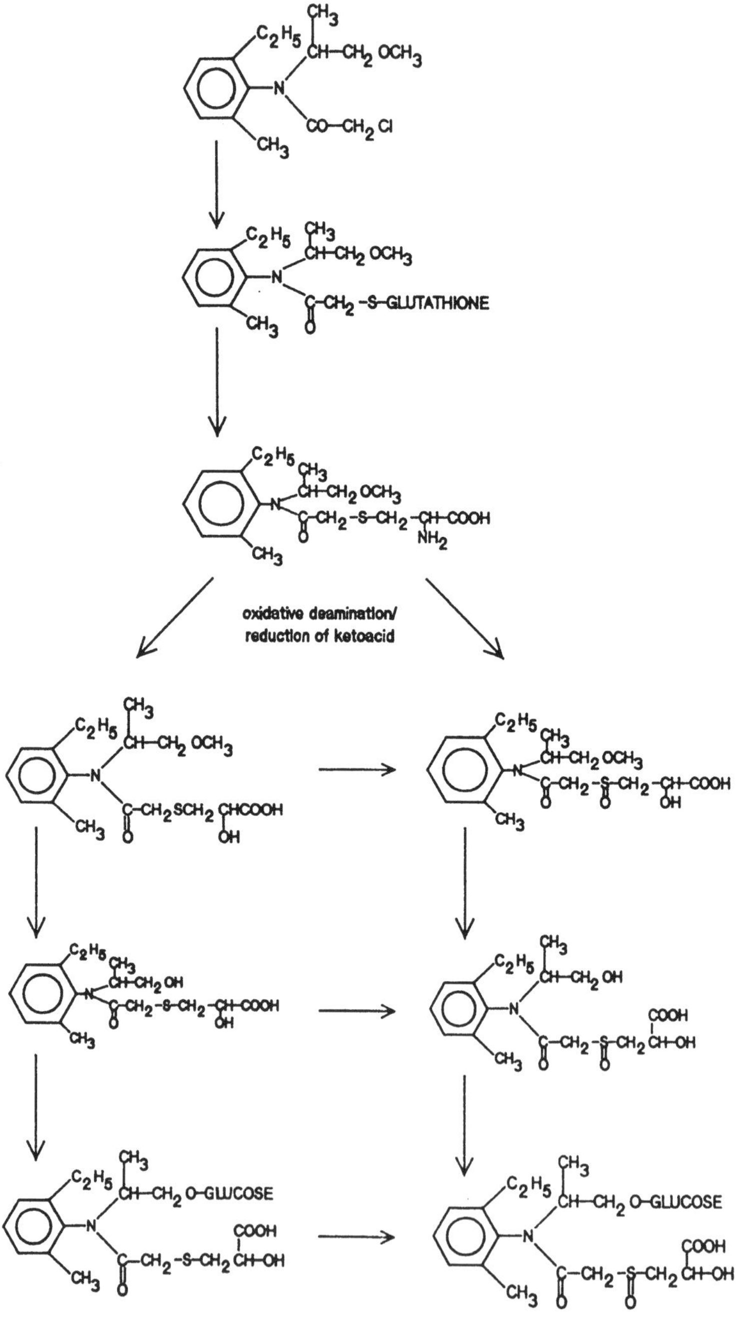

FIG. 4   Proposed metabolic pathway of metolachlor in corn.

Based on thin layer chromatography, the major metabolites in mature corn plants from the field and from corn grown for a short term in the greenhouse were the same. The thiolactic acid conjugates and their sulfoxide derivatives represented terminal metabolic products of the glutathione-dependent metabolism of metolachlor in corn [82]. Based on extraction, partition, chromatography, and hydrolysis data, the profiles of metabolites in lettuce (*Lactuca sativa*) and potatoes (*Solanum tuberosum*) are similar to those in corn [86,87]. The major metabolite in excised sorghum leaves after a 30-minute exposure was the glutathione conjugate [88].

Szolics and Cassidy [89] studied the uptake and distribution of [^{14}C]metolachlor applied preemergence at 2 lb. a.i./A to soybeans grown in a field plot. At maturity, soybean stalks and beans contained 10.8 ppm and 0.27 ppm equivalent to [^{14}C]metolachlor, respectively. Most of the radioactivity in these plant parts was extractable (65–79%), and consisted of at least eight aqueous-soluble metabolites ($\sim$60%), primarily acidic in nature. The characteristics of the extracted residues in treated soybeans indicated that the metabolic pathways of metolachlor in soybeans are similar to those in corn [90].

Other chloroacetamides are also converted to polar conjugates in soybean [91]. Breaux [81] identified the major acetochlor metabolite present in etiolated soybean seedlings 24 hours after hypocotyl treatment as the homoglutathione conjugate.

One of the initial chloroacetamide metabolites, formed in both tolerant and susceptible species, is the glutathione or homoglutathione conjugate [77,81,88]. The degradation of the chloroacetamide conjugate can proceed via the breakdown of the glutathione tripeptide to the cysteine moiety. The cysteine conjugate was the most abundant radioactive metabolite in extracts of peanut cells in suspension culture treated with metolachlor [80]. The cysteine conjugate was further metabolized to a malonylcysteine conjugate in the peanut cells.

Chloroacetamides conjugate glutathione both nonenzymatically in vitro [92,93] and enzymatically in vitro with glutathione S-transferases isolated from etiolated corn [94], corn suspension cells [95], and sorghum [93]. Edwards and Owen [95] isolated two isozymes (a major and a minor form) from untreated suspension cells of Black Mexican sweet corn. Metolachlor was rapidly metabolized to its glutathione conjugate and glutathione-related metabolites in corn cells. Although both isozymes conjugated metolachlor with glutathione, the minor form had a greater reactivity with metolachlor than the major form. Isozymes of glutathione S-transferase that vary in their reactivity to the chloroacetamides have also been isolated from etiolated corn seedlings [94].

Additional metabolism studies in a variety of species of young intact seedlings within the first hours of herbicide exposure would greatly aid the investigations of herbicide selectivity and the mode of action of the chloroacetamides.

VI. METABOLISM IN ANIMALS

Metolachlor is rapidly metabolized and almost totally eliminated in the urine and feces of rats [96-98] and ruminants (goats) [99, 100]. HambÖck [96] reported that the radioactivity in rats orally dosed with approximately 3 mg/kg of [^{14}C]metolachlor, uniformly labeled in the phenyl ring, was readily excreted in urine (35 and 44%) and feces (63 and 51%) by male and female rats, respectively. Fifty percent of the radioactivity was excreted within 19 hours by males and 33 hours by female rats. Tissue retention was low nine days after dose administration, with ^{14}C residues all below 0.05 ppm except for blood, liver, and kidney. The urine and feces of treated rats contained a complex pattern of metabolites, but no unchanged metolachlor was detected.

In an additional study, HambÖck [97] dosed 8 male rats with approximately 52 mg/kg of [^{14}C]metolachlor. The radioactivity was rapidly excreted, with a half-life of approximately 24 hours. There were no significant differences observed in the metabolite pattern in the urine and feces between the rats treated at 3 mg/kg and the rats treated at 52 mg/kg.

Roger and Cassidy [99] found that the majority of the radioactivity in a goat administered 4.7 ppm of [^{14}C]metolachlor for 10 consecutive days was excreted in the urine ($\sim$82%) and feces ($\sim$18%). Leighty et al. [100] reported that goats treated with metolachlor for four consecutive days at either 5 ppm or 25 ppm also excreted most of the radioactivity in the urine. There were no major differences in the urine metabolite patterns in goats regardless of treatment level.

Treated chickens also eliminated most of the metolachlor, and there were no detectable residues in eggs, meat, or fat samples [101].

Several urine and fecal metabolites of metolachlor in rats were isolated and purified by Mücke [102] and by Bissig et al. [103]. Identifications were confirmed by chromatography, mass spectrometry and nuclear magnetic resonance spectrometry. A comprehensive pathway for metolachlor metabolism in the rat is presented in Figure 5. The major route of degradation involves cleavage of the methyl ether, resulting in a moiety that can be further oxidized to the corresponding carboxylic acid. Substitution of the chlorine atom of the chloroacetyl group with glutathione occurs to a limited extent, giving a variety of oxidized sulfur metabolites. Minor branches of the pathway involve dealkylation to secondary amides, and oxidation

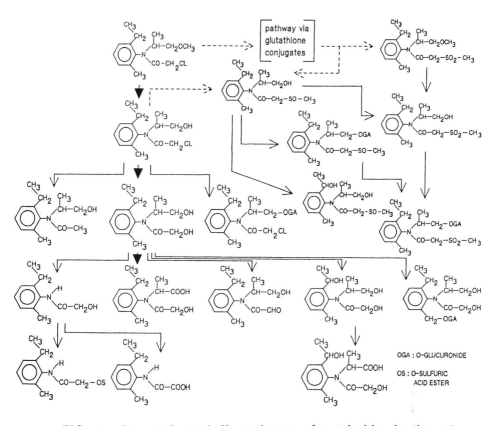

FIG. 5   Proposed metabolic pathways of metolachlor in the rat.

of the methyl and ethyl side chains on the phenyl ring.   These
metabolic processes are known detoxification mechanisms operative
in some mammals.

VII.   MODE OF ACTION

Although chloroacetamides are often classified as general plant
growth inhibitors [104,105], their primary biochemical mechanism
of action is unknown.   Metolachlor inhibits the early development of
susceptible weed species.   The treated seeds of susceptible species
usually germinate, but the seedlings either do not emerge from the
soil or emerge and exhibit stunted or abnormal growth.   General
reviews on the modes of action and physiological effects of the
chloroacetamides in plants have been reported by Fedtke [104],

Ashton and Crafts [105], and Jaworski [6]. No single model system has been used to research the modes of action of the various chloro-acetamide herbicides and the experimental conditions have varied widely from study to study. The physiological effects of chloro-acetamides on plants are dependent on the herbicide used, plant species, tissue type, herbicide concentration, and duration of treat-ment. Some of the groundwork for the experimental approaches to study the modes of action of chloroacetamides has been based on results from previous research on the structurally related chloro-acetamide, CDAA, and propachlor, the first chloroacetamide mar-keted as a herbicide. Investigations on all the chloroacetanilide herbicides, as well as on the chloroacetamide, CDAA, are included in this section in order to compare modes of action of herbicides with similar chemical structures.

Common modes of action between the chloroacetamides are usually presumed because the growth responses of susceptible plants are often similar. Most of the research on the modes of action of the chloroacetamides in plants have concentrated on the effects of these herbicides on growth inhibition, protein synthesis inhibition, inter-actions with hormones, interactions with lipids, and effects on what are often referred to as secondary metabolic pathways. Roots, shoots, or both have been reported as the sites of uptake, and the growth of roots and/or shoots of susceptible species can be inhibited by the chloroacetamides [76,88,106–114]. Detailed studies of the morpho-logic and histologic symptoms of plants treated with metolachlor have been reported by Ebert [115,116] and Paradies et al. [117] using sorghum treated with metolachlor. The first signs of metolachlor in-fluence in sorghum are at the coleoptile node, and lateral growth is favored over longitudinal growth. The sorghum coleoptile opens and releases the first leaf, which is deformed and often does not unfold. After metolachlor treatment, starch accumulates in the sorghum chloroplasts and plastids, and other membranous structures are either not formed or appear abnormal. The lignification of the bundle sheath cells also appears to be inhibited.

Deal and Hess [110] found that oats growing in sand saturated with a 1×10^{-4} M solution of metolachlor in 1% dimethyl sulfoxide (DMSO) for 6 hours showed a 47% root growth inhibition. After 24 hours, root growth was inhibited more than 80%. Both cell divi-sion in oat roots and cell enlargement in excised oat coleoptiles were inhibited within 24 hours of growth in 10^{-4} M metolachlor. They concluded that both cell division and cell enlargement must be inhibited to produce the growth reduction caused by chloro-acetamides.

Metolachlor decreased the frequency of mitosis and prolonged in-terphase in roots of broadbean (*Vicia faba* L.) treated for 24 hours [118]. Fedtke [119] found that cell division was also inhibited in the algae, *Chylamydomonas rheinhardii*, growing in 37 µ M alachlor, and he suggested that chloroacetamides might specifically interfere with the regulation of cell division. Dhillon and Anderson [120] also found that the chloroacetamides inhibited cell division in onion (*Allium Cepa* L.) root tips.

The effects of chloroacetamides on protein synthesis were initially reported in 1965 by Mann et al. [121], using excised barley (*Hordenum vulgare* L.) coleoptiles, and excised sesbania (*Sesbania exaltata*) hypocotyls. CDAA at 4 ppm decreased leucine incorporation into protein by 70% in barley, and 87% in sesbania within two hours. Duke [122] found that both auxin-induced cell elongation and protein synthesis in excised cucumber (*Cucumis sativus* L.) hypocotyls were inhibited by propachlor at 5 ppm.

Jaworski [123] and Moreland et al. [124] found that 10^{-4} M CDAA or propachlor inhibited GA_3-induced α-amalayse production. Propachlor also inhibited the incorporation of [3H]leucine into proteins of soybean hypocotyls [124]. Alachlor and propachlor inhibited both α-amalase in de-embryonated barley seed [125,126] and GA_3-induced protease [126].

Protein synthesis was inhibited in oat seedlings grown in sand saturated with 10^{-4} M metolachlor and 10^{-4} M alachlor with 1% DMSO [127]. However, protein synthesis was not inhibited in vitro with chloroacetamide herbicides in a cell-free protein synthesizing system using oat root polyribosomes, leading to the conclusion that translation of messenger RNAs is not a direct site of action [127]. Butachlor at 50 µM also inhibited protein synthesis in excised root tips of rice and barnyardgrass (*Echinochloa crusgalli* L.) by 81–90% and by 55–65%, respectively, in excised shoot segments [128].

Duke [108,122] observed that propachlor inhibited protein synthesis in cucumber roots before an effect on root growth was observed. He suggested that propachlor might be preventing the activation of amino acids, aminoacyl RNA formation or transfer, and subsequently the inhibition of nascent protein synthesis. Jaworski [123] expanded Duke's hypothesis [122], and suggested that specific aminoacyl tRNAs could be alkylated by propachlor and subsequently inhibit nascent protein synthesis.

Jaworski originally postulated the alkylation hypothesis in 1956, when he was studying the effects of CDAA on the respiration of germinating seeds [129]. He found a partial reversal of the respiration inhibition when reduced glutathione (GSH) was included, and he therefore concluded that CDAA might inhibit certain sulfhydryl-containing enzymes involved in respiration. Whether the GSH was actually protecting sulfhydryl enzymes or was conjugating with CDAA is not clear, since GSH has since been shown to conjugate chloro-acetamides in vitro [92].

Chupp and Olin [29] and Chupp et al. [30] found that the two rotamers of the 2'6'-dialkyl-2-halo-N-methylacetanilides which are similar in structure to some of the marketed chloroacetamides have two vastly different alkylation potentials, with the minor rotamer having a relative alkylation rate up to 100 times greater than the major rotamer. Hamm [31] established that the minor rotamer of a brominated acetanilide (α-bromo-2,6-di-t-butyl-N-methylacetanilide), with the greater ability to alkylate piperidine, was herbicidally active, whereas the major rotamer, with the low alkylation potential, was relatively inactive. Molin et al. [32], in their investigations on the effects of the chloroacetanilides on anthocyanin synthesis, also found that the rotamers with the greater ability to alkylate inhibited anthocyanin synthesis in excised sorghum mesocotyls, whereas the rotamers with the low alkylation potentials were relatively inactive in their system.

The chloroacetamide moiety alone is known to be an alkylating agent [130,131]. During an alkylation reaction, the halogen of a haloacetamide is displaced, and a covalent bond is formed between the acetamide moiety and the attacking nucleophile. Alachlor, metolachlor, and propachlor bind to glutathione nonenzymatically in vitro [92], and these herbicides also alkylate cysteine in vitro [34,92]. The detoxification of chloroacetamide herbicides in plants involves conjugation with the nucleophile glutathione [6,77,81,82,85].

The chloroacetamides also covalently bind to a number of unidentified proteins, both in vivo (intact oat seedlings) and in vitro [26]. The overall quantities of oat root proteins alkylated by the different chloroacetamides did not necessarily correlate with the amount of growth inhibition observed, however, alkylation to specific nucleophiles could possibly be involved in the modes of action of the chloroacetamides. Chloroacetamides are not considered general, nonspecific inhibitors of sulfhydryl enzymes, because they did not inhibit certain sulfhydryl enzymes in vitro [132]. Wilkinson [133] suggested that the inhibition of kaurene oxidation caused by CDAA could be due to the inhibition of the enzymes required in this pathway that are sensitive to sulfhydryl inhibitors. In a plant system where herbicidal activities of different chloroacetamides are similar, specific nucleophiles associated with growth may be alkylated by the chloroacetamide herbicides at similar rates.

The morphological symptoms of susceptible plants treated with chloroacetamides are often similar to a variety of hormonal effects. Dhillon and Anderson [120] found that senescence in excised squash (*Cucurbita maxima*) cotyledons was retarded by propachlor, and the degradation of storage proteins was also inhibited. Certain concentrations of gibberellic acid (GA3) reversed the inhibition of GA3-induced α-amylase and protease in de-embryonated barley seed treated with alachlor and propachlor, whereas in intact barley seedlings, GA_3 did not overcome the inhibition [126]. Chang et al. [134]

found that pretreating oats with 10^{-3} M gibberellic acid prevented 5×10^{-5} M alachlor from inhibiting shoot growth. Alachlor-induced growth inhibition of excised corn epicotyls in liquid solution was also attenuated by GA3, and resulted in growth comparable to epicotyls that were not treated with either alachlor or exogenous GA3 [135]. GA3 can overcome the inhibition of stem elongation in grain sorghum treated with metolachlor. However, the morphological symptoms of the leaves of treated sorghum cannot be reversed by GA3 [136].

In a series of papers, Wilkinson [133, 137–140] reported that chloroacetamide herbicides inhibited mevalonate incorporation into various terpenoids. Terpenes are precursors to gibberellic acid, and Wilkinson [137, 138] speculated that the morphological and some of the biochemical responses in plants treated with chloroacetamides could be partially caused by the inhibition of gibberellic acid synthesis. Metolachlor at 10^{-5} M caused a 50% inhibition of the incorporation of mevalonic acid into carotene in an in vitro carrot (*Daucus carota* var. *sativa*) disc assay [137]. The addition of metolachlor to a cell-free in vitro system containing pulverized sorghum coleoptiles (<1 cm in length) resulted in an inhibition of incorporation of mevalonic acid into kaurene. When the sorghum coleoptiles were obtained from sorghum seeds protected with the safener, cyometrinil, the kaurene synthesis was not inhibited [139]. Metolachlor did not inhibit kaurene oxidation in an in vitro system of cucurbit endosperm [141].

Ebert [115] found that germinating sorghum seeds treated with metolachlor produced more ethylene than untreated sorghum seedlings. However, when sorghum seedlings were also treated with aminoethoxyvinyl glycine, a specific inhibitor of ethylene formation, ethylene production remained normal in seedlings treated with metolachlor, but the growth of the seedlings remained inhibited and abnormal [117]. Therefore, the increased production of ethylene was a symptom, and not the direct cause of the morphological effects of metolachlor or sorghum seedlings. Ethylene production is increased 200–400 fold in sorghum, *Setaria* sp., and barnyardgrass treated with 5 and 10 ppm metolachlor or pretilachlor [142]. However, ethylene production was inhibited in treated wheat, while no effect on ethylene was observed in corn.

Lipid synthesis continues to be targeted as a possible mode of action of the chloroacetamides [143]. Sorghum treated with metolachlor contained reduced amounts of monogalactosyl diglyceride and glycerophosphatidyl glycerine. In the fatty acid fraction of treated sorghum, the concentration of linolenic acid (18:3) is reduced, while the concentration of linoleic acid (18:2) is increased (M. W. Tevini, University of Karlsruhe, Germany, 1983, personal communication). The incorporation of sodium acetate into lipids was inhibited in isolated leaf cells of red kidney beans (*Phaseolus vulgaris*) treated with butachlor [128], and the inhibition of lipid synthesis has also been reported with CDAA in excised epicotyls [144] and isolated leaf mesophyll cells [145] from sesbania. Weisshaar and Böger [146] found that metazachlor severely inhibited the incorporation of [^{14}C]acetate into polar lipids in the green algae, *Scenedesmus acutus*.

Ebert and Ramsteiner [147] found that metolachlor inhibited the synthesis of the C_{28}, C_{30}, and C_{32} long-chain alcohols and fatty acids in epicuticular wax of sorghum. The synthesis of the major plant sterols (sitosterol, stigmasterol, and campesterol) in sorghum was not influenced by metolachlor. Metolachlor also inhibited alkane-1-ol production in cucumber seedlings and Tevini and Steinmüller [148] propose that metolachlor inhibits the enzyme elongase II.

Membrane leakage occurred in roots of the susceptible species, onion and cucumber after liquid solution treatments of 1×10^{-4} M and 1×10^{-5} M metolachlor [149]. No significant membrane leakage occurred in two tolerant species, soybean and corn. Membrane disruption has also been reported in excised ion root tips [150], and sorghum leaves treated with chloroacetamides [115]. The effects of chloroacetamides on the biosynthesis of components of membrane lipids continue to be investigated as a possible site of action of the chloroacetamide herbicides.

Some attention on the modes of action of the chloroacetamides has been directed to their effects on secondary plant products. Ebert [115] observed that after metolachlor treatment, bundle sheath cells in sorghum did not have any evidence of lignification. Hickey and Krueger [151] found that lignin synthesis was inhibited in corn treated with alachlor. Molin et al. [152] reported that alachlor inhibited anthocyanin and lignin accumulation in sorghum. They studied the effects of the chloroacetamides on anthocyanin synthesis, using excised sorghum mesocotyls which were illuminated for 24 hours in the presence of liquid solutions of 20 µM alachlor, acetochlor, butachlor, metolachlor, and propachlor. Alachlor had the greatest effect on anthocyanin synthesis, causing a 64% inhibition. While acetochlor and butachlor caused a 50% inhibition, metolachlor caused only a 23% inhibition and propachlor had the least effect, causing a 21% inhibition. Lignin synthesis was inhibited 50% with 28 µM alachlor.

Lignin precursors, anthocyanin, terpenoids, and lipids are synthesized via coenzyme A (CoA) intermediates, and Molin et al. [152] suggested that some aspect of CoA metabolism may be the site of action of the chloroacetamides. Alachlor can alkylate CoA in vitro [92], but whether the conjugation occurs in vivo is unknown.

In analyzing the effects of chloroacetanilides on metabolic pathways, Molin et al. [152] noted that excised sorghum mesocotyls treated with alachlor contained 2.5X more glutamate than untreated controls, while other free amino acid levels were not dramatically affected. The significance of the increased glutamate in treated mesocotyls is unknown. Metolachlor had no effect on the amino acid composition of 13-day-old sorghum plants which had been grown for 6 days in Hewitt's nutrient solution and metolachlor at concentrations of 0.01–10 ppm [153]. Metolachlor also had no effect on electron transport in Photosystem II, CO_2-compensation point, respiration, catalase, cutinase, or pectinase activities [154].

The chloroacetamides inhibit major metabolic processes in plants, but the primary site(s) of action remain unknown. Because of the reactive halogen on the chloroacetamides, Jaworski in 1969 [123] theorized that the chloroacetamides may have multiple sites of action, which when combined, result in growth inhibition.

The use of in vitro assays are valuable and often the only feasible technique available to analyze metabolic pathways in plants. However, the concentration of herbicide exposed to a site of action in an in vitro assay or in excised tissues is often extremely differerent from the herbicide concentration that would actually reach that site of action in a plant growing in the field. Detailed accounts of the location and quantities of parent herbicide and its metabolites throughout the developmental stages of a plant in vivo are critical in order to confirm the initial or primary mode of action.

The use of rotameric forms of chloroacetamides that differ in herbicidal activity, diastereomers that differ in activity, and herbicide antidotes will continue to be valuable probes in future mode of action studies. The technical tools are now readily available to gain new scientific information that will complement past results on the modes of action of metolachlor, and will greatly contribute to our knowledge of the physiological and biochemical pathways involved in the germination and growth of plants.

VIII. INTERACTIONS WITH ANTIDOTES

Herbicide safeners or antidotes are chemicals that enhance the tolerance of crops to herbicides. Their development originated mainly from the research investigations of Hoffman [155] that began in the late 1940s. The results of these studies led to commercial development of naphthalic anhydride to improve the selectivity of thiocarbamate herbicides in corn. Subsequent research led to the development of dichlormid, designated R-25788 (N,N,diallyl-2-2-dichloroacetamide), which could be tank-mixed with the thiocarbamate herbicides and protect corn from thiocarbamate injury without protecting weed species.

The more recent development of chemicals that are able to counteract or prevent crop injury caused by chloroacetamide herbicides has greatly expanded the use of metolachlor, especially for the control of problem grasses and other weeds in sorghum. Chloroacetamides are not sufficiently selective in sorghum to permit their commercial use without herbicide safeners.

The first marketed chloroacetamide safener, cyometrinil ([(cyanomethoxy)imino]benzeneacetonitrile, or CGA-43089), was discovered in 1974, and introduced in 1980 by the Ciba-Geigy Corporation [156, 157]. Cyometrinil was marketed for use as a seed treatment to protect sorghum from metolachlor injury. The Monsanto Company introduced flurazole (5-thiazolecarboxylic acid-benzyl ester-2-chlor-4-trifluoromethyl, or MON-4606), in 1981 as a seed treatment to protect sorghum from alachlor injury [158]. In 1982, the Ciba-Geigy Corporation introduced oxabetrinil (α[1,2-dioxolan-2-yl-methoxyimino]benzeneacetonitrile, or CGA-92194) as a safener that has wider use applications than cyometrinil, by protecting high- and low-quality sorghum, yellow endosperm, and Sudan grass [159].

Fenclorim, (4,6-dichloro-2-phenyl pyrimidine, or CGA-123407) was developed by the Ciba-Geigy Corporation as a safener to be used with pretilachlor in rice [160].

Ciba-Geigy has also recently developed CGA-133205 [0-(1,3-dioxolan-2-yl-methyl)-2,2,2-trifluoro-4'-chloroacetophenone-oxime], which is a new experimental antidote applied to sorghum seed for use with metolachlor [161].

These successful antidote breakthroughs led to an increased effort to discover additional applications for existing antidotes, as well as new antidotes for use with other herbicide groups. Some of the marketed safeners can provide limited protection to plants from injury caused by several herbicide groups [162,163]. To date, however, the chloroacetamides and the thiocarbamates are the only two major groups of herbicides for which commercial safeners have been successfully developed.

Most of the marketed antidotes, with the exception of fenclorim (CGA-123407), prevent herbicide phytotoxicity to the crop but are not able to reverse crop injury. However, rice can be safened against injury by pretilachlor through application of fenclorim one day prior to, concurrent with, or up to 3 days after herbicide application [164].

Research on the development, chemistry, physiology, and mechanism of action of herbicide safeners has been reviewed by Pallos and Casida [165], Parker [162], Hatzios [163], and Hatzios and Hoagland [166]. Herbicide antidotes selectively reduce the biological activity of a herbicide which normally would result in crop phytotoxicity. This reduction in crop injury can be the result of either reducing or eliminating the herbicide from its site of action, or by an induction of an alternate pathway that will compensate for, or override, the effects of the herbicide. The biochemical antagonism can be the result of one or more factors, including reduced herbicide uptake, reduced herbicide translocation, enhanced herbicide metabolism, herbicide compartmentalization, or the induction of alternate plant metabolic pathways.

The effects of safeners on herbicide uptake vary with different experimental conditions. Christ [167] concluded that cyometrinil does not interfere with the uptake of metolachlor in sorghum. Ebert [168] found that neither oxabetrinil or fenclorim influenced herbicide uptake in hydroponically grown grain sorghum, shattercane (*Sorghum bicolor*), barnyardgrass, or rice. Metolachlor uptake in hydroponically grown sorghum was reduced when naphthalic anhydride was included in the treatment solution [169] and increased with oxabetrinil [170]. Excised sorghum coleoptiles and first leaves of seedlings treated with both metolachlor and cyometrinil after excision took up less metolachlor than seedlings treated with metolachlor alone [116]. However, excised coleoptiles from sorghum seedlings treated with oxabetrinil prior to excision absorbed twice as much metolachlor as shoots from untreated seedlings [88]. Metolachlor uptake was reduced in soil-grown sorghum treated with cyometrinil [171]. Sorghum leaf protoplasts absorbed more metolachlor when treated with oxabetrinil than when treated with metolachlor alone [172]. Obviously, no generalization can currently be made about the contribution of the effects of safeners on herbicide uptake to the safeners' mode of action.

The effects of safeners on the physiological and biochemical symptoms that result from herbicide treatment have been examined in many experimental systems. The chloroacetamide antidotes can inhibit sorghum growth when applied alone, but the injury does not effect sorghum yield [173, 174]. The incorporation of [14C]acetate by enzymatically isolated sorghum leaf protoplasts was inhibited less by metolachlor plus oxabetrinil than metolachlor alone [172]. Sorghum treated with metolachlor increased ethylene production that accompanied injury symptoms, such as leaf deformations and twisting, caused by metolachlor [115]. Sorghum seeds pretreated with cyometrinil did not produce the elevated levels of ethylene found in sorghum treated with metolachlor alone, but produced levels of ethylene similar to those of untreated controls [117].

Using scanning electron microscopy, Ebert analyzed the surfaces of coleoptiles and first leaves of sorghum seedlings grown in the presence of either metolachlor and cyometrinil or metolachlor alone [116]. Although there were no obvious differences in the surfaces of coleoptiles of plants treated with metolachlor and untreated plants, there were great differences in the surface structures of the first leaf [Fig. 6]. The epidermis of the first leaf of untreated control plants was covered with a cuticle on which epicuticular waxes were present as dense scales. The first leaf of seedlings treated with metolachlor alone did not have the waxy scales, whereas when the safener was present alone or when the safener was present with

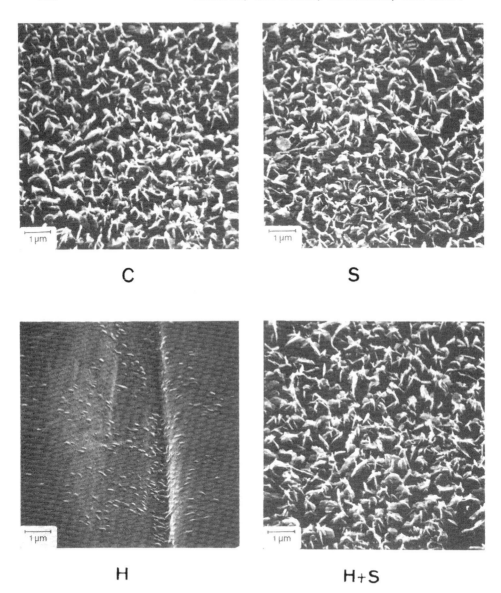

FIG. 6   Scanning electron micrographs (SEM) of the lower
surface of sorghum first leaves of 4-day-old plants (X6000).   (C)
Leaf from control plant.   (H) Leaf from plant treated with metolachlor.   (H&S) Leaf from plant treated with metolachlor and CGA-
43089.   (S) Leaf from plant treated with CGA-43089.   From Ref. 116.

metolachlor, the cuticle appeared similar to the untreated control plant. Pretilachlor also causes a decrease in the epicuticular waxes on rice leaves, and the safener fenclorim protects the rice surface waxes from pretilachlor damage [168]. However, under the same conditions, the surface waxes of barnyardgrass are not protected from pretilachlor injury by fenclorim.

An investigation of the inhibition of epicuticular wax formation by metolachlor resulted in the discovery that the synthesis of C_{28}, C_{30}, and C_{32} alcohols, as well as C_{30} and C_{32} fatty acids, was inhibited [147]. However, epicuticular wax synthesis of the dominating C_{28} to C_{30} chain length constituents was not inhibited in sorghum seedlings that had been seed-treated with cyometrinil prior to treatment with metolachlor. A reduced wax layer could have a number of physiological ramifications, including effects on the uptake of xenobiotics into the plant and the loss of water through transpiration.

Fedtke [175] speculates that the oxidation enzyme systems involved in wax biosynthesis and affected by herbicide safeners also contribute to the metabolism of the chloroacetamides. Hatzios [176] found that piperonylbutoxide, an inhibitor of mixed function oxidases, resulted in increased herbicidal activity with metolachlor. Cole and Owen [177] reported that the glutathione conjugation of metolachlor was inhibited by 100 μM 3(2,4-dichlorophenoxy)-1-propyne which is an inhibitor of mammalian P-450 oxidases.

By far the largest amount of information available on the interactions of chloroacetamide antidotes with chloroacetamides involves the effects of antidotes on herbicide metabolism. Investigations have concentrated on the ability of the safeners to increase the amounts of polar conjugates. One of the major metabolites formed in susceptible and tolerant species treated with chloroacetamide herbicides is the glutathione conjugate [77,80–82,85,178], which can occur nonenzymatically in vitro [92]. In addition, glutathione S-transferases in etiolated corn [94], sorghum shoots [93], and corn cell culture [95], are capable of enzymatically catalyzing the formation of a glutathione-chloroacetamide conjugate in vitro.

Fuerst and Gronwald [88] found that treatment of sorghum seeds with naphthalic anhydride, dichlormid, oxabetrinil, cyometrinil, and flurazole resulted in accelerated metabolism of metolachlor in etiolated sorghum shoots. Paralleling the decrease in metolachlor was an increase in the metolachlor-glutathione conjugate. Zama and Hatzios [170] also found that oxabetrinil stimulated metolachlor conjugation to glutathione in sorghum seedlings.

Sorghum and shattercane (*Sorghum bicolor* L. Moench) that were seed treated with oxabetrinil degraded metolachlor in the roots faster than untreated seedlings [168]. Oxabetrinil did not appreciably influence the degradation of metolachlor in rice, whereas the rice safener fenclorim greatly increased the degradation of metolachlor in rice.

The levels of reduced glutathione are elevated in corn and sorghum treated with flurazole [93,179]. However, the effects of safeners on glutathione levels in plants vary with different safeners. Both flurazole and dichlormid elevated the reduced glutathione levels in sorghum while naphthalic anhydride and oxabetrinil had no effect on glutathione levels [93].

Glutathione S-transferase activity increased 5- to 30-fold in sorghum treated with dichlormid, flurazole, oxabetrinil, or naphthalic anhydride when metolachlor was used as the enzyme substrate [93]. In addition, Gronwald et al. [93] reported that the enhanced glutathione S-transferase activity strongly correlated with the amount of protection the safeners conferred. Lay and Casida [180] also showed that dichlormid increased the activity of corn root glutathione S-transferases, which mediated glutathione conjugation to EPTC sulfoxide.

Mozer et al. [94] identified and purified two glutathione S-transferases in etiolated corn. One of these transferases designated GST I was constitutively present, whereas the other, designated GST II, was present in tissue that had been treated with the antidote, flurazole. Both GST I and GST II were capable of catalyzing the conjugation of alachlor with glutathione in vitro. The GST II enzyme, however, had a greater specific activity with alachlor than the constitutively expressed GST I. The herbicide antidotes, flurazole, dichlormid, and cyometrinil raised the glutathione S-transferase activity levels between 1.5 and 2.5 times in both roots and shoots of corn. Wiegand et al. [181] reported that etiolated corn treated with dichlormid contained three- to fourfold higher mRNA levels that coded for a GST subunit. They postulated that safeners may either increase the half-life of the mRNA or transcriptionally regulate a gene coding for a GST subunit. The first report on an amino acid sequence of a glutathione S-transferase in plants has been published recently [182]. Great progress is being made on the effects of herbicide antidotes on glutathione S-transferases and the differences in the isozyme populations in plants.

Safeners such as cyometrinil, flurazole, dichlormid, oxabetrinil, fenclorim, and naphthalic anhydride have extremely dissimilar chemical structures, yet all protect certain crop species from chloroacetamide injury. Recent developments support the hypothesis that the protection is a result of the stimulation in herbicide metabolism. In spite of recent progress, additional studies are needed to monitor the quantities of both parent herbicide and herbicide metabolites throughout the early stages of plant development in order to gain more information on the interactions of antidotes with the chloroacetamide herbicides.

IX. CONCLUSIONS

The very favorable biological characteristics of metolachlor have
been responsible for it becoming one of the most widely used and
valuable herbicides thorughout the world. Metolachlor can effectively
control annual grasses, many broadleaf weeds, and yellow nutsedge
in corn, soybeans, and numerous other crops. Metolachlor's versatility for weed control under most soil types and tillage conditions
has been responsible, in large measure, for the rapid increase and
successful use of conservation tillage practices, which have been responsible for preserving great quantities of essential top soil, as
well as improving the efficiency of water, energy, and labor. Further progress toward these vital agricultural and environmental objectives will require the continuing research and use of metolachlor
and other essential herbicides.

Metolachlor provides major improvements in chemical stability in
soils (slower dissipation and leaching), formulations, and other handling properties compared to earlier acetanilide herbicides, most of which
have been largely replaced. Yet, it has almost ideal soil dissipation
rate, with little or no carryover or phytotoxicity to rotational crops.
The half-life of metolachlor in most soils ranges from three weeks to
two months, with most of the loss due to microbial degradation.
Only minute quantities of this herbicide are lost through volatilization, photodecomposition, surface runoff, or leaching. Even the
metolachlor that finds its way to surface or groundwater is subject
to biodegradation. There is no evidence that either metolachlor or
its metabolites are magnified in organisms of the food chain.

In field soils, metolachlor is readily biodegraded both under
aerobic and anaerobic conditions. It can apparently be metabolized
by a variety of soil fungi, bacteria, and actinomycetes, but there
is no evidence that the microorganisms or their enzymes can be enhanced or adapted to cause a more rapid degradation of metolachlor
after repeat applications. While the major end product of soil degradation of metolachlor is CO_2, a number of intermediates are
formed, mainly products of oxidation (e.g., oxalic acid derivatives)
and hydrolysis (dechlorinated derivatives). There is evidence that
limited chemical degradation of metolachlor may also occur.

All plants, including susceptible crops and weeds, are apparently
capable of metabolizing metolachlor. Plant selectivity seems to be
based on how soon after uptake metabolism begins or how rapidly it
takes place. The major route of metolachlor degradation in plants
is by conjugation of the chloracetyl side chain with glutathione,
followed by breakdown to the cysteine conjugate, oxidative deamination, and reduction to the thiolactic acid conjugate, oxidation to

sulfoxide derivatives, cleavage of side-chain groups, and subsequent conjugation with glucose.

Metolachlor is rapidly and completely metabolized and excreted in mammals. Very little residue remains in animal tissues, including liver and kidneys. In animals, metolachlor is mainly metabolized via cleavage of the methyl ether and further oxidation and to a lesser extent oxidation of the methyl and ethyl side chains, as well as conjugation of the chloroacetyl group with glutathione.

Metolachlor inhibits major metabolic processes in plants, but the primary site of action is unknown. The chloroacetamides inhibit growth of susceptible species, and depending on the experimental conditions, these herbicides can also inhibit the synthesis of proteins, hormones, lipids and secondary metabolites in susceptible plants. The use of chloroacetamide isomers and rotamers that differ in herbicidal activity, and herbicide antidotes which protect plants from injury caused by chloroacetamides, will be valuable tools to further investigate their modes of action. Detailed information on the location and quantities of metolachlor and its metabolites in plants, correlated with its herbicidal activity, is critical for the elucidation of modes of action of metolachlor and its antidotes. This, in turn will contribute greatly to our understanding of the cellular processes involved in plant growth.

REFERENCES

1. P. C. Hamm, *Weed Sci.*, *22*, 541 (1974).
2. P. C. Hamm and A. J. Speziale, *J. Agr. Food Chem.*, *4*, 518 (1956).
3. *Herbicide Handbook*, 5th Ed., Weed Science Society of America, 1983.
4. *The Pesticide Manual*, 8th Ed., British Crop Protection Council, 1987.
5. E. G. Jaworski, in *Degradation of Herbicides* (P. C. Kearney and D. D. Kaufman, eds.), Marcel Dekker, Inc., New York, 1969.
6. E. G. Jaworski, in *Herbicides: Chemistry, Degradation and Mode of Action*, 2nd Ed. (P. C. Kearney and D. D. Kaufman, eds.), Marcel Dekker, New York, 1975.
7. O. C. Burnside and M. E. Schultz, *Weed Sci.*, *26*, 108 (1978).
8. H. D. Skipper, B. J. Gossett, and G. W. Smith, *Proc. Southern Weed Sci. Soc.*, *29*, 418 (1976).
9. H. J. Strek and J. B. Weber, *Proc. Southern Weed Sci. Soc.*, *35*, 332 (1982).
10. A. Walker, H. A. Roberts, P. A. Brown, and W. Bond, *Ann. Appl. Biol.*, *102*, 155 (1983).

11. R. J. Zimdahl and S. K. Clary, *Weed Sci.*, *30*, 545 (1982).
12. J. B. Siezcka, J. F. Creighton, and W. J. Sanok, *Proc. Northeastern Weed Sci. Soc.*, *35*, 136 (1981).
13. J. B. Weber, H. D. Coble, T. J. Monaco, A. D. Worsham, A. Mehlich, A. Hatfield, D. W. Eaddy, and C. J. Peter, *Proc. Southern Weed Sci. Soc.*, *34*, 265 (1981).
14. C. Buchholz, E. R. Higgins, M. G. Schnappinger, and S. W. Pruss, *Proc. Northeastern Weed Sci. Soc.*, *35*, 124 (1981).
15. M. G. Schnappinger, J. R. Hensley, W. C. Bay, D. L. Greene, and S. W. Pruss, *Proc. Northeastern Weed Sci. Soc.*, *35*, 36 (1981).
16. H. J. Strek and J. B. Weber, *Proc. Southern Weed Sci. Soc.*, *34*, 33 (1981).
17. T. Obrigawitch, J. R. Abernathy, and J. R. Gibson, *Weed Sci. 28*, 708 (1980).
18. J. F. Ahrens, *Proc. Northeastern Weed Sci. Soc.*, *34*, 324 (1980).
19. G. W. Selleck and R. S. Greider, *Proc. Northeastern Weed Sci. Soc.*, *34*, 250 (1980).
20. N. Burkhard, unpublished data, SPR 2/76, Ciba-Geigy Limited, Basel, Switzerland (1975).
21. N. Burkhard and J. Guth, *Pesticide Sci. 12*, 37 (1981).
22. N. Burkhard, unpublished report, Volatilization of CGA-24705 from soil under laboratory conditions, 2/77, Ciba-Geigy Limited, Basel, Switzerland (1977).
23. J. V. Parochetti, *Weed Sci. Soc. Am.*, Abstr. #17 (1978).
24. S. A. Aziz, unpublished report, Photolysis of CGA-27405 on soil slides under natural and artificial sunlight conditions, GAAC-74102, Ciba-Geigy Corporation, Greensboro, NC (1974).
25. S. A. Aziz and R. A. Kahrs, unpublished report, Photolysis of CGA-24705 in aqueous solution under natural and artificial sunlight conditions, GAAC-74041, Ciba-Geigy Corporation, Greensboro, NC (1974).
26. J. E. McFarland and F. D. Hess, *Weed Sci. Soc. of Am.* (Abstr) *25*, 72 (1985).
27. J. E. McFarland, Ph.D. dissertation, Purdue University (1986).
28. L. C. Davidse, O. C. M. Gerritsma, and G. C. M. Velthuis, *Pest. Biochem. Physiol.*, *21*, 301 (1984).
29. J. P. Chupp and J. F. Olin, *J. Org. Chem.*, *32*, 2297 (1967).
30. J. P. Chupp, J. F. Olin, and H. K. Landwehr, *J. Org. Chem.*, *34*, 1192 (1969).
31. P. C. Hamm, in *Herbicides, Fungicides, Formulation Chemistry*, Proc. 2, Int. Congr. Pestic. Chem. (IUPAC), Volume V (A. C. Tahori, ed.), Gordon and Breach, New York, 1972, pp. 41–64.
32. W. T. Molin, K. M. Naylor, J. P. Chupp, and C. A. Porter, *Weed Sci. Soc. of Am.* (Abstr.), *27*, 62 (1987).

33. H. Moser, G. Rihs, and H. Sauter, *Z. Naturforsch.*, *87b*, 451 (1982).

34. J. E. McFarland and F. D. Hess, *Weed Sci. Soc. of Am.* (Abstr) *26*, 81 (1986).

35. G. L. Jordan, Ph.D. thesis, University of Wisconsin, Madison, WI (1978).

36. P. J. Ballerstedt and P. A. Banks, *Proc. Southern Weed Sci. Soc.*, *35*, 331 (1982).

37. C. J. Peter and J. B. Weber, *Weed Sci.*, *33*, 874 (1985).

38. H. R. Gerber, G. Muller, and L. Ebner, Proc. 12th British Weed Control Conf., 1193 (1974).

39. J. B. Weber and C. J. Peter, *Weed Sci.*, *30*, 14 (1982).

40. N. Burkhard, unpublished report, Adsorption and desorption of metolachlor in various soil types, 45/78, Ciba-Geigy Limited, Basel, Switzerland (1978).

41. N. Burkhard, unpublished report, Leaching characteristics of aged ^{14}C-metolachlor residues in two standard soils, 6/80, Ciba-Geigy Limited, Basel, Switzerland (1980).

42. J. A. Guth, unpublished report, Leaching model study with the herbicide CGA-24705 in four standard soils, SPR 3/75, Ciba-Geigy Limited, Basel, Switzerland (1975).

43. T. Obrigawitch, F. M. Hons, J. R. Abernathy, and J. R. Gipson, *Weed Sci.*, *29*, 332 (1981).

44. H. D. Scott and R. E. Phillips, *Proc. Soil Sci. Soc. Am.*, *36*, 714 (1972).

45. D. L. Schertz, *J. Soil Water Conserva.*, in press (1987).

46. R. S. Fawcett, Extension Pub. Pm-1176, Iowa State University, Ames, IA (1985).

47. H. J. Strek, Ph.D. thesis, North Carolina State University, Raleigh, NC (1984).

48. D. A. Crutchfield, G. A. Wicks, and O. C. Burnside, *Weed Sci.*, *34*, 110 (1986).

49. P. A. Banks and E. L. Robinson, *Weed Sci.*, *34*, 607 (1986).

50. R. G. Harvey, A. E. Peterson, R. L. Higgs, and W. H. Paulson, *Weed Sci. Soc. Am.* Abstr. #10 (1976).

51. M. P. Braverman, T. L. Lavy, and R. E. Talbert, *Weed Sci.*, *33*, 819 (1985).

52. D. C. Bouchard, T. L. Lavy, and D. B. Marx, *Weed Sci.*, *30*, 629 (1982).

53. D. D. Sumner, unpublished data, Ciba-Geigy Corporation, Greensboro, NC (1987).

54. M. P. Braverman, T. L. Lavy, and C. J. Barnes, *Weed Sci.*, *34*, 479 (1986).

55. A. Walker and R. L. Zimdahl, *Weed Res.*, *21*, 255 (1981).

56. A. Walker and P. A. Brown, *Bull. Environ. Contam. Toxicol.*, *34*, 143 (1985).

57. G. A. Dixon, E. W. Stoller, and M. D. McGlamery, *Proc. North Central Weed Control Conf.*, *31*, 33 (1976).

58. A. S. Cornelius, W. F. Meggitt, and D. Penner, *Proc. North Central Weed Control Conf.*, *31*, 33 (1976).

59. G. B. Beestman and J. M. Deming, *Argon J.*, *66*, 308 (1974).

60. J. L. Ballard and P. W. Santelman, *Proc. Southern Weed Sci. Soc.*, *26*, 385 (1973).

61. C. Yu, G. M. Booth, D. J. Hansen, and J. R. Larsen, *J. Agric. Food Chem.*, *23*, 877 (1975).

62. L. L. McGahen and J. M. Tiedje, *J. Agric. Food Chem.*, *26*, 14 (1978).

63. J. M. Tiedje and M. L. Hagedorn, *J. Agric. Food Chem.*, *23*, 77 (1975).

64. L. L. McGahen and J. M. Tiedje, *Agron. Abstr.*, 142 (1980).

65. H. Ellgehausen, unpublished report, Degradation of CGA-24705 in aerobic, anaerobic, and autoclaved soil, 4/76, Ciba-Geigy Limited, Basel, Switzerland (1976).

66. H. Ellgehausen, unpublished report, Addendum to Project Report 4/76, 5/76, Ciba-Geigy Limited, Basel, Switzerland (1976).

67. F. J. Stevenson, *Methods of Soil Analysis* (C. A. Black, ed.), American Society of Agronomy, Madison, Wisconsin, 1965, pp. 1409–1421.

68. J. A. Guth, unpublished report, Degradation of metolachlor in aerobic soil, 41/81, Ciba-Geigy Limited, Basel, Switzerland (1981).

69. A. Krause, W. Hancock, R. Minard, A. Freyer, R. Honeycutt, H. LeBaron, D. Paulson, S. Liu, and J. Bollag, *J. Agri. Food Chem.*, *33*, 584 (1985).

70. Y. L. Chen and T. C. Wu, *J. Pestic. Sci.*, *3*, 411 (1978).

71. D. D. Chahal, I. S. Bans, and S. L. Chopra, *Plant Soil*, *45*, 689 (1976).

72. A. E. Smith and D. V. Phillips, *Agron. J.*, *67*, 347 (1975).

73. A. Saxena, R. Zhang, and J. Bollag, *Appl. Environ. Micro.*, *53*, 390 (1987).

74. G. R. Smith, C. A. Porter, and E. G. Jaworski, Am. Chem. Soc., 152nd Meeting, New York, 1966.

75. T. Obrigawitch, J. R. Abernathy, and J. R. Gipson, *Weed Sci.*, *28*, 708 (1980).

76. A. Dixon and E. W. Stoller, *Weed Sci.*, *30*, 225 (1982).

77. G. L. Lamoureux, L. E. Stafford, and F. S. Tanaka, *J. Agr. Food Chem.*, *19*, 346 (1971).

78. R. H. Shimabukuro, G. L. Lamoureux, and D. S. Frear, in *Chemistry and Action of Herbicide Antidotes* (F. M. Pallos and J. E. Casida, eds.), Academic Press, New York, 1978, pp. 133–149.

79. K. K. Hatzios and D. Penner, *Metabolism of Herbicides in Higher Plants*, Burgess Publishing Co., Minneapolis, 1982.

80. G. L. Lamoureux and D. G. Rusness, in *Pesticide Chemistry: Human Welfare and the Environment*, Vol. 3 (J. Miyamoto and P. C. Kearney, eds.), Pergamon Press, New York, 1982, pp. 295–300.

81. E. J. Breaux, *J. Agri. Food Chem.*, *34*, 884 (1986).

82. P. Blattmann, D. Gross, H. P. Kriemler, and K. Ramsteiner, Identification of thiolactic acid type conjugates as major degradation products in the glutathione dependent metabolism of the α-chloroacetamide herbicides metolachlor, dimethachlor and pretilachlor, 6th International Congress of Pesticide Chemistry (IUPAC), Ottawa, August 10–15, Abstract 7A-02 (1986).

83. D. Gross, unpublished report, Uptake, translocation and degradation of CGA-24705 in corn grown under controlled conditions, 8/74, Ciba-Geigy Limited, Basel, Switzerland (1974).

84. D. D. Sumner and J. E. Cassidy, unpublished report, Uptake and distribution of phenyl-^{14}C-CGA-24705 in field grown corn, GAAC-70422, Ciba-Geigy, Corporation, Greensboro, NC (1974).

85. D. Gross, Unpublished report, Identification of degradation products of CGA-24705 in corn plants, 28/80, Ciba-Geigy Limited, Basel, Switzerland (1980).

86. I. M. Szolics, B. J. Simoneaux, and J. E. Cassidy, unpublished report, The uptake and distribution of phenyl-^{14}C-metolachlor from soil in greenhouse grown lettuce, ABR-81021, Ciba-Geigy Corporation, Greensboro, NC (1981).

87. I. M. Szolics, B. J. Simoneaux, and J. E. Cassidy, unpublished report, Evaluation of proposed pathways for metabolism of metolachlor in lettuce and potatoes, ABR-81045, Ciba-Geigy Corporation, Greensboro, NC (1981).

88. E. P. Fuerst and J. W. Gronwald, *Weed Sci.*, *34*, 354 (1986).

89. I. M. Szolics and J. E. Cassidy, unpublished report, The uptake and balance of phenyl-^{14}C-metolachlor in field grown soybeans, ABR-78082, Ciba-Geigy Corporation, Greensboro, NC (1982).

90. D. D. Sumner and J. E. Cassidy, unpublished report, The uptake and distribution of phenyl-^{14}C-CGA-24705 from soil in greenhouse grown soybeans, GAAC-75039, Ciba-Geigy Corporation, Greensboro, NC (1975) .

91. M. Hussain, I. P. Kapoor, C. C. Ku, and S. Stout, *J. Agri. Food Chem.*, *31*, 232 (1983).

92. J. R. C. Leavitt and D. Penner, *J. Agri. Food Chem.*, *27*, 533 (1979).

93. J. W. Gronwald, E. P. Fuerst, C. V. Eberlein, and M. A. Egli, *Pest. Biochem. Physiol.*, *29*, 66 (1987).

94. T. J. Mozer, D. C. Tiemeier, and E. G. Jaworski, *Biochemistry 22*, 1068 (1983).

95. R. Edwards and W. J. Owen, *Planta, 169*, 208 (1986).

96. H. Hamböck, unpublished report, Distribution degradation and excretion of CGA-24705 in the rat, 1/74, Ciba-Geigy Limited, Basel, Switzerland (1974).

97. H. Hamböck, unpublished report, Metabolism of CGA-24705 in the rat, 7/74, Ciba-Geigy Limited, Basel, Switzerland (1974).

98. H. Hamböck, unpublished report, Addendum to Project Report 7/74, 12/74, Ciba-Geigy Limited, Basel, Switzerland (1975).

99. J-C. Roger and J. E. Cassidy, unpublished report, Metabolism and balance study of phenyl-^{14}C-CGA-24705 in a lactating goat, GAAC-74020, Ciba-Geigy Corporation, Greensboro, NC (1974).

100. E. G. Leighty, R. L. Foltz, and R. J. Jakobsen, unpublished report, Identification of metabolites in urine from goats administered CGA-24705, Battelle Columbus Laboratories, Columbus OH (1975).

101. Metolachlor Pesticide Registration Standard, Environmental Protection Agency, Washington, DC. 56 (1980).

102. W. Mücke, unpublished report, New feces metabolites of CGA-24705 in the rat, 26/81, Ciba-Geigy Limited, Basel, Switzerland (1981).

103. R. Bissig, H. Hamböck, and W. Mücke, unpublished report, Metabolism of metolachlor in the rat, Ciba-Geigy Limited, Basel, Switzerland (1987).

104. C. Fedtke, *Biochemistry and Physiology of Herbicide Action*, Springer-Verlag, New York, 1982, pp. 148–158.

105. F. M. Ashton and A. S. Crafts, *Mode of Action of Herbicides*, 2nd ed., John Wiley & Sons, New York, 1981, pp. 91–117.

106. P. E. Keeley, C. H. Carter, and Y. H. Miller, *Weed Sci.*, *20*, 71 (1972).

107. I. O. Akobundu, R. D. Sweet, W. B. Duke, and P. L. Minotti, *Weed Sci.*, *23*, 67 (1975).

108. W. B. Duke, F. W. Slife, J. B. Hanson, and H. S. Butler, *Weed Sci.*, *23*, 142 (1975).

109. P. Pillai, D. E. Davis, and B. Truelove, *Weed Sci.*, *27*, 634 (1979).

110. L. M. Deal and F. D. Hess, *Weed Sci.*, *28*, 168 (1980).

111. A. Borowy and G. F. Warren, *Weed Sci. Soc. Am.* Abstr. #158 (1978).

112. Y. Eshel, *Weed Sci.*, *17*, 441 (1969).

113. E. L. Knake and L. M. Wax, *Weed Sci.*, *16*, 393 (1968).

114. G. L. Jordon and R. G. Harvey, *Weed Sci.*, *28*, 589 (1986).

115. E. Ebert, *Pest. Biochem. Physiol.*, *13*, 227 (1980).

116. E. Ebert, *Weed Res.*, *22*, 305 (1982).

117. I. Paradies, E. Ebert, and E. F. Elstner, *Pest. Biochem. Physiol.*, *15*, 209 (1981).

118. U. Leuthold, unpublished data, Ciba-Geigy Limited, Basel, Switzerland (1975).

119. C. Fedtke, in *Biochemical Responses Induced by Herbicides*, Am. Chem. Soc. Symposium Series (D. E. Moreland, J. B. St. John, and F. D. Hess, eds.), ACS, Washington, DC., 1982, pp. 231–250.

120. N. S. Dhillon and J. L. Anderson, *Weed Res.*, *12*, 182 (1982).

121. J. D. Mann, L. S. Jordan, and B. E. Day, *Plant Phys.*, *46*, 840 (1965).

122. W. B. Duke, Ph.D. dissertation, University of Illinois (1967).

123. E. G. Jaworski, *J. Agri. Food Chem.*, *17*, 165 (1969).

124. D. E. Moreland, S. S. Malhotra, R. D. Gruenhagen, and E. H. Shokrah, *Weed Sci.*, *17*, 556 (1969).

125. R. M. Devlin and R. P. Cunningham, *Proc. Northeast Weed Control Conf.*, *24*, 149 (1970).

126. V. S. Rao and W. B. Duke, *Weed Sci.*, *24*, 616 (1976).

127. L. M. Deal, J. T. Reeves, B. A. Larkins, and F. D. Hess, *Weed Sci.*, *28*, 334 (1980).

128. S. Chang, F. M. Ashton, and D. E. Bayer, *J. Plant Growth Reg.*, *4*, 1 (1985).

129. E. G. Jaworski, *Science*, *123*, 847 (1956).

130. H. Lindley, *Biochem. J.*, *74*, 577 (1960).

131. H. Lindley, *Biochem. J.*, *82*, 418 (1962).

132. H. V. Marsh, Jr., J. Bates, and P. Trudeau, *Weed Sci. Soc. Am.*, Abstr. #176 (1975).

133. R. E. Wilkinson, *Pest. Biochem. Physiol.*, *23*, 19 (1985).

134. T. C. Chang, H. V. Marsh, and P. H. Jennings, *Pest. Biochem. Physiol.*, *5*, 323 (1975).

135. D. B. Narsaiah and R. G. Harvey, *Weed Res.*, *25*, 197 (1977).

136. A. Nyffeler, unpublished data, Ciba-Geigy Limited, Basel, Switzerland (1978).

137. R. E. Wilkinson, *Pest. Biochem. Physiol.*, *16*, 63 (1981).

138. R. E. Wilkinson, *Pest. Biochem. Physiol.*, *16*, 199 (1981).

139. R. E. Wilkinson, *Pest. Biochem. Physiol.*, *17*, 177 (1982).

140. R. E. Wilkinson, *Pest. Biochem. Physiol.*, *23*, 19 (1984).

141. E. Ebert, unpublished data, Ciba-Geigy Limited, Basel, Switzerland (1985).

142. E. Ebert and J. Gaudin, unpublished data, Ciba-Geigy Limited, Basel, Switzerland (1978).

143. B. Truelove, A. Diner, D. Davis, and J. Weete, Abstr., *Weed Sci., Soc. Am.*, 99 (1979).

144. J. D. Mann and M. Pu, *Weed Sci.*, *16*, 197 (1968).

145. F. M. Ashton, O. T. de Villiers, R. K. Glenn, and W. B. Duke, *Pest. Biochem. Physiol.*, *7*, 122 (1977).

146. H. Weisshaar and P. Böger, *Pest. Biochem. Phys.*, *28*, 286 (1987).

147. E. Ebert and K. Ramsteiner, *Weeds Res.*, *24*, 383 (1984).

148. M. Tevini and D. Steinmüller, *J. Plant Phys.*, *131*, 111 (1987).

149. J. M. Mellis, P. Pillai, D. E. Davis, and B. Truelove, *Weed Sci.*, *30*, 399 (1982).

150. M. C. Watson, P. G. Bartels, and K. C. Hamilton, *Weed Sci.*, *28*, 122 (1980).

151. J. S. Hickey and W. A. Krueger, *Weed Sci.*, *22*, 250 (1974).

152. W. T. Molin, E. J. Anderson, and C. A. Porter, *Pest. Biochem. Physiol.*, *25*, 105 (1986).

153. E. Ebert and J. Gaudin, unpublished data, Ciba-Geigy Limited, Basel, Switzerland (1986).

154. E. Ebert, unpublished data, Ciba-Geigy, Limited, Basel, Switzerland (1979).

155. O. L. Hoffman, Plant Phys., *28*, 622 (1953).

156. A. Nyffeler, H. R. Gerber, and J. R. Hensley, *Weed Sci.*, *28*, 6 (1980).

157. J. F. Ellis, J. W. Peek, J. Boehle, Jr., and G. Muller, *Weed Sci.*, *28*, 1 (1980).

158. D. E. Schafer, R. J. Brinker, and R. O. Radke, *Proc. North Central Weed Control Conf.*, *35*, 67 (1980).

159. W. E. Turner, D. R. Clark, N. Helseth, T. R. Dill, J. King, and E. B. Seifried, *Proc. Southern Weed Sci. Soc.*, *35*, 388 (1982).

160. J. Rufener and M. Quadranti, Proc. 10th Internat. Congr. Plant Prot., 332 (1983).

161. N. T. Helseth and T. R. Dill, *Proc. Southern Weed Sci. Soc.*, *39*, 35 (1986).

162. C. Parker, *Pest. Sci.*, *14*, 40 (1983).

163. K. K. Hatzois, Adv. Agron., *36*, 265 (1984).

164. R. A. Christ, *Weed Res.*, *25*, 193 (1985).

165. F. M. Pallos and J. E. Casida (eds.), *Chemistry and Action of Herbicide Antidotes*, Academic Press, New York, 1978.

166. K. K. Hatzios and R. E. Hoagland, Herbicide safeners: Development, Uses, and Mechanism of Action, in press (1987).

167. R. A. Christ, *Weed Res.*, *21*, 1 (1981).

168. E. Ebert, in *Herbicide Safeners; Development, Uses and Mechanism of Action* (K. K. Hatzios and R. E. Hoagland, eds.) in press (1987).

169. W. H. Ahrens and D. E. Davis, *Proc. Southern Weed Sci. Soc.* *31*, 19 (1978).

170. P. Zama and K. K. Hatzois, *Weed Sci.*, *34*, 834 (1986).

171. M. L. Ketchersid, D. M. Vietor, and M. G. Merkle, *J. Plant Growth Reg.*, *1*, 285 (1982).

172. P. Zama and K. K. Hatzios, *Pest. Biochem. Phyisol.*, *27*, 86 (1987).

173. M. L. Ketchersid, F. W. Plapp, and M. G. Merkle, *Weed Sci.*, *33*, 774 (1985).

174. D. L. Devlin, L. J. Moshier, O. G. Russ, and P. W. Stahlman, *Weed Sci.*, *31*, 790 (1983).

175. C. Fedtke, *Z. Pflanzenkrankheiten und Pflanzenschutz*, *92*, 654 (1985).

176. K. K. Hatzios, *Weed Sci.*, *31*, 280 (1983).

177. D. J. Cole and W. J. Owen, *Plant Sci.*, *50*, 13 (1987).

178. R. H. Shimabukuro, *Pesticide Chemistry: Human Welfare and the Environment*, Volume 3, *Mode of Action, Metabolism and Toxicology* (J. Miyamoto and P. C. Kearney, eds.), Pergamon Press, New York, 1983, p. 140.

179. E. J. Breaux, J. E. Patanella, and E. F. Sanders, *J. Agri. Food Chem.*, *35*, 474 (1987).

180. M. M. Lay and J. E. Casida, *Pest. Biochem. Physiol.*, *6*, 442 (1976).

181. R. C. Wiegand, D. M. Shah, T. J. Mozer, E. I. Harding, J. Diaz-Collier, C. Saunders, E. G. Jaworski, and D. C. Tiemeier, *Plant Mol. Biol.*, *7*, 235 (1986).

182. D. M. Shah, C. M. Hironaka, R. C. Wiegand, E. I. Harding, G. G. Krivi, and D. C. Tiemeier, *Plant Mol. Biol.*, *6*, 203 (1986).

INDEX

Printed and bound by CPI Group (UK) Ltd, Croydon, CR0 4YY

17/10/2024

01775692-0005